PHYSICAL ACTIVITY AND OBESITY

Claude Bouchard, PhD

Executive Director and George A. Bray Chair in Nutrition
Pennington Biomedical Research Center
Louisiana State University
Baton Rouge, Louisiana
U.S.A.
Editor

Human Kinetics

Library of Congress Cataloging-in-Publication Data

Physical activity and obesity / Claude Bouchard, editor.
   p. cm.
  Includes bibliographical references and index.
  ISBN 0-88011-909-8
   1. Obesity. 2. Exercise--Physiological aspects. I. Bouchard, Claude.

  RC628 .P496 2000
  616.3'98--dc21
                                                  99-088071
ISBN: 0-88011-909-8

Acquisitions Editor: Michael S. Bahrke, PhD; **Developmental Editor:** Spencer J. Cotkin, PhD; **Assistant Editors:** Laurie Stokoe, Mark E. Zulauf; **Copyeditor:** Arlene Miller; **Proofreader:** Jim Burns; **Indexer:** Vita Richman; **Permission Manager:** Heather Munson; **Graphic Designer:** Stuart Cartwright; **Graphic Artist:** Dawn Sills; **Cover Designer:** Jack W. Davis; **Printer:** Versa Press; **Binder:** Dekker Bookbinding; **Illustrator:** Accurate Art

Printed in the United States of America   10 9 8 7 6 5 4 3 2 1

Human Kinetics
Web site: http://www.humankinetics.com

United States: Human Kinetics, P.O. Box 5076, Champaign, IL 61825-5076
1-800-747-4457
e-mail: humank@hkusa.com

Canada: Human Kinetics, 475 Devonshire Road Unit 100, Windsor, ON N8Y 2L5
1-800-465-7301 (in Canada only)
e-mail: humank@hkcanada.com

Europe: Human Kinetics, P.O. Box IW14, Leeds LS16 6TR, United Kingdom
+44 (0)113-278 1708
e-mail: humank@hkeurope.com

Australia: Human Kinetics, 57A Price Avenue, Lower Mitcham, South Australia 5062
(08) 82771555
e-mail: liahka@senet.com.au

New Zealand: Human Kinetics, P.O. Box 105-231, Auckland Central
09-523-3462
e-mail: humank@hknewz.com

Contents

List of Contributors

Baker, Christina Wood, MS

Department of Psychology, Yale University, New Haven, Connecticut, USA

Barbeau, Paule, PhD

Georgia Prevention Institute, Medical College of Georgia, Augusta, Georgia, USA

Blair, Steven N., PED

Cooper Institute for Aerobics Research, Dallas, Texas, USA

Bouchard, Claude, PhD

Pennington Biomedical Research Center, Louisiana State University, Baton Rouge, Louisiana, USA

Bray, George A., MD

Pennington Biomedical Research Center, Louisiana State University, Baton Rouge, Louisiana, USA

Brodney, Suzanne, MS, RD

Cooper Institute for Aerobics Research, Dallas, Texas, USA

Brownell, Kelly D., PhD

Department of Psychology, Yale University, New Haven, Connecticut, USA

Colditz, Graham A., MD, DrPH

Channing Laboratory, Department of Medicine, Brigham and Women's Hospital, Harvard Medical School, Boston, Massachusetts, USA

Després, Jean-Pierre, PhD

Department of Food Sciences and Nutrition, Laval University and the Lipid Research Center, Laval University Research Center, Ste-Foy, Québec, Canada

Dionne, Isabelle, PhD

Division of Endocrinology, Metabolism, and Nutrition, Department of Medicine, University of Vermont, Burlington, Vermont, USA

Eckel, Robert H., MD

University of Colorado Health Sciences Center, Department of Medicine, Division of Endocrinology, Metabolism and Diabetes, Denver, Colorado, USA

Foreyt, John P., PhD

Behavioral Medicine Research Center, Department of Medicine, Baylor College of Medicine, Houston, Texas, USA

Gutin, Bernard, PhD

Georgia Prevention Institute, Medical College of Georgia, Augusta, Georgia, USA

Hill, James O., PhD

Center for Human Nutrition, University of Colorado Health Sciences Center, Denver, Colorado, USA

Ho, Richard C., MS

Department of Food Science and Human Nutrition, Colorado State University, Fort Collins, Colorado, USA

Jakicic, John M., PhD

University of Pittsburgh School of Medicine, Western Psychiatric Institute and Clinic, Pittsburgh, Pennsylvania, USA

Jebb, Susan A., PhD

MRC Human Nutrition Research, Cambridge, UK

Kelley, David E., MD

Division of Endocrinology and Metabolism, Department of Medicine, University of Pittsburgh School of Medicine and Department of Veterans Affairs Medical Center, Pittsburgh, Pennsylvania, USA

Lamarche, Benoît, PhD

Department of Food Sciences and Nutrition, Laval University and the Lipid Research Center, Laval University Research Center, Ste-Foy, Québec, Canada

Lee, Chong Do, EdD

Cooper Institute for Aerobics Research, Dallas, Texas, USA

Mariani, Anna, MPH

Channing Laboratory, Department of Medicine, Brigham and Women's Hospital, Harvard Medical School, Boston, Massachusetts, USA

Melby, Christopher L., DrPH

Department of Food Science and Human Nutrition, Colorado State University, Fort Collins, Colorado, USA

Poehlman, Eric T., PhD

Department of Medicine, University of Vermont, Burlington, Vermont, USA

Poirier, Paul, MD

Québec Heart Institute, Ste-Foy, Québec, Canada

Poston, Walker S. Carlos II, MPH, PhD

Department of Psychology, University of Missouri-Kansas City and the Mid America Heart Institute, St. Luke's Hospital, Kansas City, Missouri, USA

Prentice, Andrew M., PhD

MRC International Nutrition Group, Public Health Nutrition Unit, London School of Hygiene and Tropical Medicine, London, UK

Ravussin, Eric, PhD

National Institutes of Health, Clinical Diabetes and Nutrition Section, Phoenix, Arizona, USA

Rössner, Stephan, MD, PhD

Obesity Unit, MK-division, Huddinge University Hospital, Huddinge, Sweden

Salbe, Arline D., PhD

National Institutes of Health, Clinical Diabetes and Nutrition Section, Phoenix, Arizona, USA

Seidell, Jacob C., PhD

Department of Chronic Disease and Environmental Epidemiology, National Institute of Public Health and Environmental Protection (RIVM), Bilthoven, The Netherlands

Simoneau, Jean-Aimé, PhD

Physical Acitivity Sciences Laboratory, Division of Kinesiology, Department of Social and Preventive Medicine, Faculty of Medicine, Laval University, Ste-Foy, Québec, Canada

Suminski, Richard R., PhD

Department of Health and Human Performance, University of Houston, Houston, Texas, USA

Tremblay, Angelo, PhD

Division of Kinesiology, Physical Activity Sciences Laboratory, Department of Social and Preventive Medicine, Faculty of Medicine, Laval University, Ste-Foy, Québec, Canada

Westerterp, Klaas R., PhD

Department of Human Biology, Maastricht University, Maastricht, The Netherlands

Wing, Rena R., PhD

University of Pittsburgh School of Medicine, Western Psychiatric Institute and Clinic, Pittsburgh, Pennsylvania, USA

Preface

The suggestion for this book came from Dr. Rainer Martens, President of Human Kinetics. He felt that there was a need for a medium-size book dealing with physical activity and obesity and asked whether I would be willing to take the leadership of such a project. We both agreed that the publication should not attempt to be exhaustive, but that it should rather address the key issues and be written with the contributions of the most authoritative experts in this field. The result is *Physical Activity and Obesity*.

Even though it took a little longer than expected to produce all the manuscripts, the content of the volume is extremely up to date. This should not be surprising as the book deals with topics that are not changing as fast as some of the others pertaining to obesity. Indeed, despite the importance of physical activity in the overall energy balance equation and its influence on the health status of those affected by overweight or obesity, there is still a flagrant lack of solid data on the role of physical activity. We hope that this book will provide some encouragement to those who may be tempted by a research career in this area.

I would like to thank all the authors and coauthors of chapters of this volume for their contribution and their support. I would also like to take this opportunity to express my gratitude to Diane Drolet, MSc, for her competent help in the revision and editing of the manuscripts. As for many of our other publications, Diane has played a key role in the management of this project. Finally, a word of thanks to all those at Human Kinetics who have been associated with the production of this book.

Claude Bouchard

Pennington Biomedical Research Center
Louisiana State University
April 1999

PART I

The Obesity Epidemic

CHAPTER 1

Introduction

Claude Bouchard, PhD
Pennington Biomedical Research Center, Louisiana State University, Baton Rouge, Louisiana, U.S.A.

Physical Activity and Obesity is the result of a perceived need for such a book in the exercise science and sports medicine communities. Obesity and a physically inactive lifestyle are two of the most prevalent risk factors for common chronic diseases in the Western world. They both carry enormous health and economic costs. They are both recognized as major risk factors for cardiovascular disease, non-insulin-dependent diabetes mellitus, hypertension, and other debilitating conditions. Estimates of the number of deaths attributable in part to obesity in the United States alone reach about 300,000 per year, a figure that is comparable to the number of yearly deaths thought to be attributable to sedentarism. A physically inactive lifestyle is a risk factor for weight gain with age. Moreover, obese individuals are generally very sedentary, as their excess body mass constitutes a major obstacle in adopting a more physically active lifestyle. Furthermore, sedentarism in overweight or obese persons increases the probability that they will be affected by the common morbidities of excess weight or that they will die prematurely.

This book has been written with the collaboration of the most prominent scientists and clinicians in the field. The book is organized into 4 parts and 19 chapters. It could have been a much larger volume, but an effort was made to focus on the most important issues and a size limit was imposed on the manuscript of each contributing author.

Part I provides an overview of the current obesity epidemic, its implications for morbidity and mortality rates, and its economic burden. Part II includes chapters on the determinants of obesity and the assessments of energy expenditure components and dietary habits. It also deals with energy balance issues and the role of adipose tissue and skeletal muscle metabolism and their characteristics in the obese state. Part III provides an overview of the role of physical activity in the prevention and treatment of obesity for various population groups, including the severely obese. It also addresses the topic of weight loss maintenance. Part IV focuses on the role of physical activity in the morbidities of obesity and with respect to the health status of the obese. How to modify the physical activity habits of the obese is also discussed.

WHO and NIH Reports

Two recent reports promise to have a major influence on obesity prevention and treatment. The first was published by the World Health Organization as a consultation report in early 1998 (48). It was based largely on the deliberations and documentation of the International Obesity Task Force (IOTF) of the International Association for the Study of Obesity (IASO). The WHO Report proposed a classification of body weight based on the body mass index (BMI), defined as weight in kilograms divided by height in meters squared (kg/m^2), with the view of having it accepted internationally. With the support of the member countries represented on the IOTF and of IASO, this goal has already been achieved. The main merits of the classification are that it is simple; it is based on a large body of epidemiological and clinical data that have considered the association between BMI and morbidity and mortality rates; and it provides a useful tool for international comparisons when monitoring changes over time in a given country or region as well as changes associated with major lifestyle alterations, implementation of new public health policies, or other relevant interventions. The BMI classification scheme is reproduced in the chapter by J. Seidell (chapter 2). For the moment, suffice it to emphasize that the classification specifies that **overweight** is defined as a BMI ranging from 25 to 29.9 and that **obesity** is set at a BMI of 30 and above.

The second of these influential reports was published later in 1998 by the National Institutes of Health (29). This document was developed by a group of experts from many disciplines over a period of about two years at the initiative of the National Heart, Lung, and Blood Institute with the collaboration of several other NIH components. Its purpose was to propose guidelines for the prevention and treatment of obesity and its co-morbidities. It is an evidence-based report that considers every possible path linking behavioral characteristics to excess body mass and linking the latter to risk factors and co-morbidities. One of the lessons learned in the course of the development of these guidelines was that the body of knowledge on physical activity and relevant outcomes is extremely limited. There are few randomized clinical trials that have lasted at least six months with reasonable statistical power and with adequate monitoring of intervention protocols, high levels of compliance, and proper measurement of the outcome variables. The net result is that there is a dearth of solid research data regarding the role of physical activity in the prevention and treatment of overweight and obesity as well as their morbidities. However, the evidence-based appproach taken in the NIH effort is one that will have a growing influence on health-related issues; all the exercise specialists concerned by these issues, including physical activity and obesity, would do well to become familiar with it.

The Current Overweight and Obesity Epidemics

The data from almost all the countries of the industrialized world, and even those from the third world, reveal that a growing proportion of children and adults is overweight or frankly obese (48). The prevalence of overweight and obesity varies according to age, gender, race, and socioeconomic classes across the Western and developing worlds. About 50% of the adults in the United States, Canada, and some of the Western European countries have a BMI of 25 or more. The prevalence is slightly lower in other countries of the European Community, but it is not strikingly lower. In some population subgroups in the U.S. and elsewhere, the prevalence of those with a BMI of 25 kg/m^2 and above is more than 70% (48). Moreover, there has been a dramatic increase in the prevalence of obesity in this century, and all indications are that the problem will get even worse in the coming decades. It is striking to see that the body mass of men 1.70 m tall recruited in the United States military service increased from 66.8 kg in 1863 to 76.3 kg over a century, an increase of about 10 kg for the same stature (9). It is also commonly observed that those who are in the upper ranges of BMI, i.e., the severely obese, are heavier now than they were in the past.

The prevalence of frank obesity in childhood and adolescence has more than doubled since the early 1960s (43). For instance, by definition, 5% of the children and adolescents measured in the National Health and Examination Surveys II and III (1963-1970) had a BMI value at the 95th percentile level and above. Using the same BMI values in subsequent National Surveys revealed that a growing proportion of children and adolescents reached these high levels of body mass for stature (table 1.1). By NHANES III (1988-1991), 11% of the 6- to 11-year-old children reached the 1963-1970 95th percentile BMI value. For 12- to 17- year-old adolescents, the

Table 1.1   Trends in Prevalence of Overweight in USA Youth

	6 to 11 years	12 to 17 years
NHES II and III, 1963-1970	5	5
NHANES I, 1971-1974	5	6
NHANES II, 1976-1980	8	6
NHANES III, 1988-1991	11	11

Overweight is defined as the age- and sex-specific 95th percentile of BMI from NHES, 1963-1970 data.

Reprinted from R.P. Troiano, K.M. Flegal, R.J. Kuczmarski, S.M. Campbell, and C.L. Johnson, 1995, "Overweight prevalence and trends for children and adolescents: The national health and nutrition examination surveys, 1963 to 1991," *Archives of Pediatrics and Adolescent Medicine* 149: 1085-1091.

proportion in NHANES III attained 11% (43). The worst scenario is that these increases in childhood obesity will translate later into an even greater prevalence of adulthood obesity as that currently observed.

In this volume, Seidell reports some astounding numbers. Based on the IOTF research and the recent WHO report, he comes to the conclusion that there are currently about 250,000,000 obese adults (7% of the population) and at least 500,000,000 overweight (BMI from 25 to 29.9) people worldwide. The sad news is that these estimates are very conservative and that prevalence of both conditions is on the rise all over the world.

Important Differences Between Overweight and Obesity

Even though the BMI is a continuous variable, the WHO-NIH classification paradigm uses arbitrary cutoff points to define overweight (25.0 to 29.9) and obesity (30 and above). We are of the view that this distinction is highly justified in terms of both the etiology of the conditions and the levels of risk for morbidities and mortality rates. The levels of risk for morbidities and mortality rates is addressed in various chapters of this volume. However, the etiology of the conditions is also important in a book emphasizing the contributions and limitations of regular physical activity, and it will be briefly discussed here.

Overweight is in many ways very different from obesity. Obviously, obesity is characterized by a significantly greater excess of weight, particularly adipose tissue mass, than overweight. However, the situation is more complex. The difference between the two conditions resides mainly in the higher percentage of the body as fat in the obese. In other words, the expansion of fat-free tissue has not kept up with the growth in the adipose organ. A second difference between the two conditions is rooted in the fact that, in general, positive energy balance is likely to have been more pronounced and to have been sustained for a longer period of time in the obese than in the overweight. This is a very important notion for a proper understanding of the role of physical activity. A third difference pertains to energy expenditure: Because the obese are heavier for their stature than the overweight, they expend more energy on the average. They are characterized by a higher resting metabolic rate as a result of a greater respiring tissue mass, and they also have a higher energy expenditure above resting energy expenditure than normal-weight people. The latter is caused by the fact that it requires more energy to move around and about a bigger mass. There is no clear difference between overweight and obese when it comes to the thermic response to food. Thus, on the average, the obese have a higher resting metabolic rate and expend more energy on activities than the overweight or the normal weight.

One implication of these differences is that a sedentary lifestyle or a low level of habitual activity has the potential to account for a large proportion of the adult overweight cases. For instance, a nonresting daily energy expenditure depressed by about 300 kcal will generate chronic positive energy balance and translate into a surplus of calories consumed of more than 100,000 kcal over a year. This would undoubtedly result in body weight gain that may be as high as 50 percent of the total yearly caloric surplus caused by a sedentary lifestyle. If one assumes that one kg of body mass is, on the average, the equivalent of about 7,000 kcal, then a reduction of the daily level of habitual physical activity by 300 kcal would be associated with a body weight gain of 6 to 8 kg over a year. The weight gain would be progressively less with time as the resting metabolic rate and the energy cost of moving the body increase with body-mass accretion. After a while, energy balance would be restored but at a new body-mass level, one that is now in the overweight range. One should note that the path to overweight under these conditions of sedentarism does not require any increase in energy intake. Thus, it is possible to become overweight without being hyperphagic in comparison to other people of the same sex, age, and body build.

In contrast, it is likely that obesity, particularly the more severe types (e.g., BMIs of 35 or 40 and more), requires positive energy balance conditions that are sustained for longer periods of time, i.e., at least a few years. Such conditions will generally be achieved only when both energy intake is increased and energy expenditure is depressed, and when energy intake is further increased when body weight ceases to grow or when energy balance is nearly reached or is actually achieved for a while. Thus, one could speculate that a sedentary mode of life plays an important role in both overweight and obesity, but is, under certain circumstances, a condition sufficient in itself to lead to an overweight state.

Etiology of Overweight and Obesity

Body weight is a function of energy and nutrient balance over an extended period of time. Energy balance is determined by macronutrient intake, energy expenditure, and energy or nutrient partitioning. Positive energy balance over weeks and months will result in weight gain while negative energy balance will have the opposite effect.

The increase in the prevalence of overweight and obesity cases worldwide is occurring against a background of a progressive reduction in the energy expended for work and occupational activities as well as for the accomplishment of personal chores and daily necessities (22, 32, 45). The reduction in energy expenditure associated with physical activity brought about by automation and changing job and professional environmental circumstances has been nothing but dramatic in the second half of this century. In contrast, the energy expenditure of leisure time physical activity, the most important discretionary component of total daily energy expenditure, may have increased slightly, but not enough to keep pace with the changes brought about by urbanization and automation.

Because positive energy balance is required for weight gain to happen, dietary habits are playing a key role in the prevalence of overweight and obesity (23). In developed countries, the availability of highly palatable foods in almost unlimited abundance is undoubtedly contributing to the epidemic; some of the affected individuals eat many times a day and consume large portion sizes (18, 21). The proportion of calories derived from fats is also potentially involved, particularly in those who consume a high-fat diet while living a sedentary life (2, 11); however, the exact contribution of a high-fat diet to the current obesity epidemic remains controversial (40, 46). In addition, over that same period of time and especially over the last two or three decades, there has been a continuous increase in the prevalence of eating disorders. Restrained eating behaviors and severe dieting are two frequent occurrences. They are often followed by periods of gorging and binge eating. Such problematic eating patterns resulting in large fluctuations of caloric and nutrient intake have had influences on body weight for some individuals.

Progress has been slower in the area of energy or nutrient partitioning, probably because it was not perceived until recently as an important determinant of long-term energy balance in humans (7, 9). However, many physiological and metabolic studies on the modulation of nutrient storage into fat and protein in various organs and tissues have been conducted in the field of animal husbandry. This research has strong implications for understanding the phenotype of nutrient partitioning in humans. For instance, these studies indicate that insulin; steroid, thyroid, and growth hormones; and various growth factors all influence the fate of ingested energy. Hepatic and skeletal muscle metabolism as well as lipoprotein lipase activity in adipose tissue and skeletal muscle play an important role. The composition of ingested food, including the amino acid composition of the proteins, needs to be considered as well. This line of research suggests that being a "fat storer" as opposed to a "lean tissue storer" is a risk factor for obesity.

We have previously reviewed the correlates of overweight and obesity or of body weight gain over time (3, 4). Table 1.2 provides an updated version of these correlates and predictors. Some of these correlates are true predictors of body-fat gain and can be defined as risk factors for overweight or obesity. In some cases, it is not possible to establish whether the relationship is causal or only secondary to an obese state. However, for most cases, the associations are truly secondary and have arisen as a result of an obese state. Several of these correlates and predictors are examined in the various chapters of this volume.

Table 1.2   Correlates of Overweight and Obesity or of Body Weight and Body Fat Gain Over Time

Variable	Comment
Age	• Childhood obesity is a risk factor for adulthood obesity.
	• Body-fat content increases during adulthood.
	• Maximal rates of overweight and obesity are attained from 55 to 65 years.

Variable	Comment
Sex	• Women have more body fat. • Sex differences in prevalence of obesity vary in populations or among ethnic groups.
Socioeconomic status	• More obese in high-SES classes in poor countries. • More obese in low-SES classes in rich countries.
Energy intake	• Overfeeding causes weight gain and leads to obesity.
Dietary fat intake	• Dietary fat is related to prevalence of overweight in ecological studies. • High-fat diet causes weight gain. • Low-fat diet reduces body weight.
Resting metabolic rate	• A low body mass and composition-adjusted RMR is a risk factor for weight gain, but contradictory data have been found. • Overweight and obese people have a higher absolute RMR.
Thermic response to food	• Obese have a depressed response in some studies, but contradictory results are abundant.
Physical activity level	• A low level of PA is a risk factor for weight gain. • Level of sedentarism is higher in obese people. • Regular PA changes body composition. • High levels of PA increase SNS activity and RMR. • Regular PA contributes to weight loss and weight maintenance.
Lipid oxidation rate	• Body-fat gains decrease RQ. • A high RQ is a risk factor for weight gain, but there are contradictory results. • Ex-obese have a higher RQ than never obese.
Blood leptin level	• Low leptin levels are weakly related to weight gain, but there are contradictory results. • Most obese have high leptin levels.
SNS activity	• Low SNS activity could be a risk factor for weight gain. • SNS activity increases with overfeeding and body weight gain.
Growth hormone level	• Low GH is a risk factor for weight gain. • Most obese have low GH levels.

(continued)

Table 1.2   *(continued)*

Variable	Comment
Insulin sensitivity	• Obese are often insulin resistant and hyperinsulinemic. • Insulin resistance protects against weight gain, but there are contradictory results.
HPA axis and cortisol levels	• Obese have generally a hyper-responsive and hyperactive HPA axis. • Obese have elevated cortisol production rates, but also accelerated degradation.
Sex steroid levels	• Obese men often have low androgen levels. • Obese women often have high androgen levels with further elevation upon ACTH stimulation.
Adipose tissue metabolism	• Catecholamine-induced lipolysis is reduced in obesity. • Lipogenesis from glucose is increased in human fat cells from obese people. • Adipose tissue LPL is increased in obesity. • Elevated adipose LPL activity remains high in reduced-obese. • High adipose tissue LPL is a risk factor for weight gain.
Skeletal muscle metabolism	• SM type I fiber proportion is not affected by obesity. • SM type IIb fiber proportion is often elevated in obesity. • SM oxidative enzyme markers are inversely related to obesity. • SM LPL activity is low in obesity.
Energy and nutrient partitioning	• Under positive energy balance conditions, some people channel more food carbons into proteins. • High rates of lipid accretion could be a risk for further weight gain.
Smoking	• Smoking is associated with a lower body weight. • Cessation increases body weight in most people.

SES = Socioeconomic status; RMR = Resting metabolic rate; PA = Physical activity; SNS = Sympathetic nervous system; RQ = Respiratory quotient; GH = Growth hormone; HPA = Hypothalamic-pituitary-adrenal; ACTH = Adrenocorticotropic hormone; LPL = Lipoprotein lipase; SM = Skeletal muscle.

Expanded from Bouchard, AJCN 1991 and Bouchard, *Nutr Rev* 1996 (3, 4).

The evidence in support of these correlates can be found in several chapters of the *Handbook of Obesity* edited by Bray, Bouchard, and James 1998 (12).

Three Scenarios Behind the Epidemic

The increase in the last decades of the prevalence of overweight and obesity can be theoretically explained by one or a combination of the following three scenarios. The first scenario posits that the increase results from the fact that a large proportion of the population is consuming more calories than individuals of past generations with no change in habitual daily energy expenditure. The second scenario suggests that the cause of the increase can be found in a decrease in daily energy expenditure with no change in caloric intake. Finally, the third scenario proposes that caloric intake per capita has actually declined compared to previous generations, but daily energy expenditure has, on the average, decreased even more (6).

Although it is likely that the first scenario is applicable to a segment of the adult population of overweight and obese individuals, we believe that scenarios two and three play an even greater role. The main reason for this conclusion can be found in the nutrition surveys repeated over a few decades in several countries of the developed world. Admittedly, such data are soft, but their high degree of concordance is quite striking. For instance, such data reported for Australia (1), Japan (24), the Netherlands (28), and the United Kingdom (25, 32) are all showing a stable caloric intake per capita over time or a decrease over the last 25 years or so. The only exception to this trend comes from the comparison of the NHANES I, II, and III daily caloric intake data in which an increase of about 300 kcal was reported between NHANES II and III—that is over a period of about 15 years in the United States (15, 16). However, these data must be viewed with caution as the dietary survey methodology and the survey food coding and nutrient composition databases changed between NHANES II and NHANES III. The improvement in methodology is thought to have resulted in more complete intake information in NHANES III (15).

In addition, it is useful to note that the average proportion of calories derived from fats has been diminishing steadily during the same period of time (15, 41). Despite this trend, however, it is important to recognize that a fraction of the population of all these nations still consumes high amounts of dietary fat. The latter are likely more at risk to gain weight until the fat mass becomes large enough to restore a better coupling between lipid intake and amount of lipids oxidized.

There are several prospective studies that have demonstrated the presence of a significant and inverse relationship between the level of habitual physical activity and weight gain over a number of years (20, 26, 35). It is not infrequent in these studies to show that the level of physical activity is a better predictor of weight gain than estimates of caloric or fat intake. One of the most impressive demonstrations of the strong role of a sedentary mode of life, compared to caloric or fat intake in the growing prevalence of overweight and obesity cases, comes from the analysis of survey data in the UK collected over a period of four decades (32). The analysis of this unique material shows convincingly that energy expenditure associated with habitual physical activity is a key determinant of the population increase in the prevalence of excess body weight. See Prentice and Jebb, chapter 12, this volume.

Total daily energy expenditure measured over 24 hours in a metabolic chamber was found to be inversely correlated with the subsequent rate of weight gain in Pima Indians ($r = -0.39$) (33). Estimates of total daily energy expenditure in free-living individuals assessed with the doubly labeled water technique are not affected by the limits imposed on physical activity in the confines of a metabolic chamber. Few prospective studies have been reported to date based on doubly labeled water estimates of physical activity levels for periods of about 7 to 10 days. However, Schulz and Schoeller (39) reported a compilation from individual data points from several studies. They found that nonbasal energy expenditure (calculated as total energy expenditure minus resting metabolic rate, in MJ per kg weight per day) was inversely correlated with percentage body fat in women ($r = -0.83$) and men ($r = -0.55$). Even though these observations are cross-sectional, they suggest a strong and robust relationship between habitual level of physical activity (nonresting energy expenditure) and body weight, and body fat gained over the years. In support of this view, a study performed in elderly women was able to show that differences in body mass between sedentary and active women were already detectable at age 25 (44).

From the above studies, one can hypothesize that the contribution of a diminished energy expenditure to the current overweight and obesity epidemic is determined by the decrease in the level of habitual physical activity associated with work and chores of daily living; and by the growing amount of time spent in a very sedentary mode, such as watching TV, working on the computer, playing video games, etc. It is not associated with decreases in resting metabolic rate or in dietary-induced thermogenesis. Indeed, there is absolutely no indication that there is a downward secular trend for these two components of daily energy expenditure.

Can Overweight and Obesity be Prevented?

The various treatment modalities of obesity are characterized by only modest success rates (with the exception of bariatric surgery), and the trend often reported is for weight loss to be followed by weight regain. Thus, it would seem judicious to consider the prevention of obesity as an important agenda item (30). If severe obesity is as incurable as it seems to be based on the experience of the last 30 to 40 years, then prevention of weight gain should become a priority (37).

The rationale supporting the view that a good fraction of obesity cases could have been prevented is based on the following considerations (4). First, the level of heritability of obesity or body-fat content is only moderate. Second, most intermediate phenotypes that can be defined as determinants of body-fat content are also characterized by low to moderate levels of heritability. Third, the prevalence of overweight and obesity has been steadily increasing over the last 50 years or so, and population studies in the Western countries seem to indicate that the prevalence is still rising. This increase has occurred over a period of time that is too short to be caused by changes in the frequency of obesity genes or susceptibility alleles. Fourth, the increase in the prevalence of obesity can therefore only be due to the fact that a greater number of children and adults are in chronic positive energy balance. Fifth, the large number of people in positive energy balance for sustained periods can only be explained by the three scenarios defined earlier in the chapter.

A more physically active lifestyle is likely to be the cornerstone of a prevention strategy centered on the concept of the promotion of a healthy weight (8). However, it is equally important to recognize that energy balance will be easier to achieve in the long term if the physically active lifestyle is associated with a low-fat diet (about 30% of calories). Energy balance, and particularly balance between lipid intake and lipid oxidation, is quite difficult and perhaps impossible to sustain when dietary fat intake is high (42).

Two other lines of evidence support the concept that overweight or obesity can be prevented in a yet-undetermined fraction of the affected. The first results from the comparison of the Arizona Pima Indians and Mexican Pima Indians living in Maycoba, a poor and remote area of Mexico (19, 34). Pima Indians living in Arizona exhibit one of the highest prevalence rates of obesity and type 2 diabetes in the world. In contrast, the 208 Mexican Pima Indians measured, among a total population of about 600 according to a recent census, had mean BMIs of 25.9 for women (mean age of 34.1 years) and 23.6 for men (mean age of 39.8 years) (19), values that are markedly lower than their Arizonian kin. These data are admittedly preliminary, but they strongly suggest that environment and lifestyle have a strong impact on body mass for height.

The second observation comes from the survey of monozygotic twins discordant for BMI. Rönnemaa and collaborators surveyed monozygotic twin pairs of the Finnish Twin Cohort registry (36). A total of 1,453 such pairs born between 1932 and 1957 responded to a mail questionnaire in 1990. Among them, 50 pairs were identified as being discordant for BMI (Rönnemaa, personal communication). The latter was defined as a BMI difference between the two identical brothers or sisters of at least 4 kg/m^2, with one twin having a BMI of at least 27. Hence, about 3 percent of this sample of identical twins were markedly discordant for an indicator of overweight or obesity. In a subgroup of 23 of these pairs who were extensively studied under controlled laboratory conditions, the mean body weight difference between the overweight or obese co-twin and the lean brother or sister reached 16 kg in men and 19 kg in women (36). Such data indicate that for the same genetic characteristics, it is possible to remain normal weight or become obese. There is no doubt that dietary and physical activity habits can have a major contribution to body weight regulation over and above those imposed by the genotype.

To prevent obesity, one should target young children, adolescents, and young adults. The urgency would seem to be with children and adolescents (31). Recent studies have indicated that low levels of physical activity in preschool children, estimated from doubly labeled water, were already indicative of a higher body-fat content (13). It seems as if the stage was set in these early postnatal years for the obesity phenotype to unfold more readily with time. The risk is likely to be even more dramatic for the less-active infants and older children who happen to have inherited genes making them more susceptible to be in positive energy balance and to gain body fat.

The tools necessary to reverse this unhealthy trend are remarkably simple in appearance, as they center on the promotion of eating regular and healthy meals, avoiding snacking, drinking water instead of energy-containing beverages, keeping dietary fat at about 30% of calories, cutting down on TV viewing time, walking more, participating more in sports and other energy-consuming leisure activities, and other similar measures (4). However, it will be a daunting task to change the course of nations that have progressively become quite comfortable with an effortless lifestyle in which individual consumption is almost unlimited. It will require massive resources and an unprecedented level of concentration among all public health agencies and private organizations to begin reversing the trends that have emerged over the last decades (48).

A Gloomy Future

Kinesiologists and other exercise advocates do not see physical activity as a burden, as something that people engage in because they have no choice. This view of physical activity in the lives of human beings contrasts sharply with the struggle that went on for millennia to reduce the amount of muscular work and physical activity required in daily life (5). The war on muscular work has been a remarkable success. Thus, the amount of energy expended by individuals to ensure sustained food supply, decent housing under a variety of climatic conditions, safe and rapid transportation, personal and collective security, and diversified and abundant leisure activities has decreased.

Throughout this journey, the struggle to free human beings from muscular work and physical exertion was a constant feature. It made very impressive gains around the period of the industrial revolution and, even more, in the present century with the technological progress achieved in industrialized countries (22). Lately, however, the alarm bell was activated to propose that the reduction in the amount of physical work may have gone too far. The benefits of a physically active lifestyle have been compared with those of an inactive mode of life and while all the evidence is not in, it seems that human beings are better off when they maintain a physically active lifestyle. The health problems associated with a sedentary mode of life were collectively referred to as "hypokinetic diseases" almost 40 years ago (27). This is one of the most striking paradoxes emanating from the evolutionary and historical journey of *Homo sapiens*.

Nowhere are the consequences of the present level of low energy expenditures more obvious than on energy balance and body weight regulation. Because the health consequences of excess body fat do not become immediately manifest, the current epidemic of obesity in children, adolescents, and young adults will translate later in unprecedented numbers of cases of type 2 diabetes, hypertension, cardiovascular disease, gallbladder disease, postmenopausal breast cancers, osteoarthritis at the knees, back pain, and physical and mental disabilities. Americans alone are currently spending about $100 billion annually as a result of the direct and indirect costs of obesity (47). Foreyt and Goodrick (17) have argued that the increase in the prevalence of overweight and obesity appears to be unstoppable as a side effect of modernization. In the United States, they estimated that, at the current rate of increase, 100% of adults will be overweight or obese by the year 2230. This would no doubt constitute the "ultimate triumph of obesity" (17). The health consequences of a situation in which obesity is ever becoming more prevalent will be catastrophic. And the time bomb of obesity is already ticking (10).

Even though individuals bear responsibility for maintaining healthy weights, national surveys in developed countries and the compendium of data around the world by the International Obesity Task Force indicate that programs with a focus on individuals are not enough. Of course, one must continue and expand the education efforts aimed at teaching people about healthy diets and physically active lifestyles. Promoting healthy weights through controlled behaviors and increased awareness should remain on the agenda (25, 45). However, it is obvious that this approach will not be sufficient, as it has not succeeded in containing the present obesity epidemic (48).

What is needed is a series of major policies aimed at transforming our environment and the way we live. Indeed, nothing short of a paradigm shift has any chance of success in the efforts to curtail the increase in the number of people who are chronically in positive energy balance. City planning, building codes, mass transit systems, car use, safe footpaths and cycling paths, pedestrian-only city centers, school schedules and programs, and the media are among the areas that will require transformations if we are to attenuate the impact of the current "obesogenic" environment (14).

The challenge is enormous. Evolution has endowed *Homo sapiens* with complex regulatory systems of appetite and satiety as well as with physiological and metabolic characteristics determining basal metabolic rates and food- or cold-induced thermogenesis. The recent past in affluent societies reveals that these biological systems cannot cope well in an environment in which palatable foods are abundant and energy expenditure of activity is low. In particular, the lesson from the last decades is that it seems to be extremely difficult and perhaps impossible for a large fraction of sedentary individuals to regulate food and caloric intake to be in balance at low levels of daily energy expenditure. The energy expenditure from physical activity is thus too low for most people to be able to eat normally without having to be on caloric restriction diets from time to time or having to be constantly restraining their food intake. It has been estimated that the current deficit of energy expenditure from physical activity compared to that of the recent past ranges on average from about 300 to 800 kcal/day (25, 38). If this range of estimates is close to the true values, it implies that adults would have to add one to three hours of brisk walking every day to their current daily regimen to be in energy balance at a normal body weight level. This is a major public health challenge indeed!

References

1. Abraham, B., E.T. d'Espaignet, and C. Stevenson. 1995. *Australian health trends 1995.* Australian Institute of Health and Welfare. Canberra: Australian Government Publishing Service.
2. Astrup, A., S. Toubro, A. Raben, and A.R. Skov. 1997. The role of low-fat diets and fat substitutes in body weight management: What have we learned from clinical studies? *Journal of the American Dietetic Association* 97: S82-S87.
3. Bouchard, C. 1991. Current understanding of the etiology of obesity: genetic and nongenetic factors. *American Journal of Clinical Nutrition* 53: 1561S-1565S.
4. Bouchard, C. 1996. Can obesity be prevented? *Nutrition Reviews* 54: S125-S130.
5. Bouchard, C. 1997. Biological aspects of the active living concept. In *Physical Activity in Human Experience—Interdisciplinary Perspectives,* eds. J.E. Curtis and S.J. Russell, 11-58. Champaign, IL: Human Kinetics.
6. Bouchard, C. 1998. L'obésité est-elle une maladie génétique? *Médecine Thérapeutique* 4: 283-289.
7. Bouchard, C., A. Tremblay, J.P. Després, O. Dériaz, and F.T. Dionne. 1992. The genetics of body energy content and energy balance: an overview. In *The science of food regulation: food intake, taste, nutrient partitioning and energy expenditure,* eds. G.A. Bray and D.H. Ryan, 3-21. Baton Rouge, LA: Louisiana State University Press.
8. Bouchard, C., J.P. Després, and A. Tremblay. 1993. Exercise and obesity. *Obesity Research* 1: 133-147.
9. Bouchard, C., and G.A. Bray. 1996. Introduction. In *Regulation of body weight: Biological and behavioral mechanisms.* Report of the Dahlem Workshop. eds. C. Bouchard and G.A. Bray, 1-13. Chichester, England: John Wiley & Sons.
10. Bray, G.A. 1998. Obesity: a time bomb to be defused. *The Lancet* 352: 160-161.
11. Bray, G.A., and B.M. Popkin. 1998. Dietary fat intake does affect obesity! *American Journal of Clinical Nutrition* 68: 1157-1173.
12. Bray, G.S., C. Bouchard, and W.P.T. James, eds. 1998. *Handbook of Obesity.* New York: Dekker Inc.
13. Davies, P.S., J. Gregory, and A. White. 1995. Physical activity and body fatness in pre-school children. *International Journal of Obesity* 19: 6-10.

14. Egger, G., and B. Swinburn. 1997. An "ecological" approach to the obesity pandemic. *British Medical Journal* 315: 477-480.
15. Ernst, N.D., E. Obarzanek, M.B. Clark, R.R. Briefel, C.D. Brown, and K. Donato. 1997. Cardiovascular health risks related to overweight. *Journal of the American Dietetic Association* 97: S47-S51.
16. Ernst, N.D., C.T. Sempos, R.R. Briefel, and M.B. Clark. 1997. Consistency between US dietary fat intake and serum total cholesterol concentrations: the National Health and Nutrition Examination Surveys. *American Journal of Clinical Nutrition* 66: 965S-972S.
17. Foreyt, J., and K. Goodrick. 1995. The ultimate triumph of obesity. *The Lancet* 346: 134-135.
18. Foreyt, J.P., and W.S. Carlos Poston II. 1997. Diet, genetics, and obesity. *Food Technology* 51: 70-73.
19. Fox, C.S., J. Esparza, M. Nicolson, P.H. Bennett, L.O. Schulz, M.E. Valencia, and E. Ravussin. 1998. Is a low leptin concentration, a low resting metabolic rate, or both the expression of the "thrifty genotype"? Results from Mexican Pima Indians. *American Journal of Clinical Nutrition* 68: 1053-1057.
20. French, S.A., R.W. Jeffery, J.L. Forster, P.G. McGovern, S.H. Kelder, and J.E. Baxter. 1994. Predictors of weight change over two years among a population of working adults: the healthy worker project. *International Journal of Obesity* 18: 145-154.
21. Grundy, S.M. 1998. Multifactorial causation of obesity: implications for prevention. *American Journal of Clinical Nutrition* 67: 563S-572S.
22. Haskell, W.L. 1996. Physical activity, sport, and health: Toward the next century. *Research Quarterly for Exercise and Sport* 67: S37-S47.
23. Heymsfield, S.B., P.C. Darby, L.S. Muhlheim, D. Gallagher, C. Wolper, and D.B. Allison. 1995. The calorie: myth, measurement, and reality. *American Journal of Clinical Nutrition* 62: 1034S-1041S.
24. Iwane, H. Physical activity and health—Japanese experiences. In: Health Promotion and Physical Activity. 1996. *Proceedings of the Joint Meeting WHO & FIMS, Cologne*, April 1994, pp. 79-95.
25. James, W.P.T. 1995. A public health approach to the problem of obesity. *International Journal of Obesity* 19: S37-S45.
26. Klesges, R.C., L.M. Klesges, C.K. Haddock, and L.H. Eck. 1992. A longitudinal analysis of the impact of dietary intake and physical activity on weight change in adults. *American Journal of Clinical Nutrition* 55: 818-822.
27. Krauss, H., and W. Raab. 1961. *Hypokinetic disease: diseases produced by lack of exercise.* Springfield, IL: Charles C. Thomas.
28. Kromhout, D., C. de Lezenne Coulander, G.L. Obermann-de Boer, M. van Kampen-Donker, E. Goddijn, and B.P.M. Bloembert. 1990. Changes in food and nutrient intake in middle-aged men from 1960 to 1985 (the Zutphen Study). *American Journal of Clinical Nutrition* 51: 123-129.

29. National Institutes of Health and National Heart, Lung, and Blood Institute. 1998. Clinical guidelines on the identification, evaluation, and treatment of overweight and obesity in adults. The Evidence Report. *Obesity Research* 6, Suppl 2.

30. National Task Force on Prevention and Treatment of Obesity. 1994. Towards prevention of obesity: research directions. *Obesity Research* 2: 571-584.

31. Poskitt, E.M.E. 1994. The prevention of childhood obesity. In *Obesity in Europe 93: Proceedings of the 5th European Congress on Obesity.* eds. H. Ditschuneit, F.A. Gries, H. Hauner, V. Schusdziarra, and J.G. Wechsler, 141-145. Paris: John Libbey.

32. Prentice, A.M., and S.A. Jebb. 1995. Obesity in Britain: gluttony or sloth? *British Medical Journal* 311: 437-439.

33. Ravussin, E., S. Lillioja, W.C. Knowler, L. Christin, D. Freymond, W.G. Abbott, V. Boyce, B.V. Howard, and C. Bogardus. 1988. Reduced rate of energy expenditure as a risk factor for body weight gain. *New England Journal of Medicine* 318: 467-472.

34. Ravussin, E., M.E. Valencia, J. Esparza, P.H. Bennett, and L.O. Schulz. 1994. Effects of a traditional lifestyle on obesity in Pima Indians. *Diabetes Care* 17: 1067-1074.

35. Rissanen, A.M., M. Heliovaara, P. Knekt, A. Reunanen, and A. Aromaa. 1991. Determinants of weight gain and overweight in adult Finns. *European Journal of Clinical Nutrition* 45: 419-430.

36. Rönnemaa, T., M. Koskenvuo, J. Marniemi, T. Koivunen, A. Sajantila, A. Rissanen, M. Kaitsaari, C. Bouchard, and J. Kaprio. 1997. Glucose metabolism in identical twins discordant for obesity. The critical role of visceral fat. *Journal of Clinical Endocrinology and Metabolism* 82: 383-387.

37. Rössner, S. 1994. Is obesity incurable? In *Obesity in Europe 93: Proceedings of the 5th European Congress on Obesity.* eds. H. Ditschuneit, F.A. Gries, H. Hauner, V. Schusdziarra, and J.G. Wechsler, 203-208. Paris: John Libbey.

38. Schoeller, D.A., K. Shay, and R.F. Kushner. 1997. How much physical activity is needed to minimize weight gain in previously obese women? *American Journal of Clinical Nutrition* 66: 551-556.

39. Schulz, L.O., and D.A. Schoeller. 1994. A compilation of total daily energy expenditures and body weights in healthy adults. *American Journal of Clinical Nutrition* 60: 676-681.

40. Seidell, J.C. 1998. Dietary fat and obesity: an epidemiologic perspective. *American Journal of Clinical Nutrition* 67: 546S-550S.

41. Stephen, A.M., and G.M. Sieber. 1994. Trends in individual fat consumption in the UK 1900-1985. *British Journal of Nutrition* 71: 775-788.

42. Stubbs, R.J., P. Ritz, W.A. Coward, and A.M. Prentice. 1995. Covert manipulation of the ratio of dietary fat to carbohydrate and energy density: effect on food intake and energy balance in free-living men eating ad libitum. *American Journal of Clinical Nutrition* 62: 330-337.

43. Troiano, R.P., K.M. Flegal, R.J. Kuczmarski, S.M. Campbell, and C.L. Johnson. 1995. Overweight prevalence and trends for children and adolescents. The National Health and Nutrition Examination Surveys, 1963 to 1991. *Archives of Pediatrics and Adolescent Medicine* 149: 1085-1091.
44. Voorrips, L.E., J.H.H. Meijers, P. Sol, J.C. Seidell, and W.A. van Staveren. 1992. History of body weight and physical activity of elderly women differing in current physical activity. *International Journal of Obesity* 16: 199-205.
45. Weinsier, R.L., G.R. Hunter, A.F. Heini, M.I. Goran, and S.M. Sell. 1998. The etiology of obesity: Relative contribution of metabolic factors, diet, and physical activity. *The American Journal of Medicine* 105: 145-150.
46. Willett, W.C. 1998. Is dietary fat a major determinant of body fat? *American Journal of Clinical Nutrition* 67: 556S-562S.
47. Wolf, A.M., and G.A. Colditz. 1998. Current estimates of the economic cost of obesity in the United States. *Obesity Research* 6: 97-106.
48. World Health Organization. 1998. Obesity—preventing and managing the global epidemic. Report of a WHO consultation on obesity. Geneva: World Health Organization.

CHAPTER 2

The Current Epidemic of Obesity

Jacob C. Seidell, PhD

Department of Chronic Disease and Environmental Epidemiology, National Institute of Public Health and Environmental Protection (RIVM), Bilthoven, The Netherlands

Obesity is now considered to be one of the major health threats in the developed world. The scope of the problem is the result of the prevalence of a combination of health hazards. This chapter will review the classification of obesity, which is primarily based on data from studies of white men living in North America and Europe.

Classification of Obesity and Fat Distribution

The World Health Organization (WHO) has provided the most recent classification of adults based on weight relative to height as a gauge of obesity (table 2.1) (27).

Table 2.1   World Health Organization Classification of Adults by Body Mass Index

Classification	BMI (kg/m^2)	Associated health risks
Underweight	< 18.5	Low (but risk of other clinical problems increased)
Normal range	18.5-24.9	Average
Overweight	25.0 or higher	
Pre-obese	25.0-29.9	Increased
Obese class I	30.0-34.9	Moderately increased
Obese class II	35.0-39.9	Severely increased
Obese class III	40 or higher	Very severely increased

In many community studies in affluent societies this classification scheme has been simplified, and boundaries of 25 and 30 kg/m² are used for descriptive purposes. Both a very low BMI (< 18.5 kg/m²) and a very high BMI (40 kg/m² or higher) usually occur in only 1-2% or less of the population. The subdivision of obese classes that is provided in the World Health Organization classification (table 2.1) is particularly relevant for the clinical study of obesity and in weight management.

Much research during the last decade has suggested that abdominal obesity needs to be considered in order to accurately classify overweight subjects with respect to their health risks. Traditionally, abdominal obesity has been indicated by a relatively high waist-to-hip circumference ratio. It has become increasingly clear that the complex classification of body mass index and waist-hip ratio is not a powerful tool in health promotion and that it should be replaced by a classification based on waist circumference alone (11, 5). A tentative classification scheme is given in table 2.2.

Table 2.2   Sex-Specific Cutoff Points for Waist Circumference

	Level 1 ("alerting zone")	Prevalence	Level 2 ("action level")	Prevalence
Men	≥ 94 cm (~37 inches)	24.1%	≥ 102 cm (~40 inches)	18.0%
Women	≥ 80 cm (~32 inches)	24.4%	≥ 88 cm (~35 inches)	23.9%

Level 1 was initially based on replacing the classification of overweight (BMI ≥ 25 kg/m²) in combination with high waist-hip ratio (WHR ≥ 0.95 in men and ≥ 0.80 in women). Level 2 was based on classification of obesity (BMI ≥ 30 kg/m²)˙in combination with high waist-hip ratio (5).

In June 1998 the National Institutes of Health (National Heart, Lung, and Blood Institute) adopted the BMI classification and combined this with waist cutoff points (15). In this particular classification the combination of overweight (BMI between 25 and 30 kg/m²) or moderate obesity (BMI between 30 and 35 kg/m²) and a large waist circumference (≥ 102 cm in men or ≥ 88 cm in women) is proposed to carry additional health risk.

Global Prevalence of Obesity and Time Trends

Many reviews have shown that obesity (defined here as a body mass index of 30 kg/m² or higher) is a prevalent condition in most countries with established market economies (20). However, there is a wide variation in the prevalence of obesity within and among these countries. Some countries display at least a twofold difference in the prevalence of obesity from one area to another. For example, in Toulouse, France, 9% of men are obese and 11% of women are obese; whereas in Strasbourg, 22% of men are obese and 23% of women are obese. Usually, obesity is more frequent among those with a relatively low socioeconomic status. Also, the prevalence of obesity increases with age until about 60-70 years of age, after which it declines (21). In most of these established market economies, the prevalence of obesity is also increasing with time (21). Figure 2.1 illustrates this trend using data from the United States, United Kingdom, The Netherlands, Brazil, and Japan. The magnitude of the increase in obesity varies among countries, and in many affluent countries the increase seems to be stronger in men than in women.

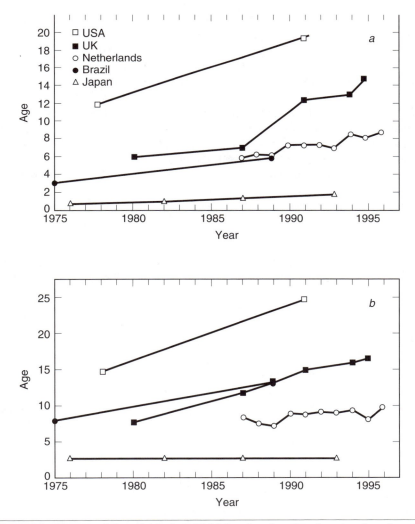

Figure 2.1   Percent of obese (a) males and (b) females plotted against time for selected countries (adapted from reference 1). Obesity is defined as a body mass index of 30 kg/m² or greater. The increase in the prevalence of obesity is clearly illustrated.

In other parts of the world, obesity is also common. Regions with a relatively high incidence of obesity are the formerly socialist economies of Europe, Latin America, the Caribbean, and the Middle Eastern Crescent (including north African countries) (22). Some recent data (Sans et al., unpublished) suggest that the very high prevalence of obesity in some former socialist economies of Europe (as high as 45% in women living in parts of Russia and Lithuania) may be declining. But other reports show that obesity is still on the increase in Latin America, the Caribbean, and the Middle Eastern Crescent.

Obesity is uncommon in sub-Saharan Africa, China, and India. However, even in these regions the prevalence of obesity seems to be increasing, particularly among the affluent parts of the population in the large cities (22). In these countries the paradoxical states of increasing undernutrition and overnutrition are present. This is related to growing inequities in income and access to food in these regions.

Table 2.3 attempts to quantify the approximate number of obese adults in the world. A lack of reliable data, large differences among countries within a region, and secular trends, such as the rapid increase in the prevalence of obesity, make this estimate uncertain. Summation of the midpoints of these estimates yields a total of about 250 million obese adults worldwide. This equals approximately 7% of the world population of adults and appears to be a reasonable estimate. However, in most countries the prevalence of overweight (BMI between 25 and 30 kg/m^2) is about two to three times as great as the prevalence of obesity, which

Table 2.3   Estimated World Prevalence of Obesity

Region*	Population aged 15+ in millions	Prevalence of obesity percentage	Approximate estimate of number of obese subjects in millions (midpoint)	
Established market economies	640	15-20%	96-128	(112)
Former socialist economies	330	20-25%	66-83	(75)
India	535	0.5-1.0%	3-7	(5)
China	825	0.5-1.0%	4-8	(6)
Other Asia and islands	430	1-3%	4-12	(8)
Sub-Saharan Africa	276	0.5-1.0%	1-3	(2)
Latin America and Caribbean	280	5-10%	14-28	(21)
Middle Eastern Crescent	300	5-10%	15-30	(22)
World	3616			(251)

*Population size and regions taken from *The Global Burden of Disease* by C.J.L. Murray and A.D. Lopez, 1996 (WHO/World Bank).

implies that as many as one billion people are either overweight or obese. If the trends shown in figure 2.1 continue, if the projected increase in the world's population is accurate, and if the developments of economic transition develop as currently predicted, we may see a dramatic increase in the prevalence of obesity in the coming decades.

Prevalence of a Large Waist Circumference

The data of the WHO MONICA obesity study (second survey carried out between 1987 and 1992) have recently been analyzed with respect to waist cutoff points (14) (see table 2.4). From this analysis it is clear that the prevalence of obesity varies

Table 2.4   Prevalence of a Large Waist Circumference and of Obesity in the WHO MONICA Study

Population (countries by alphabetical order)	Men Large waist %	Obesity %	Women Large waist %	Obesity %
Australia (Newcastle)	24	20	31	19
Australia (Perth)	15	14	17	14
China (Beijing)	4	4	21	8
Czech Republic (Rural)	32	25	48	30
Denmark (Glostrup)	18	11	17	8
Finland (Kuopio Province)	16	16	23	20
Finland (N. Karelia)	19	19	23	21
Finland (Turku/Loimaa)	18	16	21	19
France (Toulouse)	19	12	–	–
Germany (Augsburg Rural)	21	18	26	21
Germany (Augsburg Urban)	19	15	25	15
Germany (Halle County)	29	18	40	26
Italy (Area Brianza)	11	12	23	17
Italy (Friuli)	15	15	26	15
Spain (Catalonia)	23	17	33	22
Sweden (Gothenburg)	12	9	14	10
Sweden (Northern Sweden)	13	11	18	11
United Kingdom (Glasgow)	22	14	27	19
Yugoslavia (Novi Sad)	17	15	32	26

Adapted from Molerius et al. (14)

greatly from country to country. The prevalence of a large waist circumference ($\geq$102 cm in men and $\geq$ 88 cm in women) and of obesity (BMI $\geq$ 30 kg/m^2) are shown in table 2.4. In general, the prevalence of a large waist is higher than the prevalence of obesity, because the large waist circumference group also includes overweight subjects with abdominal obesity.

Obesity in Children and Adolescents

Comparison of data concerning obesity in children and adolescents around the world is difficult because of the lack of standardization of the classification of obesity and interpretation of indicators of overweight and obesity in these age groups. Usually, local or national percentile distributions for weight-for-age, weight-for-height, or body-mass-index-for-age are being used. These may not only differ among regions and nations, but are also subject to change over time. In addition, different percentile cutoff points are being used for the definition of overweight or obesity (e.g., 85th, 90th, 95th, and 97th percentiles are all being used in different countries).

Another difficulty with these criteria is that they not only are impossible to compare across populations, but when they are applied with advancing age, they do not correspond to the criteria for classification of overweight based on BMI for adults.

With respect to the interpretation of criteria for overweight in different age groups, it is also important to know whether or not these criteria are predictive of later obesity. It is now generally accepted that body weight before the age of six years has very limited predictive power for the occurrence of overweight or obesity in adulthood, regardless of the family history of obesity (26). Data at this age may, however, be predictive in another way, as suggested by Rolland-Cachera et al. (19). The BMI-for-age from infancy until adulthood has the form of a J. The nadir of this curve usually occurs in the age range of 5 to 7 years old. It has been suggested that when this nadir is at a relatively early age ("early-adiposity-rebound"), the chances of adult obesity are higher than when there is a relatively late adiposity rebound (19, 3). In addition, time trends in overweight may be sensitive indicators of secular changes in energy balance.

The World Health Organization has now tentatively recommended the use of BMI-for-age as an indicator of overweight or obesity (28). In The Netherlands, the French reference curves (> 97th percentile of BMI-for-age) were used to show that there was a slight increase in obesity during the early 1990s (23). This trend was also observed in other countries, particularly the United States (25, 13, 16) and the United Kingdom (7). Similar data were reported from Denmark (24) and Sweden (8). Military conscripts are particularly useful in giving an unbiased view of long-term national trends. Figure 2.2 shows these time trends in overweight and obesity among young Danish and Swedish men, illustrating a persistent increase in both countries.

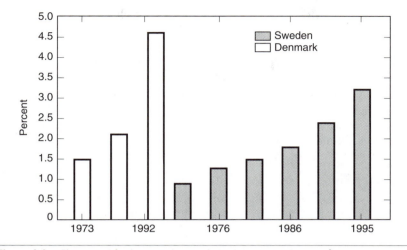

Figure 2.2   Time trend in the prevalence of obesity (BMI ≥ 30 kg/m^2) among Danish and Swedish male conscripts (adapted from references 8, 24).

Currently, a subgroup of the WHO International Obesity Task Force (IOTF) is attempting to develop international BMI-by-age standards. Ideally, these will be able to be used universally; preferably, they are based on longitudinal tracking data of BMI for children and adolescents and will match the adult classification of BMI by age 20. Although BMI may not be a very precise indicator of body fatness for an individual, many studies support the use of BMI as an indicator of fatness for a larger population (17, 2).

Interpretation of the increases in childhood and adolescent obesity rates is difficult. Accurate explanations require unbiased and precise estimates of energy intake and energy expenditure, and these are often unavailable. This is further complicated by the likelihood that reported energy intake in children is considerably underestimated (1). Small secular changes in obesity may be interpreted as the result of minute shifts in energy balance, which are all well within the margin of error of all available methods. However, the United States is among those countries in which, despite a dramatic recent increase in the prevalence of obesity, there is no good evidence for any appreciable change in energy intake over the last decades; in fact, there may even have been some improvement (10). Some crude evidence suggests that in particular the reduction in energy expenditure in children and adults is the most important determinant of overweight, and it is not difficult to see that major changes in lifestyle have occurred in youngsters over the last few decades (9). Several studies report low physical activity in obese children compared to their lean counterparts (6, 12). This may be the cause or the consequence of their obesity. Prospective studies, however, have also linked sedentary behavior such as television viewing to the development of obesity (4, 18).

Implications

The increase in the prevalence of obesity among children, adolescents, and adults in many countries around the world is alarming. Prevention of obesity should be among the high priorities in public health. This prevention should certainly include encouraging healthy lifestyles in all age groups including children and adolescents. This cannot be achieved by efforts aimed at the individual level. Communities, governments, the media, and the food industry need to work together to modify the environment so that it is less conducive to weight gain (27).

References

1. Champagne, C.M., N.B. Baker, J.P. Delany, D.W. Harsha, G.A. Bray. 1998. Assessment of energy intake underreporting by doubly labeled water and observations on reported nutrient intakes in children. *Journal of the American Dietetic Association* 98:426-33.
2. Deurenberg, P., J.A. Weststrate, J.C. Seidell. 1991. Body mass index as a measure of body fatness: age- and sex-specific prediction formulas. *British Journal of Nutrition* 65:105-14.
3. Dietz, W.H. Critical periods in childhood for the development of obesity. 1994. *American Journal of Clinical Nutrition* 59:955-9.
4. Gortmaker, S., A. Must, A. Sobel, K. Peterson, G.A. Colditz, W.H. Dietz. 1996. Television viewing as a cause of increasing obesity among children in the United States. *Archives of Pediatric Adolescent Medicine* 150:356-62.
5. Han, T.S., E.M. Van Leer, J.C. Seidell, M.E.J. Lean. 1995. Waist circumference action levels in the identification of cardiovascular risk factors: prevalence study in a random sample. *British Medical Journal.* 311:1401-5.
6. Harrell, J.S., S.A. Gansky, C.B. Bradley, R.G. McMurray. 1997. Leisure time activities of elementary school children. *Nutrition Research* 46:246-53.
7. Hughes, J.M., L. Li, S. Chinn, R.J. Rona. 1994. Trends in growth in England and Scotland, 1972 to 1994. *Archives of Diseases in Children* 76:182-9.
8. Hulthen, L., D. Höglund, L. Hallberg. 1998. Obesity and physical performance in Swedish young men—a 26-year trend in 1.1 million conscripts. *Kuopio University Publications D. Medical Series 149* 1998:16.
9. Jebb, S.A. Aetiology of obesity. 1997. *British Medical Bulletin* 53:264-85.
10. Kennedy, E., R. Powell. Changing eating patterns of American children: a view from 1996. 1997. *Journal of the American College of Nutrition* 16:524-9.
11. Lean, M.E.J., T.S. Han, J.C. Seidell. 1998. Impairment of health and quality of life in men and women with a large waist. *Lancet* 351:853-6.

12. Maffeis, C., M. Zaffanello, Y. Schutz. 1997. Relationship between physical inactivity and adiposity in prepubertal boys. *Journal of Pediatrics* 131:288-92.

13. Mei, Z., K.S. Scanlon, L.M. Grummer-Strawn, D.S. Freedman, R. Yip, F.L. Trowbridge. 1997. Increasing prevalence of overweight among US low-income preschool children: the CDC Pediatric Nutrition Surveillance 1983 to 1995. *Pediatrics* 101(1). Available: **http://www.pediatrics.org/cgi/content/full/101/1/e12**.

14. Molarius, A., J.C. Seidell, S. Sans, J. Tuomilehto, K. Kuulasma. 1999. Varying sensitivity of waist action levels to identify subjects with overweight or obesity in 19 populations of the WHO MONICA project. *Journal of Clinical Epidemiology* 52:1213-24.

15. NIH. 1998. *Clinical guidelines on the identification, evaluation, and treatment of overweight and obesity in adults. The Evidence Report*. NIH, NHLBI, June 1998.

16. Ogden, C.L., R.P. Troiano, R.R. Briefel, R.J. Kuczmarski, K.M. Flegal, C.L. Johnson. 1997. Prevalence of overweight among preschool children in the United States, 1971 through 1994. *Pediatrics* 99(4). Available: **http://www.pediatrics.org/cgi/content/full/99/4/e1**.

17. Pietrobelli, A., M.S. Faith, D.B. Allison, D. Gallagher, G. Ciumello, S.B. Heymsfield. 1998. Body mass index as a measure of adiposity among children and adolescents: a validation study. *Journal of Pediatrics* 132:204-10.

18. Robinson, T.N. 1998. Does television cause childhood obesity? *JAMA* 279:959-60.

19. Rolland-Cachera, M.F., M. Deheeger, M. Guilloud-Bataille, P. Avons, M. Sempe. 1987. Tracking the development of adiposity from one month of age to adulthood. *Annals of Human Biology* 14:219-29.

20. Seidell, J.C. 1997. Time trends in obesity: an epidemiological perspective. *Hormone and Metabolic Research* 29:155-8.

21. Seidell, J.C., K.M. Flegal. 1997. Assessing obesity: classification and epidemiology. *British Medical Bulletin* 53:238-52.

22. Seidell, J.C., A. Rissanen. 1997. World-wide prevalence of obesity and time-trends. In *Handbook of obesity*, eds. G.A. Bray, C. Bouchard, W.P.T. James, 79-91. New York: Marcel Dekker, Inc.

23. Seidell, J.C. 1999. Obesity, a growing problem. *Acta Paediatrica* (suppl. 428):46-51.

24. Sörensen, H.T., S. Sabroe, M. Gillman, K.J. Rothman, K.M. Madsen, P. Fischer, T.I.A. Sörensen. 1997. Continued increase in prevalence of obesity in Danish young men. *International Journal of Obesity* 21:712-4.

25. Troiano, R.P., K.M. Flegal, R.J. Kuczmarski, S.M. Campbell, C.L. Johnson. 1995. Overweight prevalence and trends for children and adolescents. *Archives Pediatric Adolescent Medicine* 149:1085-91.

26. Whitaker, R., J. Wright, M. Pepe, et al. 1997. Predicting adult obesity from childhood and parent obesity. *New England Journal of Medicine* 337:869-73.
27. WHO. 1998. Obesity: preventing and managing the global epidemic. WHO, Geneva WHO/NUT/NCD/98.1.
28. WHO. 1995. Physical status: the use and interpretation of anthropometry. *WHO Technical Report Series #854.* WHO, Geneva.

CHAPTER 3

Overweight, Mortality, and Morbidity

George A. Bray, MD
Pennington Biomedical Research Center, Louisiana State University, Baton Rouge, Louisiana, U.S.A.

Introduction

Overweight patients are at risk for developing a number of medical, social, and psychological disabilities. These are shown in figure 3.1.

The medical profession has known about the effects of excess weight on morbidity and mortality for more than 2,000 years. Hippocrates recognized that "sudden death is more common in those who are naturally fat than in the lean." A large volume of literature has been written on this subject over the past 50 years, and the reader is referred to additional reading for more detail (1, 7, 12, 13, 21, 24, 25).

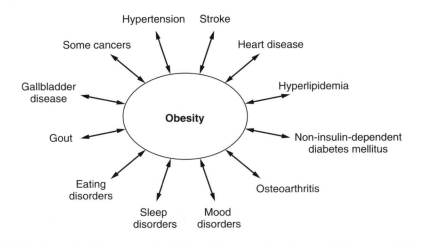

Figure 3.1   Conditions associated with obesity.

Effects of Overweight, Body-Fat Distribution, Weight Gain, and Sedentary Lifestyle on Mortality

Changes in weight, body-fat distribution, weight gain, and level of physical activity can all influence quality of life. These are detailed in the following paragraphs.

Increased Body Weight

Increased overweight is associated with increased risk of death (figure 3.2). The life insurance industry has played a key role in bringing public awareness to the relationship between increased weight and higher death rates. Four large studies, including the Life Insurance Build Study of 1979 (26), the American Cancer Society

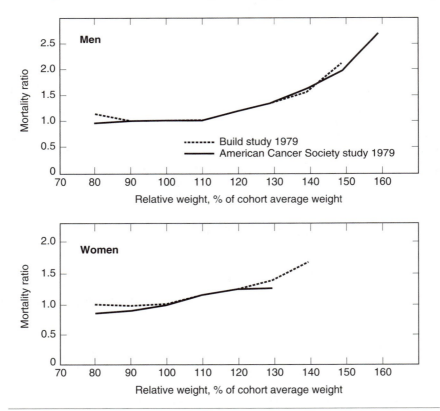

Figure 3.2   Relationship of increasing weight to excess mortality in two large studies (17). The mortality ratio is the ratio of deaths in each subgroup to the total deaths in the population. Relative weight is the weight of each group relative to the standards of the Metropolitan Life Insurance Company Table of 1959, which provides the age group (cohort) for comparison.

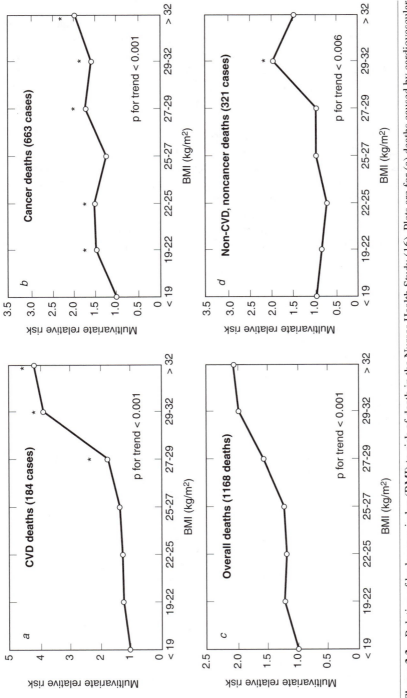

Figure 3.3 Relation of body mass index (BMI) to risk of death in the Nurses Health Study (16). Plots are for (a) deaths caused by cardiovascular disease (CVD), (b) deaths caused by cancer, (c) total deaths, and (d) noncancer and non-CVD deaths. The multivariate relative risk refers to the relative risk of death compared to the lowest BMI determined from multivariate statistical analysis. Asterisks indicate data for which p < 0.05.

33

study (15), the Norwegian population study (30), and the Nurses Health Study (16), as well as many smaller epidemiological studies, have supported these findings. The similarity of the life insurance data and the American Cancer Society study is shown in figure 3.2 (17). As noted in a critical review of the studies, there was no control for the effects of smoking or for early deaths. When smokers were eliminated in a 12-year follow-up of nurses, there was only a graded increase in mortality with increasing body mass index (BMI) that became significant at a BMI of 27 to 29.9 kg/m^2 (figure 3.3d) (16).

A very graphic demonstration of the impact of excess weight is seen in a 35-year follow-up of inductees into the Danish army. The overweight men with a BMI > 31 kg/m^2 had a shorter life span than men with a normal-range BMI (figure 3.4) (27). As the years passed, the two curves continued to separate. The obvious impact of excess weight on increased mortality has not always been found in epidemiological studies. Sjostrom (25) analyzed 40, some of which showed and some of which did not show an effect of overweight on mortality. Sjostrom found that large studies (more than 10,000 participants) and smaller studies with a long-term follow-up of

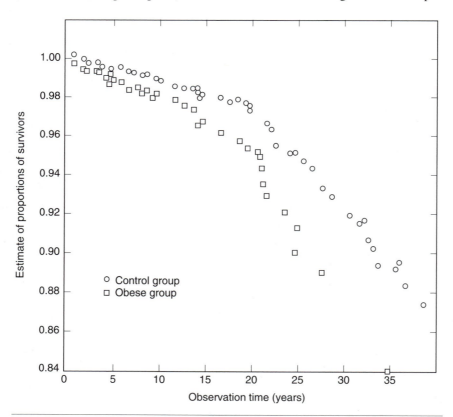

Figure 3.4   Effect of overweight on survival following examination for induction in the Danish army (27). The proportion of survivors at increasing intervals from the time of examination for the military is plotted on the x-axis.

more than 15 years were almost unanimous in finding a relationship between BMI and mortality (25).

Figure 3.5 shows the relationship between BMI and two other "risk factors"— blood pressure and cholesterol (4). Areas of low risk, moderate risk, and high risk are demarcated by vertical dashed lines. Long-term elevated levels of high blood pressure, high cholesterol, and overweight are risk factors for disease. For example, renal failure, heart failure, and stroke can result from prolonged elevations in blood pressure; atherosclerosis, with coronary and cerebrovascular occlusion, results from elevated cholesterol; and diabetes, gallstone disease, heart failure, and some kinds of cancer have the strongest association with overweight. It is the chronic exposure to the elevated risk factor that produces the disease states.

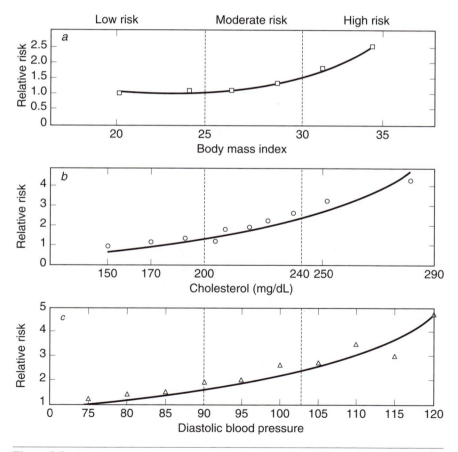

Figure 3.5   Relative risk of death, compared to that for the lowest level of (a) body mass index, (b) cholesterol, and (c) blood pressure. All three risk factors show a curvilinear increase as the value rises. Vertical dashed lines separate areas of low, moderate, and high risk.

Several other studies are also instructive. In the Nurses Health Study, deaths from cardiovascular disease and cancer were the major source of increased mortality among 114,868 nurses observed over 13 years of follow-up (figure 3.3) (16). Among 8,800 Seventh Day Adventist men followed for 26 years, those with a BMI greater than 27.5 kg/m² had a twofold risk of death from all causes and a 3.3-fold risk of dying from coronary heart disease compared to Seventh Day Adventist men with a BMI < 22.3 kg/m². Of interest was that the mean age at death of men with a BMI below 22.3 kg/m² was 80.5 years compared with 75.8 years in the Seventh Day Adventist men with BMI greater than 27.5 kg/m².

The Harvard Alumni Study also showed important effects of overweight on mortality. Among the 19,292 male Harvard alumni with an average age of 46.6 years at the time the study was done, those who had a BMI greater than 26 kg/m² showed a 1.67-fold increase in mortality compared to those who were followed for the same 27 years, but whose BMI was less than 22.5 kg/m². The lowest mortality in the Harvard Alumni Study was among men weighing an average of 20% below the mean for men of comparable age in the United States.

In 1993, there were 1.25 million deaths in the United States from natural causes in men and women who were 35 to 74 years of age and whose BMI was greater than 21 kg/m². Of 406,923 deaths from coronary heart disease and 55,110 deaths from diabetes, some 77,315 and 24,413 deaths, respectively, can be attributed to obesity (18). When the BMI is greater than 30 kg/m², more than 50% of all causes of death among the 18 million women and 16.7 million men in the United States ages 20-74 can be attributed to overweight.

Regional Fat Distribution

Regional fat distribution also plays an important role in the risk of death (7, 12). The life insurance industry first noted this at the beginning of the 20th century. This theme resumed after World War II when it was noted that obese individuals with an android, or male, distribution of body fat were at higher risk for diabetes and heart disease than those with a gynoid, or female, type of obesity. However, it was clinical and epidemiological work in the 1980s that convinced the world of the relationship between body-fat distribution and the risk of excess mortality. The data in figure 3.6 is from the Gothenburg longitudinal study (14). In this study, central obesity was evaluated using the waist-hip ratio (WHR), or the waist circumference divided by the hip circumference. Women and men with the highest degree of central obesity had a higher death rate than those with the least central fat. Women in the highest tertile of central obesity had mortality rates similar to those of men in the lowest tertile of central obesity, suggesting that the higher mortality rates in men may be related to the effects of differences in fat distribution. Differences in central fatness, as with BMI, can be used in evaluating the risk of overweight. In addition to excess mortality, the risk of

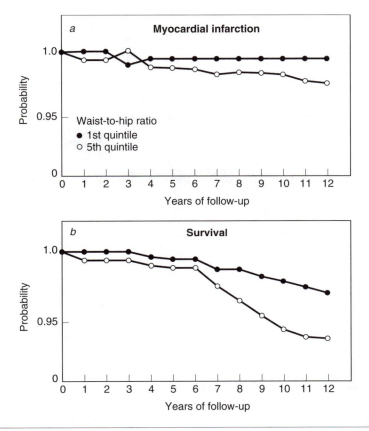

Figure 3.6   Relationship of the waist-to-hip ratio (waist circumference divided by hip circumference) to the probability of (a) remaining free of a myocardial infarction and (b) survival in the Gothenburg study (14).

heart attacks, diabetes, and some forms of cancer were increased by central fatness.

Weight Gain

In addition to overweight and central fatness, the amount of weight gain after age 18-20 also predicts mortality. This is clearly illustrated in the Nurses Health Study (figure 3.7) and the Health Professionals Study, where there is a graded increase in mortality from heart disease associated with increasing degrees of weight gain (31). A gain of more than 10 kg is the dividing line for a higher level of increased risk. Weight gain in men after age 20 in the Health Professionals Study shows the same relationship.

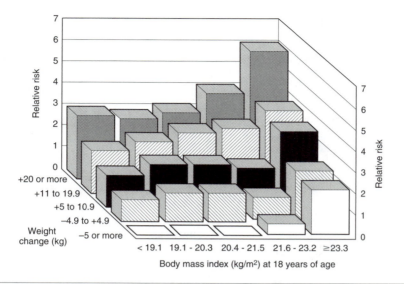

Figure 3.7   Effect of weight gain since age 18 on risk of death. The BMI at age 18, weight gain since age 18, and relative risk of mortality in subjects from the Nurses Health Study are plotted (31).

Sedentary Lifestyle

A sedentary lifestyle is the final important component in the relationship of excess mortality to overweight. A sedentary lifestyle by itself increases the mortality rates from all causes as shown in figure 3.3 and part of this may relate to increased fatness and its co-morbid correlates.

Intentional Weight Loss

If overweight increases risk of mortality, then one would anticipate that intentional weight loss would reduce it. A definitive demonstration of this prediction is not available, but several studies suggest that intentional weight loss does reduce risk. Weight loss reduces blood pressure, improves abnormal lipids, and reduces risk of diabetes (23). Patients treated for obesity with gastric operations have been reported to have lower rates of death. A follow-up of women age 40-64 in the American Cancer Society Study who intentionally lost weight found a significant reduction in all-cause mortality of 20-25% (32).

Weight loss affects a number of risk factors. The data on participants in the Swedish Obesity Study show the degree of weight loss necessary for individual risk factors to respond. Changes in blood pressure and triglyceride levels are very responsive to weight loss and show a decrease after a 5-10% weight loss. HDL cholesterol with a similar weight-related change. Cholesterol, on the other hand, does not show a sustained effect until weight loss exceeds 20%. For most co-morbidities, however, a 10% weight loss is sufficient to see significant improvement in risk factors (11, 23).

Morbidity Associated With Obesity and Increased Central Fat

Overall Morbidity

Table 3.1 shows the relative risks for a variety of conditions including hypertension, diabetes mellitus, and hypercholesterolemia among overweight individuals. For several diseases overweight has a particularly great effect, whereas for others overweight is only one of several factors. Sjostrom et al. (273) evaluated the 2-year incidence rate of new cases of several diseases in the overweight control group of the Swedish Obesity Study. In this 2-year follow-up of untreated overweight patients with a BMI averaging 38 kg/m², there was a 15% incidence of new cases of hypertension, a 7.8% incidence of new cases of diabetes, a 5.8% increase in hyperinsulinemia, a 27.8% incidence for hypertriglyceridemia, and a 15.9% incidence of increased HDL (23).

Table 3.1   Relative Risk of Health Problems Associated With Obesity in Developed Countries

Greatly increased (relative risk >> 3)	Moderately increased (relative risk *ca* 2-3)	Slightly increased (relative risk *ca* 1-2)
Diabetes	Coronary heart disease	Cancer (breast cancer in postmenopausal women, endometrial cancer, colon cancer)
Gallbladder disease	Osteoarthritis (knees)	Reproductive hormone abnormalities
Hypertension	Hyperuricemia and gout	Polycystic ovary syndrome
Dyslipidemia		Impaired fertility
Insulin resistance		Low back pain
Breathlessness		Increased anesthetic risk
Sleep apnea		Fetal defects arising from maternal obesity

Ethnic Groups and Gender

The risk of developing diabetes, hypertension, gallbladder disease, and coronary artery disease differs among ethnic groups and by gender within ethnic groups (table 3.2). These risks are particularly evident in the people with a BMI > 40kg/m^2, but are also present at BMI between 30 and 40 kg/m^2 (19). In white women the risk of non-insulin-dependent diabetes mellitus (NIDDM) was greater than hypertension, which was in turn greater than gallbladder disease. Although the risk of diabetes was also high in white men, gallbladder disease and hypertension were reversed in risk compared to white women. In African-American and Mexican-American men, the risk of hypertension was higher than the risk of diabetes. In African-American women the risk of diabetes and gallbladder disease was higher, whereas in Mexican-American women coronary heart disease, diabetes, and gallbladder disease were similar and the odds ratio less than in African-American or white women. These ethnic and gender differences undoubtedly reflect the interaction of genetic and environmental factors.

Table 3.2   Effect of Gender and Ethnic Group on Health Risks of Men and Women

| | Age-adjusted odds ratio for BMI > 40 kg/m^2 | | | | | |
| | Men | | | Women | | |
Disease	White	African-American	Mexican-American	White	African-American	Mexican-American
Diabetes (NIDDM)	24.3	5.1	12.2	15.7	20.0	4.6
Hypertension	6.8	14.4	22.5	10.4	3.5	3.1
Gallbladder disease	12.5	–	6.5	7.8	9.8	4.4
CHD	2.6	1.5	7.0	3.3	2.8	5.4
High cholesterol	2.3	1.5	0.8	1.1	1.7	0.8

Must et al. (19)

Pathophysiology From Excess Fat

Each of the disease entities whose risk is increased by overweight can be put into one of two categories from a pathophysiological perspective. The first category contains the risks that result from the metabolic changes associated with excess fat. These include diabetes mellitus, gallbladder disease, hypertension, cardiovascular disease, and some forms of cancer. The second group of disabilities arises from the increased mass of fat per se. These would include osteoarthritis, sleep apnea, and the stigmatization of being obese.

The fat cell is an endocrine organ that secretes a number of biologically active peptides and metabolites. The first of these is adipsin, which is a cellular form of the clotting factor, complement D. However, leptin is clearly the most important of these peptides, and it cements the role of the adipocyte as an endocrine cell and of fat as an endocrine organ. From the pathophysiologic point of view, however, released free fatty acids are probably the most important secreted peptide.

The distribution of fat plays an important role in the response to the endocrine products of the fat cell. The accumulation of body fat in visceral fat cells is modulated by a number of factors, some of which are presented in figure 3.8. Androgens and estrogen produced by the gonads and adrenals, as well as peripheral conversion of Δ^4-androstenedione to estrone in fat cells, play a pivotal role in body-fat distribution. Male, or android, fat distribution and female, or gynoid, fat

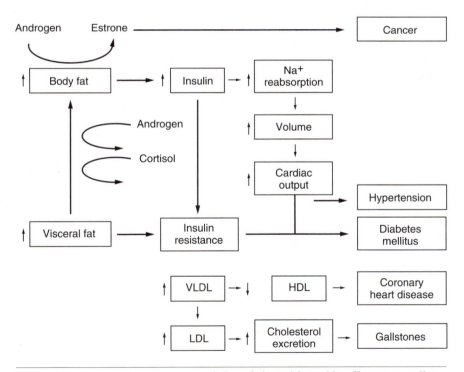

Figure 3.8   Factors influencing accumulation of visceral fat and its effect on mortality. Increasing body-fat stores increase the production of estrone from androgen and provide one cause for the increased risk of breast and uterine cancer in overweight women. The increased production of cortisol in the visceral fat, along with insulin resistance, can be helpful in explaining the genesis of hypertension, diabetes mellitus, coronary heart disease, and gallstones.

distribution develop during adolescence. The increasing accumulation of visceral fat in adult life is related to gender, but the effect of cortisol, decreasing growth hormone, and changing levels of testosterone play important roles in age-related fat accumulation. Increased visceral fat enhances the degree of insulin resistance associated with obesity and hyperinsulinemia. Together, the hyperinsulinemia and insulin resistance can be viewed as enhancing the risk of the co-morbidities described below.

Diabetes Mellitus

Type II, or non-insulin-dependent diabetes mellitus (NIDDM), is strongly associated with overweight in both genders in all ethnic groups (1) (table 3.2). The risk of NIDDM increases with the degree of overweight, with its duration, and with a more central distribution of body fat. The relationship between increasing body mass index and the risk of diabetes in the Health Professional Study is shown in figure 3.9 (5). For individuals with a body mass index below 24 kg/m², the risk of diabetes was at its lowest. As body mass index increased, the relative risk increased so that at a body mass index of 35 kg/m², the relative risk increased 40-fold, or 4,000%. In the Nurses Health Study, a similar strong curvilinear relationship is observed in women. The lowest risk in women was associated with a body mass index below 22 kg/m², which is slightly lower than in the Health Professionals Study. At a body mass index greater than 35 kg/m², the age-adjusted relative risk for diabetes in nurses had increased to 60.9, or over 6,000% (6).

Weight gain also increases the risk of diabetes. More than 80% of the NIDDM cases can be attributed to overweight. Of the 11.7 million cases of diabetes, overweight may account for two-thirds of the diabetic deaths. Using the BMI at age

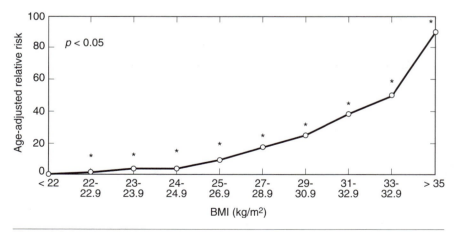

Figure 3.9   Relationship of body mass index to the risk of diabetes. For each higher level of BMI above 22 kg/m² there was a further increase in the incidence of diabetes mellitus (6). Asterisks indicate data for which $p < 0.05$, which refers to statistically significant value compared to BMI < 22 kg/m².

18, a 20-kg weight gain increased the risk for diabetes 15-fold, whereas a weight reduction of 20 kg reduced the risk to almost zero. In the Health Professional Study, weight gain was also associated with an increasing risk of NIDDM, whereas a 3-kg weight loss was associated with a reduction in relative risk (5).

Weight gain appears to precede the onset of diabetes. Among the Pima Indians, body weight showed a slow steady increase of 30 kg (from 60 kg to 90 kg) in the years preceding the diagnosis of diabetes (figure 3.10a) (20). Following the diagnosis of diabetes, there was a small decrease in body weight. In the Health Professional Study (5), there was an increased relative risk of developing diabetes with weight gain as well as with an increase in body mass index. In long-term follow-up studies, there was also a strong relationship between the duration of overweight and the change in plasma glucose during an oral glucose tolerance test. When overweight was present for less than 10 years, there was no increase in plasma glucose. With longer durations, of up to 45 years, there was a nearly linear increase in plasma glucose following an oral glucose tolerance test. Risk of diabetes is

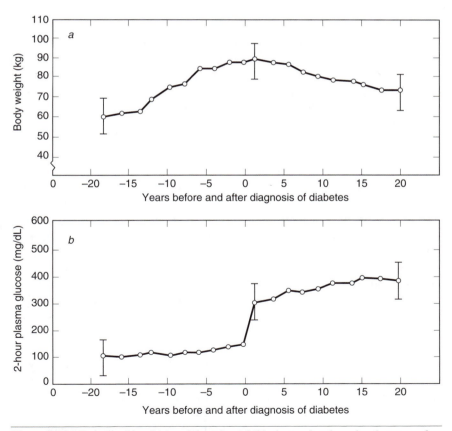

Figure 3.10   Relationship of (a) weight gain and (b) glucose levels to development of diabetes in the Pima Indians (20).

increased in hypertensive individuals treated with diuretics or β-blocking drugs, and this risk was increased in the overweight.

In the Swedish Obesity Study, Sjostrom and his colleagues observed that diabetes was present in between 13 and 16% of their obese subjects at baseline (23). Of those who underwent gastric bypass and subsequently lost weight, 69% of those who initially had diabetes were cured and only 0.5% of those who did not have diabetes at baseline developed it. In contrast, in the obese control group who lost no weight, there was a small 16% cure rate and a 7.8% incidence of new cases of diabetes.

Weight loss or moderating weight gain over years reduces the risk of developing diabetes. This is most clearly shown in the Health Professional Study where relative risk declined by nearly 50% with a weight loss of 5 to 11 kg; with a weight loss of greater than 20 kg or a BMI below 20 kg/m^2, Type II diabetes was almost nonexistent (6).

Both increased insulin secretion and insulin resistance are part of obesity. The relation of insulin secretion to BMI has already been noted. The greater the BMI, the greater the secretion of insulin. In nonhuman primates, obesity develops in over 50% of individual animals, and nearly half of these obese animals subsequently develop diabetes. The time course for the development of obesity in the nonhuman primates is spread over a number of years. The same is true for the Pima Indians. After the animals gain weight, the next demonstrable effect is an impairment in glucose removal and increased insulin resistance as measured by impaired glucose clearance with an euglycemic hyperinsulinemic clamp. The hyperinsulinemia in turn increases hepatic VLDL triglyceride synthesis and secretion, increases PAI-1 (plasminogen activator inhibitor-1) synthesis, increases sympathetic nervous system activity, and increases renal sodium reabsorption.

With insulin resistance there is a decrease in the glut-4 glucose transporter in adipose tissue and a decrease in glucokinase activity in liver. The promoter region of the genes for these two enzymes has an insulin response element and a peroxisome-proliferater activator receptor-γ (PPAR-γ) response element. Insulin resistance may reflect a reduction in the occupancy of the PPAR-γ response element with a reduced transcription of these two components of glucose metabolism. A factor from adipose tissue, such as leptin or TNF-α, or some as yet undescribed peptide or rexinoid, might impair PPAR-γ activation and thus diminish the response to insulin and lower glut-4 and glucokinase, which will produce insulin resistance. Diabetes would then develop in individuals where pancreatic β-cell production was unable to provide sufficient insulin.

Gallbladder Disease

Cholelithiasis is the primary pathology in the hepatobiliary system associated with overweight (13). The old clinical adage "fat, female, fertile, and forty" describes the epidemiologic factors often involved in the development of gallbladder disease. This is demonstrated in the Nurses Health Study (28) (figure 3.11). When the body mass index is less than 24 kg/m^2, the incidence rate of clinically symptomatic gallstones is approximately 250 per 100,000 person-years of follow-up. There is a

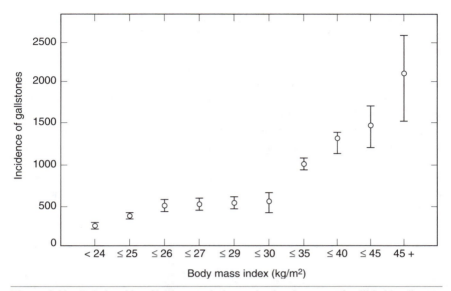

Figure 3.11   Relationship of body mass index to the development of gallbladder disease (28). Rates are given in number of clinically symptomatic gallstones per 100,000 person-years. Average and range of data are shown.

gradual increase in incidence rate with increasing body mass index to a level of 30 kg/m^2 and a very steep increase when the body mass index is higher than 30 kg/m^2. This confirms published work by many other authors.

Part of the explanation for the increased risk of gallstones is the increased cholesterol turnover related to total body fat (13). Cholesterol production is linearly related to body fat with approximately 20 mg per day of additional cholesterol synthesized for each kg of extra body fat. Thus, a 10 kg increase in body fat leads to the daily synthesis of as much cholesterol as is contained in the yolk of one egg. The increased cholesterol is in turn excreted in the bile. High concentrations of cholesterol relative to bile acids and phospholipids in bile increase the likelihood of precipitation of cholesterol gallstones within the gallbladder. Additional factors such as nidation conditions also play a role in whether or not gallstones form (13).

During weight loss, the likelihood of precipitating gallstones increases because the flux of cholesterol is increased through the biliary system. Diets with moderate levels of fat will trigger gallbladder contraction and thus empty the cholesterol content and this may reduce the risk of gallstones. Similarly, the use of bile acids, such as urodeoxycholic acid, may be advisable if the risk of gallstone formation is thought to be increased.

The second feature of the gastrointestinal (GI) system that is altered in obesity is the quantity of fat in the liver (13). Increased steatosis is characteristic of the liver of overweight people and may reflect the increased very low density lipoprotein (VLDL) production associated with hyperinsulinemia. The fact that lipid accumulates in the liver suggests that the secretion of VLDL in response to hyperinsulinemia is inadequate to keep up with the high rates of triglyceride turnover.

Bones, Joints, Muscles, Connective Tissue, and Skin

Osteoarthritis is significantly increased in overweight individuals. The osteoarthritis that develops in the knees and ankles may be directly related to the trauma associated with the degree of excess body weight (figure 3.12) (8). However, the increased osteoarthritis in other non-weight-bearing joints suggests that there are components of the overweight syndrome that alter cartilage and bone metabolism independent of weight bearing. The effect of increased weight, as shown by increasing quintile of Metropolitan Relative Weight, is greater in females than in males, although both show significant rises. The increased osteoarthritis accounts for a significant component of the cost of overweight.

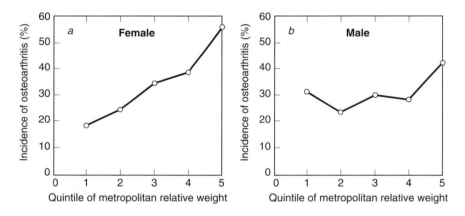

Figure 3.12   Incidence of osteoarthritis for (a) females and (b) males in relation to quintile of Metropolitan Relative Weight determined using the 1959 Metropolitan Life Insurance Weight Tables (8).

There are several changes in the skin associated with excess weight. Stretch marks, or striae, are common and reflect the pressures on the skin from expanding lobular deposits of fat. Acanthosis nigricans, with deepening pigmentation in the folds of the neck, knuckles, and extensor surfaces, occurs in many overweight individuals but is not associated with increased risk of malignancy. Hirsutism in women may reflect the altered reproductive status observed in these individuals.

Obesity and Hypertension

Blood pressure commonly increases in overweight individuals (21). In the Swedish Obesity Study, hypertension was present at baseline in between 44 to 51% of the subjects. One estimate suggests that control of overweight would eliminate 48% of the hypertension in whites and 28% in blacks. For each decline of 1 mm Hg in diastolic blood pressure, it is estimated that the risk of myocardial infarction is decreased 2-3%.

Overweight and hypertension interact on cardiac function. Hypertension in normal weight people produces concentric hypertrophy of the heart with thickening of the ventricular walls. In overweight individuals, eccentric dilatation occurs. There is increased preload and stroke work associated with hypertension. The combination of overweight and hypertension leads to a thickening of the ventricular wall, a larger heart volume, and thus to a greater likelihood of cardiac failure. Table 3.3 summarizes the changes in overweight with and without hypertension (2).

The hypertension of overweight people appears to have a strong relationship to altered sympathetic activity. During insulin infusion, overweight subjects have a much greater increase in muscle sympathetic nerve firing rate than normal-weight subjects, but the altered activity is associated with a lesser change in the vascular resistance of calf muscles.

Table 3.3   Comparison of Cardiac Structural and Hemodynamic Alterations in Patients With Obesity Alone, Systemic Hypertension Alone, and Combined Obesity and Hypertension

Variable	Obesity alone	Hypertension alone	Obesity and hypertension
Heart rate	N	N	N
Blood pressure	N	↑	↑
Stroke volume	↑	N	↑
Cardiac output	↑	N	↑
Systematic vascular resistance	↓	↑	N or ↑
LV volume	↑	N	↑
LV wall stress	N or ↑	N or ↑	↑
LV hypertrophy	Eccentric	Concentric	Hybrid
LV diastolic dysfunction	Usually present	Usually present	Usually present
LV systolic dysfunction	Occasionally present	Usually absent	Occasionally present
LV failure	Occasionally present	Occasionally present	Commonly present
RV hypertrophy	Occasionally present	Usually absent	Occasionally present
RV enlargement	Occasionally present	Usually absent	Occasionally present
RV failure	Occasionally present	Usually absent	Occasionally present

Alpert and Hashimi (2)

Hypertension has a strong association with Type II diabetes (NIDDM), impaired glucose tolerance, hypertriglyceridemia, and hypercholesterolemia. This association or symptom cluster is referred to as the insulin resistance syndrome, the metabolic syndrome, or syndrome X. In a group of 2930 subjects, these conditions individually showed a rate of 1.5% for hypertension compared to a rate of 11.1% for hypertension when the other factors were also present (9). The presence of hyperinsulinemia in the overweight and in the hypertensive patient suggests the presence of insulin resistance. An analysis of the factors that predict blood pressure and changes in peripheral vascular resistance in response to body weight showed that a key determinant of the weight-induced increases in blood pressure was a disproportionate increase in cardiac output that could not be fully accounted for by the hemodynamic contribution of new tissue. This hemodynamic change may be due to a disproportionate increase in cardiac output related to an increase in sympathetic activity.

In experimental studies where animals are force fed to produce obesity, the response to chronic insulin infusion depends on the species. In dogs made obese by a high-fat diet, there was no increased pressor effect of hyperinsulinemia. In rats, on the other hand, seven days of insulin infusion produced a sustained increase in arterial blood pressure. The key question is whether overweight human beings respond to insulin more like the rat or the dog. This question remains to be resolved.

Heart Disease
Using data from the Nurses Health Study (16) the risk for U.S. women developing coronary artery disease is increased 3.3-fold with a BMI > 29 kg/m² compared to women with a BMI < 21 kg/m². With a BMI in the range of 25 to < 29 kg/m² the relative risk is increased to 1.8 (figure 3.3a). Weight gain also has a strong effect on this risk at any initial level of BMI (31). That is, at all levels of initial BMI there was a graded increase in risk of heart disease with weight gain. This was particularly evident in the highest quintile where weight gain was more than 20 kg.

Dyslipidemia may be an important component in the relationship of BMI to increased risk of heart disease (7). A positive correlation between BMI and triglycerides has been demonstrated repeatedly. Of more importance may be the inverse relationship between HDL cholesterol and BMI, because a low HDL cholesterol carries a greater relative risk than elevated triglyceride. Central fat distribution also plays an important role in lipid abnormalities. The waist circumference alone accounted for as much or more of the variance in triglycerides and HDL cholesterol as either the waist/hip ratio (WHR) or the sagittal diameter. A positive correlation for central fat and triglyceride and the inverse relationship for HDL-cholesterol is evident for all measures (figure 3.13). A detailed analysis showed that waist circumference was better than sagittal diameter and both were better than the WHR. As with most other risk factors, weight loss lowers cholesterol slightly and significantly raises HDL cholesterol (figure 3.13).

Increased body weight is associated with a number of cardiovascular abnormalities (review table 3.3) (2). Cardiac weight increases with increasing body weight,

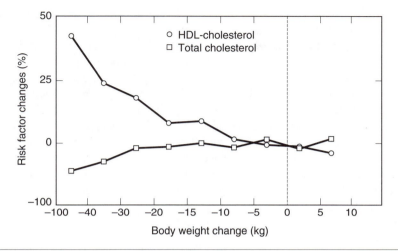

Figure 3.13   Changes in total cholesterol and HDL cholesterol with weight loss (23).

which in turn suggests increased cardiac work. The heart weight, as a percent of body weight, is lower than in a normal weight control group. The increased cardiac work associated with overweight may produce cardiomyopathy and heart failure. With weight loss, there was a decrease in heart weight, which was linearly related to the degree of weight loss in both men and women. Echocardiographic study of left ventricular mid-wall function showed that obese individuals compensated by using cardiac reserve, especially in the presence of hypertension. Heart rate, however, was well within the normal range (table 3.3).

Central fat distribution is associated with small dense low-density lipoproteins (LDL) as opposed to the large fluffy LDL (7). For a similar level of cholesterol, the risk of CHD is significantly higher in individuals with small dense LDL than large fluffy LDL. Because each LDL particle has a single molecule of apo B protein, the concentration of apo B can be used to estimate the number of LDL particles. Despres and his colleagues (7) have demonstrated that the level of apo B is a strong predictor of the risk for CHD. Based on a study of French Canadians, the authors (7) proposed that estimating apo B, the levels of fasting insulin, the concentration of triglyceride, the concentration of HDL cholesterol, and waist circumference could help identify individuals at high risk for the metabolic syndrome.

Respiratory System
Alterations in pulmonary function have been described in overweight subjects; however, in only a few of these studies have the subjects been free of other potential chronic pulmonary diseases. In studies of individuals without underlying pulmonary disease, only the presence of major degrees of increased body weight significantly affected pulmonary function. The major effect is a decrease in residual lung volume associated with increased abdominal pressure on the diaphragm (29).

Fat distribution, independent of total fat, also influences ventilatory capacity in men possibly through effects of visceral fat level.

In contrast to the relatively benign effects of excess weight on respiratory function, per se, the effects of overweight associated with sleep apnea can be severe (29). Overweight subjects with obstructive sleep apnea show a number of significant differences from overweight subjects without sleep apnea. Sleep apnea is considerably more common in men than women, and as a group those with sleep apnea were significantly taller than individuals without sleep apnea. People with sleep apnea have an increased snoring index and increased maximal nocturnal sound intensity, as would be expected. Nocturnal oxygen saturation was also significantly reduced. One interesting hypothesis is that the increased neck circumference and fat deposits in the pharyngeal area may lead to the obstructive sleep apnea of obesity.

Cancer

Certain forms of cancer are significantly increased in overweight individuals, and this contributes to the increased mortality with obesity (figure 3.3b) (15). In males, the increased risk is for neoplasms of the colon, rectum, and prostate. In women it is cancers of the reproductive system and gallbladder. One explanation for the increased risk of endometrial cancer in overweight women is the increased production of estrogens by adipose tissue stromal cells. This increased production is related to the degree of excess body fat, which accounts for a major source of estrogen production in postmenopausal women. Breast cancer is related not only to total body fat, but may have a more important relationship to central body fat (22). The increased visceral fat measured by CT scan shows an important relation to the risk of women developing breast cancer.

Endocrine Changes

A variety of endocrine changes are associated with overweight and are summarized in table 3.4. The changes in the reproductive system are among the most important. Irregular menses and frequent anovulatory cycles are common in women, and the rate of fertility may be reduced. There are reports of increased risks of toxemia, hypertension, and the use of cesarean section in the overweight.

Table 3.4   Common Hormonal Abnormalities Associated With Obesity

1. Increased cortisol production
2. Insulin resistance
3. Decreased sex hormone binding globulin (SHBG) in women
4. Decreased progesterone levels in women
5. Decreased testosterone levels in men
6. Decreased growth hormone production

Psychosocial Function

Overweight is a stigmatized condition (10). That is, overweight individuals are exposed to the consequences of public disapproval of their fatness. This stigma is seen in education, employment, health care, etc. (table 3.5). Examining education, marital status, and income level in adolescents who were overweight into adult life reveals psychosocial consequences. These effects were more evident in the females than in the males.

Table 3.5   Estimated Effect of Overweight in Adolescence on Subsequent Social and Economic Characteristics and Self-Esteem Among Women

| | Observed Value | | |
Variable	Overweight (N = 195)	Nonoverweight (N = 4943)	p Value
Married (%)(n = 4922)	28%	56%	< 0.001
Household income ($) (n = 4286)	$18,372	$30,586	< 0.001
Income below poverty level (%)[†] (n = 4286)	32%	13%	< 0.001
Education (yr) (n = 4881)	12.1 yr	13.1 yr	0.009
Completed college (%) (n = 4881)	9%	21%	0.21
Self-esteem in 1987 (n = 5138)	32.4	33.6	0.38

[†] Household poverty was defined according to federal poverty guidelines.
Gortmaker et al. (32).

References

1. Albu, J., and F.X. Pi-Sunyer. 1997. Obesity and diabetes. In *Handbook of Obesity*. eds. G.A. Bray, C. Bouchard, and W.P.T. James, 697-707. New York: Marcel Dekker.
2. Alpert, M.A., and M.W. Hashimi. 1993. Obesity and the heart. *American Journal of Medical Sciences* 306: 117-123.
3. Blair, S.N., H.W. Kohl, C.E. Barlow, R.S. Paffenbarger, L.W. Gibbons, and C.A. Macera. 1995. Changes in physical fitness and all-cause mortality. *JAMA* 273: 1093-1098.
4. Bray, G.A. 1996. Coherent preventive and management strategies for obesity. In *The origins and consequences of obesity*. eds. D.J. Chadwick and G. Cardew, 228-246. Ciba Foundation Symposium 201. London: John Wiley.

5. Chan, J.M., E.B. Rimm, G.A. Colditz, M.J. Stampfer, and W.C. Willett. 1994. Obesity, fat distribution, and weight gain as risk factors for clinical diabetes in men. *Diabetes Care* 17: 961-969.
6. Colditz, G.A., W.C. Willett, A. Rotnitzky, and J.E. Manson. 1995. Weight gain as a risk factor for clinical diabetes mellitus in women. *Annals of Internal Medicine* 122: 481-486.
7. Després, J-P., and R.M. Krauss. 1997. Obesity and Lipoprotein metabolism. In *Handbook of Obesity*. eds. G.A. Bray, C. Bouchard, and W.P.T. James, 651-675. New York: Marcel Dekker.
8. Felson, D.T., J.J. Anderson, A. Naimark, A.M. Walker, and R.F. Meenan. 1988. Obesity and knee osteoarthritis. The Framingham Study. *Annals of Internal Medicine* 109: 18-24.
9. Ferrannini, E., S.M. Haffner, B.D. Mitchell, and M.P. Stern. 1991. Hyperinsulinemia: the key feature of a cardiovascular and metabolic syndrome. *Diabetologia* 34: 416-422.
10. Gortmaker, S.L., A. Must, J.M. Perrin, A.M. Sobol, and W.H. Dietz. 1993. Social and economic consequences of overweight in adolescence and young adulthood. *New England Journal of Medicine* 329: 1008-1012.
11. Institute of Medicine (IOM). 1995. In *Weighing the options: Criteria for evaluating weight-management programs*. ed. Thomas, P.R. Washington D.C.: National Academy Press.
12. Kissebah, A.H., and G.R. Krakower. 1994. Regional adiposity and morbidity. *Physiological Reviews* 74: 761-811.
13. Ko, C.W., and S.P. Lee. 1997. Obesity and gallbladder disease. In *Handbook of Obesity,* eds. G.A. Bray, C. Bouchard, and W.P.T. James, 709-724. New York: Marcel Dekker.
14. Lapidus, L., C. Bengtsson, B. Larsson, K. Pennert, E. Rybo, and L. Sjostrom. 1984. Distribution of adipose tissue and risk of cardiovascular disease and death: a 12 year follow up of participants in the population study of women in Gothenburg, Sweden. *British Medical Journal* 289: 1257-1261.
15. Lew, E.A. 1985. Mortality and weight: insured lives and the American Cancer Society studies. *Annals of Internal Medicine* 103: 1024-1029.
16. Manson, J.E., M.J. Stamfer, C.H. Hennekens, and W.C. Willett. 1987. Body weight and longevity. A reassessment. *JAMA* 257: 353-358.
17. Manson, J.E., W.C. Willett, M.J. Stamfer, G.A. Colditz, D.J Hunter, S.E. Hankinson, C.H. Hennekens, and F.E. Speizer. 1995. Body weight and mortality among women. *New England Journal of Medicine* 333: 677-685.
18. McGinnis, J.M., and W.H. Foege. 1993. Actual causes of death in the United States. *JAMA* 270: 2207-2212.
19. Must, A. J. Spadano, and W.H. Dietz. 1997 Prevalence of obesity-related morbidity in relation to adult overweight in the United States. *Obesity Research* 5(Suppl. 1): 27.
20. Ravussin, E. 1993. Energy metabolism in obesity. Studies in the Pima Indians. *Diabetes Care* 16(Suppl. 1): 232-238.

21. Rocchini, A.P. 1997. Obesity and blood pressure regulation. In *Handbook of Obesity*. eds. G.A. Bray, C. Bouchard, and W.P.T. James, 677-695. New York: Marcel Dekker.

22. Schapira, D.V., R.A. Clark, P.A. Wolff, A.R. Jarrett, N.B. Kumar, and N.M. Aziz. 1994. Visceral obesity and breast cancer risk. *Cancer* 74: 632-639.

23. Sjostrom, C.J., L. Lissner, and L. Sjostrom. 1997. Relationships between changes in body composition and changes in cardiovascular risk factors: The SOS Intervention Study. *Obesity Research* 5(6): 519-530.

24. Sjostrom, L. 1992. Morbidity in severely obese subjects. *American Journal of Clinical Nutrition* 55 (Suppl. 2): 508S-515S.

25. Sjostrom, L. 1992. Mortality in severely obese subjects. *American Journal of Clinical Nutrition* 55 (Suppl. 2): 516S-523S.

26. Society of Actuaries. 1980. *Build Study of 1979*. Chicago: Society of Actuaries/Association of Life Insurance Medical Directors of America.

27. Sonne-Holm, S., and T.I.A. Sorensen. 1977. Post-war course of the prevalence of extreme overweight among Danish young men. *Journal of Chronic Disorders* 30: 351-358.

28. Stampfer, M.J., K.M. Maclure, G.A. Colditz, J.E. Manson, and W.C. Willett. 1992. Risk of symptomatic gallstones in women with severe obesity. *American Journal of Clinical Nutrition* 55: 652-658.

29. Strohl, K.P., R.J. Strobel, and R.A. Parisi. 1997. Obesity and Pulmonary Function. In *Handbook of Obesity*. eds. G.A. Bray, C. Bouchard, and W.P.T. James, 725-739. New York: Marcel Dekker.

30. Waaler, H.T. 1984. Height, weight and mortality: The Norwegian experience. *Acta Medica Scandinavica* 679: 1-56.

31. Willett, W.C., J.E. Manson, M.J. Stampfer, G.A. Colditz, B. Rosner, F.E. Speizer, and C.H. Henneken. 1995. Weight, weight change and coronary heart disease in women: Risk within the normal weight range. *JAMA* 273: 461-465.

32. Williamson, D.E., E. Pamuk, M. Thun, D. Flanders, T. Byers, and C. Heath. 1995. Prospective study of intentional weight loss and mortality in never smoking overweight white women aged 40 to 64 years. *American Journal of Epidemiology* 141: 1128-1141.

CHAPTER 4

The Cost of Obesity and Sedentarism in the United States

Graham A. Colditz, MD, DrPH, and Anna Mariani, MPH
Channing Laboratory, Harvard Medical School, Boston, Massachusetts, U.S.A.

To estimate the economic impact of obesity and sedentarism in the United States it is important to look at the full range of diseases that are related to these lifestyle markers. This chapter presents a summary of the diseases related to being obese or sedentary, which collectively serve as a measure of sedentary lifestyle; the increased relative risk of these diseases for level of obesity and inactivity; and the proportion of cases of these diseases attributable to inactivity. In addition, we present the direct medical cost to society for those conditions that are attributable to obesity and sedentarism in the United States. This cost represents the proportion of total United States health care expenditures that would be avoided if obesity and sedentarism did not exist.

One proven way to combat obesity and sedentarism is to engage in regular lifelong physical activity. Regular physical activity done three times a week for 30 minutes at a time for moderate levels of activity (35) greatly reduces the risk of developing and/or dying from many of the leading causes of illness and death in the United States. As summarized in the comprehensive U.S. Surgeon General's Report on Physical Activity (36), regular physical activity will improve health because it

- reduces the risk of dying prematurely,
- reduces the risk of developing heart disease,
- reduces the risk of developing cardiovascular disease,
- reduces the risk of developing diabetes,
- reduces the risk of developing high blood pressure,
- reduces the risk of developing colon cancer, and
- reduces feelings of depression and anxiety.

In addition to improving health in the above conditions, regular physical activity will also

- help control weight;
- help to reduce blood pressure in people who already have high blood pressure;

- help to build and maintain healthy bones, muscles, and joints (thereby reducing osteoporosis and osteoporotic hip fractures);
- help older adults become stronger and better able to move about without falling; and
- promote psychological well-being (35).

Millions of Americans suffer from illnesses that could have been prevented if they had engaged in regular physical activity. Currently, 60 million people are overweight; that is approximately one-third of the population (25). Furthermore, the prevalence of overweight and obesity has been increasing (8). The proportion of the U.S. adult population aged > 20 years old exceeding the healthy weight ranges (generally BMI > 25.0) is high, including 59.4% of men, 50.7% of women, and 54.9% of the total population (8).

United States Prevalence Estimates

Despite the documented benefits to health and well-being, many people do not engage in regular physical activity. The Behavioral Risk Factor Surveillance System 1994 and 1995 (BRFSS) reported that in 1995, on average, 28.8% of the total adult U.S. population reported no leisure-time physical activity. The prevalence of no leisure time physical activity was as high as 48% in at least one state (22). Lack of physical activity is related to a large number of chronic diseases. There is strong evidence to support a causal relationship between participation in regular physical activity and lower risk of disease. Such a causal relationship is supported for cardiovascular disease (CVD), coronary heart disease (CHD), colon cancer, non-insulin-dependent diabetes mellitus, obesity, high blood pressure, and depression and anxiety.

In 1990, the Department of Health and Human Services (DHHS) estimated that as much as 70% of premature mortality might be attributed to lifestyle (32). McGinnis and Foege conducted an extensive review of the causes of mortality and estimate that 14% of total deaths in the U.S. are attributable to diet and activity patterns (20). Prospective studies that have recorded lifestyle habits for individuals and followed them over time support the summary estimate reported by McGinnis and Foege. For example, data from the follow-up of Iowa women indicate a strong inverse relationship between physical activity and total mortality among women (17); the Harvard Alumni Study shows a similar relationship among men (28).

Cardiovascular disease is the leading cause of all deaths in the United States. Approximately 58 million persons in the United States (20% of the total population) have one or more types of CVD (23). These types of CVD include high blood pressure, coronary heart disease, stroke, rheumatic heart disease, and others. Behavioral risk factors for CVD include both physical inactivity and overweight (35). In 1994, CVD accounted for 45.2% of all deaths in the United States.

Higher levels of physical activity are inversely related to risk of coronary heart disease (1). In the United States 13.5 million people have coronary heart disease (CHD), which is the leading cause of CVD mortality in the United States. Each year CHD is newly diagnosed in approximately 1.5 million persons. Of the many risk factors associated with CHD, uncontrolled hypertension, obesity, and physical activity are included. A meta-analysis of prospective studies of physical activity and CHD showed twice the risk of CHD among sedentary as compared with active occupations. Similar results were observed for leisure-time activity (1).

Physical activity is inversely related to the risk of Type 2 diabetes. About 16 million Americans have diabetes, but only about 10 million have been diagnosed. The prevalence of diagnosed diabetes in the U.S. is approximately 3%, and some 798,000 new cases are diagnosed annually. Diabetes is the nation's seventh leading killer and contributed to about 187,800 deaths in 1995. The majority (90-95%) of those individuals with diabetes have type 2 diabetes. Diabetes accounts for approximately 15% of total U.S. health care expenditures annually (21).

The Behavioral Risk Factor Survey reported that in 1995, 22.0% of the total U.S. population reported that they had been told by a health professional that they had high blood pressure (22). Physical activity is inversely related to the risk of developing high blood pressure.

Colon cancer is the second leading cause of cancer mortality in the United States. The risk of colon cancer can be significantly reduced if a pattern of regular physical activity is followed. In a review of epidemiological studies published through 1996 Colditz et al. noted that a consistent inverse relationship was observed between physical activity and risk of colon cancer. About 50% reduction in incidence was observed among those with the highest level of activity (6, 7).

The relationship between physical activity and depression or anxiety is less clearly documented than several of the other major conditions described above. The available research shows that physical activity relieves and reduces symptoms of depression and anxiety (35).

Regular physical activity reduces the risk of osteoporosis and osteoporotic fracture of the hip, a major burden among the elderly population (13). Physical activity helps maintain healthy bones, muscles, and joints, thereby reducing osteoporosis and osteoporotic hip fractures. An estimated 850,000 fractures occur annually in the United States among persons 65 years of age or older. Approximately 25 million persons may be at increased risk for fracture because of low bone mass. Physical activity can also help to increase muscle strength, which reduces the impact of low bone density and helps to prevent falls. Randomized trials of muscle strengthening activity show lower risk of falling with strength training (13, 29, 34). Finally, the Study of Osteoporotic Fractures showed decreased risk of hip fracture with higher levels of physical activity (12), which supports evidence that activity patterns relate to preservation of bone mineral density of the proximal femur (26).

Physical activity can greatly improve people's ability to control their weight. Regular lifelong physical activity can help to prevent weight gain and obesity.

Obesity is an independent risk factor for coronary heart disease, hypertension, non-insulin-dependent diabetes mellitus, breast cancer (among postmenopausal women), endometrial cancer, colon cancer, and certain musculoskeletal disorders (knee, osteoarthritis, low back pain).

Analysis of Costs

We use a prevalence-based approach to estimate the direct costs of the medical conditions that can be attributed to lack of physical activity. This follows the same methods used by Wolf and Colditz to estimate the economic costs of obesity (37). Direct medical costs include personal health care, hospital care, physician services, allied health services, and medications. We do not estimate the indirect costs (lost output as a result of a reduction or cessation of productivity due to morbidity or mortality) of these illnesses, but rather focus on direct costs that would be avoided if none of the population were inactive.

To estimate the proportion of disease attributable to being sedentary, the population attributable risk (PAR%), we use the standard formula $PAR = P(RR-1)/1 + P(RR-1)$, where P is the prevalence of lack of activity in the population, and RR is the relative risk for contracting the disease comparing the active to the inactive. For relative risk we use the increased risk associated with being sedentary as compared to being *moderately* active as reported in the literature. These relative risks are summarized in table 4.1. The prevalence of sedentary behavior is taken from the

Table 4.1   Costs of Inactivity (billion $) in the United States, 1995

Condition	Relative risk (RR)	PAR%	Direct costs
Type 2 diabetes	1.5	12%	6.4
Coronary heart disease	2	22%	8.9
Hypertension	1.5	12%	2.3
Gallbladder disease	2	22%	1.9
Breast cancer	1.2	5%	0.38
Colon cancer	2	22%	2.0
Osteoporotic fractures	2	18%	2.4
Total			24.3 billion

RR = relative risk of disease among inactive compared with active population.

PAR% population attributable risk of disease: the proportion of total disease in society due to obesity.

1995 BRFS report noted earlier. We use the median for total adults reporting no leisure-time physical activity in 1995, which was 28.8% (22).

The procedures used to estimate the costs of diabetes and gallstones are presented elsewhere (5, 15). Annual direct costs of hypertension and coronary heart disease were extrapolated from Hodgson and Kopstein (14). All costs are inflated to 1995 dollars using the medical component of the consumer price index. We multiply the total direct cost of care for each illness by the PAR% and estimate the annual expenditure due to being sedentary.

Results

Costs of Inactivity

We estimate that 22% of CHD, colon cancer, and gallbladder disease; 18% of osteoporotic fractures; 12% of diabetes and hypertension; and perhaps 5% of breast cancer are attributable to lack of physical activity (see table 4.1).

Overall, sedentary behavior and lack of physical activity costs the U.S. a conservative total of 24.3 billion dollars per year for direct health care delivery costs. Thus, approximately 2.4% of all health care costs in 1995 are due to lack of physical activity.

In a sensitivity analysis to address the robustness of the results, we modified the prevalence of inactivity. We estimate the costs due to inactivity using the upper bound of state-level reports of no physical activity (48%). This may more closely approximate the proportion of the population with insufficient activity to avoid the health consequences of inactivity. Using this prevalence and retaining the relative risk for the association between activity and the major health outcomes, we estimate the costs of inactivity as 37.2 billion dollars (3.7% of direct health care costs).

Costs of Obesity

Using a prevalence-based approach to estimate the costs of obesity in the United States, Wolf and Colditz (37) estimated the proportion of disease attributable to obesity and the associated costs. Conditions included in their analysis were type 2 diabetes, coronary heart disease, hypertension, gallbladder disease, postmeno-pausal breast cancer, endometrial cancer, colon cancer, and osteoarthritis. That analysis used a definition of obesity as BMI greater than 27.8 for men and 27.3 for women. Here, we use the current recommended definition of obesity at BMI of 30. Using prevalence estimates for obesity from NHANES III (22.4% overall and 24.9% for breast and endometrial cancers), we updated the costs attributable to obesity. Overall, the direct health care costs of obesity were approximately 70 billion dollars, or 7% of total health care cost (see table 4.2).

Indirect Costs

Early retirement and increased risk of disability pensions together add indirectly to the direct costs summarized above. Narbro et al. (24) estimate that in Sweden obese subjects are 1.5 to 1.9 times more likely to take sick leave and that 12% of obese women had disability pensions attributable to obesity, costing some 300 million US dollars for 1 million females in the adult population. Overall, approximately 10% of sick leave and disability pensions in women may be related to obesity and obesity-related conditions (24).

Total Direct Costs of Inactivity

The sum of obesity (7% of health care costs) and of inactivity (2.4% of health care costs) is here used to estimate the total direct costs of inactivity. Overall, a minimum 9.4% of all direct costs incurred in delivering health care in the U.S. is attributable to insufficient energy expenditure. As noted in the discussion section, these two

Table 4.2   Costs (billion $) of Obesity (BMI > 30) in the United States, 1995

Condition	Relative risk (RR)	PAR%	Direct costs
Type 2 diabetes	11	69%	36.6
Coronary heart disease	4	40%	16.2
Hypertension	4	40%	7.6
Gallbladder disease	5.5	50%	4.3
Breast cancer	1.3	7%	0.53
Endometrial cancer	2.5	27%	0.23
Colon cancer	1.5	10%	0.89
Osteoporotic fractures	2.1	20%	3.6
Total			70 billion

Uses prevalence of obesity = 22.5% as reported in NHANES III; for breast and endometrial cancer uses prevalence of 24.9% as reported by Flegal et al. 1998.

RR = relative risk of disease among obese compared with leaner population.

PAR% population attributable risk of disease: the proportion of total disease in society due to obesity.

indicators of sedentary lifestyle are independent risk factors in epidemiological studies, supporting the combination of the results. A sedentary lifestyle directly leads to medical conditions, or alternatively, the accumulation of adiposity, which in turn contributes to excess morbidity and mortality. A more reasonable estimate of the proportion of the population that lacks sufficient physical activity gives some 4% of health care costs due to inactivity.

Note that these are conservative estimates, as the costs of obesity are estimated for those with body mass index of 30 or more kg/m^2. However, adverse health effects are present at levels of obesity below a BMI of 30 (27, 35), and higher levels of activity are associated with even lower risk of many chronic conditions (36). There are substantial additional costs incurred among those who are overweight (BMI 25 to 29.9; 32% of the U.S. population). Likewise for estimates of inactivity, we use the prevalence of approximately 28% of the U.S. population being inactive. As there are clear health benefits with increasing levels of physical activity, the definition of inactivity is somewhat arbitrary and could be raised so that more of the population is considered inactive and more of the chronic conditions considered here might be attributable to lack of activity. Such changes would substantially increase the economic burden of lack of activity and of obesity or excess adiposity.

Discussion

McGinnis and Foege (20) estimated that 14 percent of U.S. deaths in 1992 were attributable to inactivity and diet. The burden of inactivity is not only substantial for mortality, but also for chronic conditions that negatively impact quality of life and life expectancy.

The costs of inactivity and obesity compare with the total estimated impact of cigarette smoking in the United States. In 1990 the lost productivity of persons disabled by disease attributable to cigarette smoking and forgone earnings of those dying prematurely totaled $47 billion (18). These data indicate that excess body weight or adiposity places a substantial annual burden on the health care system as measured by the economic impact of a range of obesity-related diseases. These costs in large part reflect the impact of weight gain in adult life. These costs could be avoided if the population maintained a healthy weight. This is now a priority recommendation for the United States Department of Health and Human Services dietary and weight guidelines (35).

The addition of costs attributable to inactivity to those attributable to obesity is reasonable, given the growing evidence that Western lifestyle is associated with excess weight gain and lack of energy expenditure through physical activity. For example, in a prospective study among women, weight gain after cessation of smoking was attenuated among those women who increased their level of physical activity (16). The health consequences of sedentary lifestyle are independent of those associated with obesity as indicated in several major studies that have simultaneously evaluated physical activity and obesity in relation to colon cancer, diabetes, and total mortality. For example, evaluating relationships between physical activity and colon cancer, Giovannucci (10) and Martinez (19) found that the inverse associations with level of activity were independent of body mass index. Likewise, Blair has shown independent effects of fitness and activity on mortality (2, 3). These and other data support the assumption of independence and hence allow us to add the economic costs of obesity to the costs of a sedentary lifestyle. We, therefore, add the direct costs of obesity to those for a sedentary lifestyle.

An alternative approach to estimating the costs of obesity and lack of physical activity is to use direct evaluation of medical expenses as may be possible through access to computerized health data records. Quesenberry et al. (30) used such an approach to document the direct relationship between obesity and use of services in a health maintenance organization. Hospitalizations, laboratory services, outpatient visits, and outpatient pharmacy and radiologic services were included in an estimate of direct costs of health care. Total costs increased with rising levels of obesity, and were 25% higher for those with BMI of 30 to 34.9 compared to those with BMI less than 25, and 44% higher for those with BMI of 35 or greater.

Gorsky et al. (11) simulate three hypothetical cohorts to estimate the costs of health care according to level of obesity over a 25-year period, discounting future costs at 3% per year. They estimate that 16 billion additional dollars will be spent over the next 25 years treating health outcomes associated with obesity among middle-aged women. Using an incidence-based approach to cost of illness, Thompson et al. (33) estimated the excess costs of health services according to level of obesity. Using a conservative approach that does not include any future weight gain, and starting with NHANES III population estimates for BMI, cholesterol, hypertension, and diabetes, they estimated the lifetime future costs per person as comparable to that of smoking.

A social cost that is not considered in any of the economic summaries to date is that of reduced physical functioning that is associated with higher levels of obesity (4, 9). Research is needed to quantify these relationships and incorporate the deterioration in quality of life into summaries of the burden of inactivity. With this added understanding of the magnitude of the burden of inactivity, we might then better motivate the political will to support the social changes necessary to increase energy expenditure in our society and avoid the excess weight gain that is associated with our Western ways of living.

While benign prostatic hyperplasia was not considered here, it is noteworthy that this common condition may be but one of many complications of obesity and inactivity that has to date been overlooked. Likewise there is a direct relationship between increasing adiposity and the diagnosis of infertility among women (31). Recent evidence also suggests that obesity is related to asthma. This common condition, which is rising in prevalence around the world, is related to adiposity among children and also adults. Likewise, the costs to society of higher levels of depression and other chronic conditions attributable to inactivity, have not been quantified.

Summary

Growing levels of both inactivity and obesity pose major health problems in Western society. These preventable causes of morbidity and mortality require focused strategies to increase the level of energy expenditure in our lifestyles. Substantial benefits will likely accrue through reduced health care costs, as well as reduced indirect costs and gains in the quality of life.

References

1. Berlin, J., and G. Colditz. 1990. A meta-analysis of physical activity in the prevention of coronary heart disease. *American Journal of Epidemiology* 132: 612-628.
2. Blair, S., H.I. Kohl, C. Barlow, R.J. Paffenbarger, L. Gibbons, and C. Macera. 1995. Changes in physical fitness and all-cause mortality: a prospective study of healthy and unhealthy men. *JAMA* 273: 1093-1098.
3. Blair, S., H.I. Kohl, R.J. Paffenbarger, D. Clark, K. Cooper, and L. Gibbons. 1989. Physical fitness and all-cause mortality: a prospective study of healthy men and women. *JAMA* 262: 2395-2401.
4. Coakley, E., I. Kawachi, J. Manson, F. Speizer, W. Willett, and G. Colditz. 1998. Lower levels of physical functioning are associated with higher body weight among middle-aged and older women. *International Journal of Obesity* 22: 958-996.
5. Colditz, G. 1992. The economic costs of obesity. *American Journal of Clinical Nutrition* 55: 503S-507S.
6. Colditz, G., C. Cannuscio, and A. Frazier. 1997. Physical activity and colon cancer. *Cancer Causes Control* 8: 649-667.
7. Colditz, G.A., D. DeJong, D.J. Hunter, D. Trichopoulos, and W.C. Willett. 1996. Harvard Report on Cancer Prevention. Volume 1. Causes of Human Cancer. *Cancer Causes Control* 7: 1-59.
8. Flegal, K., M. Carroll, R. Kuczmarski, and C. Johnson. 1998. Overweight and obesity in the United States: prevalence and trends, 1960-1994. *International Journal of Obesity* 22: 39-47.
9. Fontaine, K., L. Cheskin, and I. Barofsky. 1996. Health-related quality of life in obese persons seeking treatment. *Journal of Family Practice* 43: 265-270.
10. Giovannucci, E., A. Ascherio, E.B. Rimm, G.A. Colditz, M.J. Stampfer, and W.C. Willett. 1995. Physical activity, obesity, and risk for colon cancer and adenoma in men. *Annals of Internal Medicine* 122: 327-334.
11. Gorsky, R., E. Pamuk, D. Williamson, P. Shaffer, and J. Koplan. 1996. The 25-year health care costs of women who remain overweight after 40 years of age. *American Journal of Preventive Medicine* 12: 388-394.

12. Gregg, E., J. Cauley, D. Seeley, K. Ensrud, and D. Bauer. 1998. Physical activity and osteoporotic fracture risk in older women. Study of Osteoporotic Fractures Research Group. *Annals of Internal Medicine* 129: 81-88.

13. Henderson, N., C. White, and J. Eisman. 1998. The role of exercise and fall risk reduction in the prevention of osteoporosis. *Endocrinology and Metabolism Clinics of North America* 27: 369-387.

14. Hodgson, T., and A. Kopstein. 1984. Health care expenditures for major diseases in 1980. *Health Care Financial Review* 5: 1-12.

15. Huse, D., G. Oster, A. Killen, M. Lacey, and G. Colditz. 1989. The economic costs of non-insulin-dependent diabetes mellitus. *JAMA* 262: 2708-2713.

16. Kawachi, I., R. Troisi, R. Rotnitzky, E. Coakley, and G. Colditz. 1996. Can physical activity minimize weight gain in women after smoking cessation? *American Journal of Public Health* 86: 999-1004.

17. Kushi, L., R. Fee, A. Folsom, P. Mink, K. Anderson, and T. Sellers. 1997. Physical activity and mortality in postmenopausal women. *JAMA* 277: 1287-1292.

18. MacKenzie, T., C. Bartecchi, and R. Schrier. 1994. The human costs of tobacco use. *New England Journal of Medicine* 330: 975-980.

19. Martinez, M.E., E. Giovannucci, D. Spiegelman, D.J. Hunter, W.C. Willett, and G.A. Colditz. 1997. Leisure-time physical activity, body size, and colon cancer in women. Nurses' Health Study Research Group. *Journal of the National Cancer Institute* 89: 948-955.

20. McGinnis, J.M., and W.H. Foege. 1993. Actual causes of death in the United States. *JAMA* 270: 2207-2212.

21. MMWR. 1997. Diabetes-specific preventive-care practices among adults in a managed care population. Colorado, Behavioral Risk Factor Surveillance System, 1995. *Morbidity and Mortality Weekly Reports* 46.

22. MMWR. 1997. State- and sex- specific prevalence of selected characteristics. Behavioral Risk Factor Surveillance System, 1994 and 1995. *Morbidity and Mortality Weekly Report* 46: 1-29.

23. MMWR. 1998. Missed opportunities in preventive counseling for cardiovascular disease. *Morbidity and Mortality Weekly Report* 47.

24. Narbro, K., E. Jonsson, B. Larsson, H. Wedel, and L. Sjostrom. 1996. Economic consequences of sick-leave and early retirement on obese Swedish women. *International Journal of Obesity* 20: 895-903.

25. National Center for Health Statistics. 1996. Monitoring health care in America, Quarterly Fact Sheet. June: National Center for Health Statistics.

26. Nguyen, T., P. Sambrook, and J. Eisman. 1998. Bone loss, physical activity, and weight change in elderly women: the Dubbo Osteoporosis Epidemiology Study. *Journal of Bone and Mineral Research* 13: 1458-1467.

27. NHLBI Obesity Initiative Expert Panel. 1998. Clinical guidelines on the identification, evaluation, and treatment of overweight and obesity in adults—The evidence report. *Obesity Research* 6: 51s-209s.

28. Paffenbarger, R.J., R. Hyde, A. Wing, I.M. Lee, D. Jung, and J. Kampert. 1993. The association of changes in physical activity level and other lifestyle characteristics with mortality among men. *New England Journal of Medicine* 328: 538-545.

29. Province, M.A., E.C. Hadley, M.C. Hornbrook, L.A. Lipsitz, J.P. Miller, C.D. Mulrow, M.G. Ory, R.W. Sattin, M.E. Tinetti, and S.L. Wolf. 1995. The effects of exercise on falls in elderly patients. A preplanned meta-analysis of the FICSIT trials. Frailty and injuries: cooperative studies of intervention techniques. *JAMA* 273: 1341-1347.

30. Quesenberry, C.J., B. Caan, and A. Jacobson. 1998. Obesity, health services use, and health care costs among members of a health maintenance organization. *Archives of Internal Medicine* 158: 466-472.

31. Rich-Edwards, J.W., M.B. Goldman, W.C. Willett, D.J. Hunter, M.J. Stampfer, G. A. Colditz, and J.E. Manson. 1994. Adolescent body mass index and ovulatory infertility. *American Journal of Obstetrics and Gynecology* 171: 171-177.

32. Sullivan, L. 1990. Healthy People 2000. *New England Journal of Medicine* 323: 1065-1067.

33. Thompson, D., J. Edelsberg, G.A. Colditz, A.P. Bird, and G. Oster. 1999. Lifetime health and economic consequences of inactivity. *Archives of Internal Medicine* 159(18): 2177-2183.

34. Tinetti, M.E., D.I. Baker, G. McAvay, E.B. Claus, P. Garrett, M. Gottschalk, M.L. Koch, K. Trainor, and R.I. Horwitz. 1994. A multifactorial intervention to reduce the risk of falling among elderly people living in the community. *New England Journal of Medicine* 331: 821-827.

35. U. S. Department of Agriculture and U. S. Department of Health and Human Services. 1995. Nutrition and Your Health: Dietary Guidelines for Americans. Washington, DC: U.S. Government Printing Office.

36. U.S. Department of Health and Human Services. 1996. Physical activity and health: A Report of the Surgeon General. Atlanta, GA: US Department of Health and Human Services, Centers for Disease Control and Prevention, National Center for Chronic Disease Prevention and Health Promotion.

37. Wolf, A., and G. Colditz. 1998. Current estimates of the economic cost of obesity in the United States. *Obesity Research* 6: 97-106.

PART II

The Biological and Behavioral Determinants of Obesity

CHAPTER 5

The Determinants of Obesity

Arline D. Salbe, PhD, and Eric Ravussin, PhD

National Institutes of Health, Clinical Diabetes and Nutrition Section, Phoenix, Arizona, U.S.A.

Obesity is highly prevalent in the industrialized world, and the chronic diseases associated with it are the major killers in these countries. In the United States, obesity and overweight affect 55% of adults (58). Of even greater concern is the high prevalence of obesity in youths 6 to 17 years of age (174) and in preschool children (117). The prevalence of obesity appears to be increasing despite numerous public health campaigns to reduce it, including the best efforts of many governmental agencies as well as the medical, nutritional, agricultural, and pharmaceutical industries. In addition, alarming figures on the increasing prevalence of obesity-related diseases are emerging from many developing countries in Africa, Asia, and South America (195). Why is it that we are experiencing a worldwide epidemic of obesity at this point in history, and what can be done to stem the tide? Although this epidemic is likely the result of changes in environmental conditions, within any given environment, the large variability in body size and body composition seen is probably due to inherited metabolic characteristics. For the tide of this epidemic to be stemmed, scientists and public health workers will need to uncover the metabolic and environmental triggers leading to the development of obesity.

Energy Balance Equations

Under conditions of weight maintenance, the simple energy balance equation:

$$\text{Energy Intake} = \text{Energy Expenditure} \qquad \text{(Eq. 5.1)}$$

is quite valid because only limited changes in body composition are possible without changing body weight. This form of the energy balance equation has been extremely useful because of the light it has shed on the nature of reported energy intake. When assessing intakes of obese individuals, an apparent conflict was noted between reported energy intake and energy expenditure (8, 12). After careful investigation (82), it was found that this discrepancy is largely due to underreporting of food intake and overreporting of physical activity by overweight individuals.

When energy stores are changing, the following static energy balance equation is most often used in discussions and calculations of energy balance:

$$\text{Change in Energy Stores} = \text{Energy Intake} - \text{Energy Expenditure} \quad \text{(Eq 5.2)}.$$

Intuitively, it seems valid, but Alpert (5) has elegantly demonstrated that this equation is static and mathematically unbounded; thus it is invalid for calculations on living organisms for whom energy balance is dynamic and bounded. Indeed, a small initial increase in energy intake sustained over a number of years does not lead to a large weight increase, as is often claimed, because in response to weight gain, energy expenditure increases and a new equilibrium is reached. A consequence of using the static equation has been the search for small "defects" in energy expenditure using cross-sectional studies to explain significant obesity (7, 100, 192). With the dynamic energy balance equation (see equation 5.3) as the underlying tenet, the search moves away from looking for small initial "defects" in energy intake or expenditure toward finding chronic states of imbalance between the two.

The dynamic energy balance equation uses "rates" to introduce time dependency, thereby allowing the effect of changing energy stores (especially fat-free mass and weight) on energy expenditure to enter into the calculations (5):

$$\text{Rate of Change of Energy Stores} =$$
$$\text{Rate of Energy Intake} - \text{Rate of Energy Expenditure} \quad \text{(Eq 5.3)}.$$

As a consequence, after a short period of positive energy balance, the energy stores (fat-free mass and fat mass) will increase and cause an increase in energy expenditure that will balance the increased energy intake. The individual will then once again be in energy balance, but with a higher energy intake, a higher energy expenditure, and higher energy stores. Weight gain, therefore, can be viewed not only as the consequence of an initial positive energy balance but also as the mechanism by which energy balance is eventually restored.

The variability in energy intake requirements is related to the variability in energy expended for the three major components of daily expenditure: resting metabolic rate, thermogenesis, and physical activity. The *resting metabolic rate* (RMR) is the energy expended by a subject resting in bed in the morning in the fasting state under comfortable, ambient conditions. RMR includes the cost of maintaining the integrated systems of the body and the homeothermic temperature at rest. In most sedentary adults, RMR accounts for approximately 60-70% of daily energy expenditure (89, 127). *Thermogenesis* can be defined as an increase in metabolic rate in response to stimuli such as food intake, exposure to high or low temperatures, to psychological influences such as fear or stress, or as the result of administration of drugs or hormones that mimic the physiological response to such stimuli. The thermic effect of food (the major form of thermogenesis) accounts for approximately 5-15% of the daily energy expenditure (151). Lastly, *physical activity,* the most variable component of daily energy expenditure, can account for a significant number of calories in very active people. However, sedentary adult individuals exhibit a range of physical activity that represents only 20-30% of total energy expenditure.

We know that significant weight gain results from a sustained imbalance between energy intake and energy expenditure; however, this simple statement belies the complex, multifactorial nature of obesity and the numerous biological and behavioral factors that can affect both sides of the equation (figure 5.1). This chapter will review the biological and behavioral factors influencing the development of obesity.

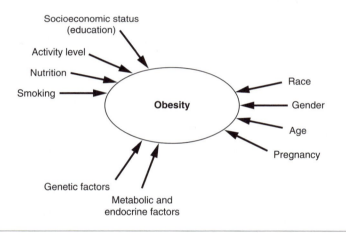

Figure 5.1   The various causes of obesity grouped according to the behavioral (activity level, nutrition, smoking status, socioeconomic status), metabolic (genetic and metabolic/endocrine factors), and biological (race, gender, age, pregnancy status) influences that predispose the individual to the development of obesity.

Metabolic Predictors of the Development of Obesity

Cross-sectional studies that compare lean and obese individuals have found that obesity is associated with high absolute energy expenditure, low respiratory quotient, insulin resistance, high sympathetic nervous system activity, and elevated plasma leptin concentrations (106, 131). In contrast, longitudinal studies that follow the same individuals over time indicate that relative to body size, a low metabolic rate, a high respiratory quotient, insulin sensitivity, low sympathetic nervous system activity, and low plasma leptin concentrations predict weight gain. Upon gaining weight, the original "abnormal" metabolic state becomes "normalized." Such "normalization" indicates why cross-sectional studies have not led to the identification of metabolic risk factors for obesity (table 5.1). Weight gain thus causes an increase in metabolic rate (126), a decrease in respiratory quotient (202), a decrease in insulin sensitivity (166), an increase in sympathetic nervous system activity (169), and an increase in plasma leptin concentrations (128, 139), all of which are much greater than the cross-sectional data would predict (130), and all of which serve to counteract further weight gain. The changes in metabolic factors that are not explained by changes in body weight and body composition are the hallmark of adaptation in response to a transient energy imbalance between intake and expenditure.

Table 5.1   Metabolic Characteristics of Obese and Pre-Obese Individuals

	Obese (factors associated with obesity)	Pre- or post-obese (factors predicting weight gain)
Relative resting metabolic rate	High or normal	Low
Energy cost of activity	Normal	Low
Fat oxidation	High	Low
Sympathetic nervous system activity	High	Low
Insulin sensitivity	Low	High
Relative leptin concentration	High	Low

Energy Intake

Hill et al. (83) consider that eating behaviors bridge the gap between the nutritional environment and the biological mechanisms of weight control. The quantity and quality of food consumed, the frequency of meal consumption, and the factors motivating one to eat are important aspects of food intake regulation. Using twin studies, de Castro (40) has shown that daily intake, including macronutrient intake and meal patterns, is genetically influenced. He found that heredity accounts for 50-64% of the variance in carbohydrate, fat, protein, alcohol, and water intake; 44% of the variance in meal frequency; and 76% of intake expressed in grams. These effects were independent of body size (41) as well as environmental factors.

Many studies have shown that dietary fat intake is a major determinant of body fat (10, 49, 132, 186), and fat intake is significantly associated with an increase in body mass index in women (81). However, there is considerable evidence showing that fat intake has actually been decreasing as the prevalence of obesity has been rising (58, 98, 191), which suggests that differences among individuals in fat metabolism may offer protection from or susceptibility to obesity; it is likely that these differences are under genetic control (132).

Several recent studies have shown that energy density is perhaps more important than dietary fat per se in influencing the number of calories consumed (14, 121, 164). High-fat diets are by definition energy dense, because fat contains more than twice the number of calories per gram as carbohydrate, and low energy-dense diets are likely to contain considerable amounts of water. People appear to eat for volume; i.e., they consume a given amount of food without adjustment for energy density (14). Thus, consumption of the same volume of a high-carbohydrate/low-fat diet might yield a considerably lower total caloric intake than consumption of a similar quantity of a high-fat/low-carbohydrate diet. Indeed, Cooling and Blundell (36) found that appetite control in habitual high-fat consumers is regulated differently than in habitual high-carbohydrate consumers; high-fat consumers ate a constant weight of food, whereas high-carbohydrate consumers ate a constant level of energy.

Long-term (over a 24-hour period) adaptation to energy density can take place in three ways: by modification of subsequent portion sizes; via adjustment of the energy density of the subsequent meal; or through alterations in meal frequency (187). Normal-weight women adapt to an energy-dense diet by increasing subsequent consumption of foods containing low energy density and decreasing consumption of foods containing high energy density (187). In contrast, obese women adapt by eating equal amounts of high and low energy-dense foods.

Appetite control, and hence energy intake, may be influenced by many different metabolic factors, such as neuropeptide activity and nutrient metabolism, and anatomic factors, such as adipose tissue storage ability and gastrointestinal tract morphology. Several of these factors will be discussed later in this chapter.

Energy Expenditure

The following sections discuss the major components of energy expenditure and their determinants (see figure 5.2).

Resting Metabolic Rate

The close correlation between resting metabolic rate (RMR) and body size has been known for some time. In earlier years, RMR was essentially considered constant for a given body size, and this led to the development of equations, now widely used, to predict RMR based on height and weight (37, 80, 141, 149). More recent studies (64, 127), however, have shown that at any given body size, and body composition, RMR can be quite different among individuals. A great deal of our effort over the past several years has been to assess the underlying mechanisms determining the variance in metabolic rate (167). We recently reported that the sleeping metabolic rate (SMR, which differs only slightly from RMR due to the energy cost of arousal), measured in 916 individuals after at least a week on a weight-maintaining diet, ranged from 876 to 3728 kcal/day with a standard deviation of 315 kcal/day. Furthermore, we found that SMR correlated best with fat-free body mass, which explained 66% of its variance (188). However, the metabolic rate still varies considerably among individuals, even after accounting for differences in fat-free body mass, age, sex, and methodological variability. Part of the unexplained variance in RMR is accounted for by family membership, suggesting that RMR is at least partially genetically determined (17). Further support for a genetic determinant of RMR comes from studies of twins by Bouchard et al. (19). These authors showed convincingly that the RMRs of monozygotic twins were more alike than the RMRs of dizygotic twins, even after adjustment for individual differences in body size and body composition.

In our search for the possible mechanisms underlying the intersubject variability in RMR, we have explored the impact of gender, physical training, age, muscle metabolism, sympathetic nervous system activity, and body temperature on RMR (129). In a large number of Caucasian volunteers, we found that females had a lower RMR compared to males (~100 kcal/d less), independent of differences in fat-free mass, fat mass, and age (54). This difference can be most likely attributed to the

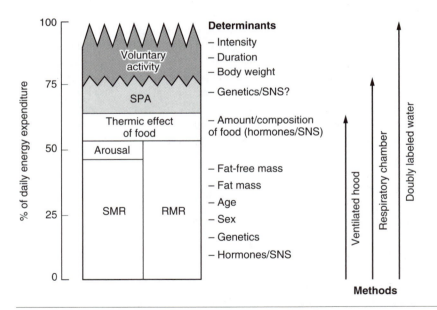

Figure 5.2   Components of daily energy expenditure in humans. Daily energy expenditure can be divided into three major components: the basal metabolic rate (sum of the sleeping metabolic rate and the energy cost of arousal), which represents 50-70% of total daily energy expenditure; the thermic effect of food, which represents ~10% of total daily energy expenditure; and the energy cost of physical activity (sum of spontaneous physical activity and unrestricted/voluntary physical activity), which represents 20-40% of total daily energy expenditure. The major determinants of the different components of total daily energy expenditure and the methods to measure them are presented.
Adapted from reference 129.

effect of sex hormones on metabolic rate and may explain part of the high susceptibility of females to obesity.

Cross-sectional studies of RMR indicate significant age-related declines (18, 156). Based on longitudinal data, Keys et al. (92) estimated that the decline in RMR was less than 1-2% per decade from the second to the seventh decade of life. Subsequent work (25, 31, 176) has supported Keys' conclusion that the decrease in RMR seen in elderly people can be explained largely by decreases in fat-free mass. In our laboratory (180), we found a small but significant negative impact of age on RMR independent of fat-free mass, fat mass, and sex. Saltzman and Roberts (147) performed over- and underfeeding experiments in older and younger men and found that aging is associated with a significantly diminished RMR response to energy imbalance and a significantly delayed thermic response to a standard meal.

Reports have shown that the RMR adjusted for fat-free mass is lower in African-Americans compared to Caucasians (4, 63, 87); this difference is especially pronounced in females (28) and among both normal weight and overweight subjects (28). This racial difference has also been noted in children (91, 199).

The *absolute* metabolic rate, expressed in kcal/day, is positively correlated with fat-free mass (17). The relationship holds for either basal or 24-hour metabolic rate versus fat-free mass or body weight (127), so that obesity clearly is associated with a high *absolute* metabolic rate (133). However, the scatter about the regression line means that at any given body size some individuals have a high, a normal, or a low *relative* metabolic rate (see figure 5.3). Several studies have examined this question

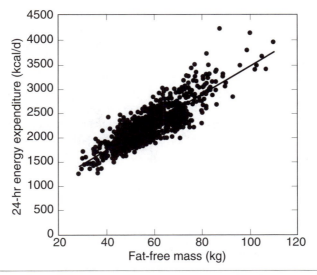

Figure 5.3 Relationship between 24-h energy expenditure (24-EE) measured in a respiratory chamber and fat-free mass in 916 subjects of different age, body weight and composition, sex, ethnicity, and glucose tolerance. Twenty-four–hour EE was significantly related to fat-free mass (r = 0.84, p < 0.0001), but at any given body size, there was a high inter-individual variability, corresponding to a standard deviation of 422 kcal/day (188).

in different populations (74, 126, 135). Roberts et al. (135) measured energy expenditure by the doubly labeled water technique in 18 infants at three months of age and then divided them according to whether they were overweight or normal weight at 12 months of age. Energy expenditure (mostly activity) was, on average, 20% lower in those who became overweight at 12 months compared to those who did not become overweight. Similarly, Griffiths et al. (74) reported that a lower energy expenditure (assessed by measured energy intake and expressed per kg body weight) in 3- to 4-year-old girls correlated negatively with BMI at adolescence.

From our own studies in adult Pima Indians (126), we have found that low *relative* metabolic rates (resting and 24-hour, normalized for fat-free mass, fat mass, age, and sex) were risk factors for body weight gain. After four years of follow-up, the risk of gaining 10 kg was approximately seven times greater in those subjects with the lowest RMR (lower tertile) compared with those with the highest RMR (higher tertile). In 95 subjects, the rate of 24-hour energy expenditure, adjusted for fat-free mass, fat mass, age, and sex, correlated negatively with the rate of weight change ($r = -0.39$, $p < 0.001$). The low *relative* metabolic rate could account for no more than 40% of the weight gain; upon gaining the weight the metabolic rate "normalized" and, indeed, even went higher than normal (126). Recently, Astrup et al. (11) reported that RMR was 8% lower in weight-stable individuals after weight loss than in matched never-obese control subjects. These investigators also found that plasma-free triiodothyronine concentrations were lower in the post-obese subjects, suggesting a possible, but currently unproved, mechanism.

Thermic Effect of Food
Many factors influence the thermic effect of food: the meal size and composition, the palatability of the food, and the time of the meal, as well as the subject's genetic background, age, physical fitness, and sensitivity to insulin. These influences plus the technical aspects of thermogenesis assessment, such as the position of the subject and the duration of measurement, mean that the thermic effect of food is the most difficult and the least reproducible component of daily energy expenditure to measure (127). All of these physiological and technical factors might explain the inconsistency and variability of the thermic effect of food as reviewed by D'Alessio et al. (38). However, in a recent review evaluating 49 studies measuring the thermic effect of food, de Jonge and Bray (43) found that 22 of 29 studies in lean and obese subjects reported a statistically significant impaired thermic effect of food among the obese, possibly related to the degree of insulin resistance among these individuals.

Although there is disagreement in the literature regarding the role of an impaired thermic effect of food in the pathogenesis of obesity, one can safely state that individual differences in the thermic effect of food can only account for small differences in daily energy expenditure. This implies that a minimal weight gain will increase energy expenditure (mostly in relationship to resting metabolic rate and the energy cost of physical activity) and will, therefore, be sufficient to offset any initial "impairment" in the thermic effect of food. Recently, Tataranni et al. (168) found no relationship between the thermic effect of food and weight change in more than 100 subjects. In contrast to rodent studies, decreased thermogenesis provides a very unlikely explanation for significant degrees of obesity in humans.

Spontaneous Physical Activity

Another important component of the 24-hour metabolic rate is the energy cost of physical activity; this will be extensively discussed in the following chapters. Physical activity can be divided into voluntary and involuntary (spontaneous) physical activity. Spontaneous physical activity (fidgeting) can account for between 8-10% of total daily energy expenditure as measured in the confines of a respiratory chamber (201). In a study conducted in Pima Indians, we found that spontaneous physical activity is a familial trait (201). Most important, in this study we also found that a low level of spontaneous physical activity was associated with subsequent weight gain in males, but not in females.

Respiratory Quotient and Low Rates of Fat Oxidation

The composition of nutrient intake has been shown to be an important factor in the genesis of obesity, and consequently, one might expect that the composition of nutrient oxidation would also play a role. Although this topic will be addressed in greater detail in chapter 7, it is essential to make mention of it here because of the impact of the respiratory quotient on the etiology of obesity. The nonprotein respiratory quotient, which is an index of the ratio of carbohydrate to fat oxidation, ranges from a value of about 0.80 after an overnight fast, where fat is the main oxidative substrate (108), to values close to 1.00 after a carbohydrate meal where glucose is the major substrate (1). Apart from the effects of diet composition, the respiratory quotient is also influenced by recent energy balance (negative balance causing more fat oxidation), sex (females tending toward carbohydrate oxidation and fat storage), adiposity (higher fat mass means higher fat oxidation), and family membership, suggesting genetic determinants. Indeed, in a recent study, Toubro et al. (171) found that substrate oxidation rates measured by respiratory quotient exhibit familial correlation after adjustment for energy balance, sex, and age.

In a longitudinal study in Pima Indians, the 24-hour respiratory quotient showed considerable interindividual variation (part of it aggregating in families) and predicted weight gain (202). Those in the 90th percentile for respiratory quotient ("low fat oxidizers") had a 2.5 times greater risk of gaining 5 kg or more body weight than those in the 10th percentile ("high fat oxidizers"). Similar results were found in Caucasian volunteers participating in the Baltimore Longitudinal Study of Aging (155). In support of these observations, recent studies have demonstrated that post-obese volunteers have a consistently high respiratory quotient, i.e., a low rate of fat oxidation (9, 99); and those who were able to maintain their weight loss during follow-up were found to have a lower respiratory quotient compared to those experiencing weight relapse (65, 178).

Insulin Sensitivity

Insulin sensitivity is measured as the total glucose disposal stimulated by insulin and is the sum of carbohydrate oxidation and nonoxidative glucose disposal (mainly storage). Because higher rates of carbohydrate oxidation (as measured by the 24-hour respiratory quotient) predict weight gain, we investigated whether insulin sensitivity would also predict weight gain in Pima Indians. Subjects in the 90th percentile (most insulin sensitive) were 3 to 4 times more likely to gain 10 kg or more body weight compared to those in the 10th percentile (most insulin resistant), and this effect was more closely related to the oxidative than the nonoxidative component (166). Thus insulin sensitivity predicts weight gain, or conversely, insulin resistance is associated with significantly lower rates of weight gain. Similar observations were reported in Mexican Americans (177) and Caucasians from the San Luis Valley study (85). Recently, Yost et al. (200) reported that increased insulin sensitivity predicts weight regain following sustained weight loss.

It appears that, at least in adults, insulin resistance seems to be a mechanism counteracting further weight gain (50). In contrast, in 5- to 10-year-old children a high fasting plasma insulin concentration (reflecting insulin resistance) seemed to predict overweight 7-8 years later, especially in boys (116). Taken together, these results indicate that hyperinsulinemia (possibly due to the effect of insulin resistance) promotes excess weight gain during growth in childhood (antilipolytic effect of insulin) and that long-lasting insulin resistance then protects adults against further weight gain.

Sympathetic Nervous System Activity

The sympathetic nervous system (SNS) is believed to play a role in the etiology of obesity through its effects on both energy expenditure and food intake. Among Pima Indians, low activity of the SNS at baseline has been found to predict weight gain and is also associated with increased accumulation of abdominal adipose tissue (169). SNS activity seems to be involved in the regulation of the RMR under eucaloric conditions (142, 152). Using microneurography to assess muscle sympathetic nervous system activity (MSNA), we found that variability in energy expenditure was related to variability in sympathetic activity (162). A low level of SNS activity, therefore, may cause weight gain via a decreased metabolic rate. However, low SNS activity has also been shown to be associated with hyperphagia in both animals (21) and humans (123). Furthermore, a recent study in our laboratory indicates that low levels of SNS activity, as measured by microneurography, are associated with low rates of fat oxidation (159). In response to weight gain, SNS activity clearly increases and, therefore, restores energy balance via increased metabolic rate and increased fat oxidation (169).

The causes of the inverse relationship found between food intake and SNS activity are not clear. A currently popular view is that the same central pathways that regulate SNS activity also influence feeding behavior and that autonomic activity and food intake are coordinately regulated in the control of energy balance (90). Clearly a better understanding of the factors affecting SNS activity is needed.

Leptin

Leptin, a 16-kilodalton protein secreted by adipocytes, is believed to regulate food intake through a negative feedback signal between adipose tissue stores and the satiety centers in the hypothalamus, possibly mediated by decreased neuropeptide Y activity and increased corticotropin releasing hormone (CRH) activity (154). Although plasma leptin concentrations are significantly correlated to body weight and body-fat percentage (35, 106) in many diverse groups of subjects, leptin concentrations in the central nervous system remain fairly constant when plasma concentrations are greater than approximately 25 ng/ml (26), suggesting that most obesity is leptin resistant (26, 153). Leptin resistance may be due to defects in leptin transport across the blood-brain barrier (a barrier membrane between the circulating blood and the brain that prevents certain damaging substances from reaching brain tissues and cerebrospinal fluid), or to defects in the leptin-signaling sites and action within the central nervous system (27).

Despite the correlation between leptin concentrations and body fatness, there is substantial variation in plasma leptin concentrations among individuals with comparable obesity, suggesting that some individuals produce relatively low amounts of leptin (106). We have found that relatively low plasma leptin concentrations are a risk factor for weight gain (128). In response to weight gain, plasma leptin concentrations now become "normal" for the new body size and body composition. However, baseline leptin concentrations do not appear to predict success at weight loss or weight loss maintenance (194).

Other Hormones

Although it has long been recognized that hypothyroidism can lower the RMR, that overactivity of the adrenal gland causes central obesity, and that certain rare genetic abnormalities (such as pseudohypoparathyroidism) are associated with obesity, few other hormones seem to influence obesity. Several studies have convincingly shown that there are food intake (20, 107) and energy expenditure (161, 184) fluctuations over the course of the menstrual cycle, which may indicate that sex hormones play a role in the development of obesity (24). A recent study has found that women who consistently ate the most high-fat snacks in response to laboratory stressors secreted significantly more endogenous cortisol after stress than those who did not consistently choose high-fat snacks (51), suggesting that the role of hormones in obesity warrants greater scrutiny.

Neuropeptides

The role of the central nervous system in the regulation of food intake has long been recognized, but the exact mechanisms have not been clearly delineated. Lesions of the ventromedial hypothalamus result in hyperphagia, whereas destruction of the lateral hypothalamus decreases food intake and results in starvation, suggesting that these areas represent the "satiety" and "feeding" centers of the brain, respectively (61). Recent efforts have identified several neuropeptides that modulate food intake, appetite, and energy expenditure. The list of these neuropeptides is growing every month and was recently reviewed by Flier and Maratos-Flier (61). These neuropeptides are discussed in the following paragraphs.

Neuropeptide Y (NPY) is a member of the pancreatic polypeptide family of proteins that is widely distributed in the central and peripheral nervous systems. When administered directly into the brain, NPY is a potent stimulator of food intake and inhibitor of energy expenditure (181). NPY was considered among the most compelling targets of leptin action (163) until it was found that NPY knockout mice are not obese and respond normally to the satiety effects of leptin (53). Such results suggest that the cerebral physiological pathways regulating body weight are most likely redundant.

Agouti; Agouti Gene-Related Protein (AGRP); and Pro-Opiomelanocortin (POMC) are recently discovered proteins related to obesity. Yellow agouti mice overexpress a protein known as agouti in the central nervous system (110). These animals exhibit a phenotype consisting of obesity, insulin resistance, infertility, and tumor susceptibility (111). Subsequent studies have shown that agouti causes obesity by antagonizing the action of α-melanocyte stimulating hormone (α-MSH) at the level of the hypothalamic melanocortin-4 receptor (MC4; 66). α-MSH is derived from the gene known as pro-opiomelanocortin (POMC), which has been linked with body fat in Mexican-Americans (34), suggesting a possible role for this pathway in the pathophysiology of human obesity. Indeed, a recent study described two individuals with mutations in the POMC gene who exhibited early-onset obesity (97), confirming a role for POMC in human energy balance. Currently, both the human homologue of agouti, known as the agouti signaling protein, and the melanocortin receptors are under investigation and may provide new pathways for the treatment of human obesity. The agouti gene-related protein (AGRP) is another protein that acts like the agouti protein by blocking the anorectic effect of α-MSH on the MC4 receptor (119).

Melanin-Concentrating Hormone (MCH) neurons in the lateral hypothalamus project to the nucleus of the solitary tract as well as to the medial prefrontal cortex (16). Intracerebroventricular administration of MCH to rats appears to stimulate food intake; this action is countered by the effects of α-MSH. Recent evidence suggests that these peptides exert functional and antagonistic influences on feeding behavior (105).

Galanin is a neuropeptide that stimulates feeding when injected into the paraventricular nucleus of the rat hypothalamus. Although galanin gene expression has been found to be positively correlated with spontaneous fat intake (3) and administration of galanin results in acute hyperphagia, compensatory hypophagia appears to follow shortly thereafter (158). The physiological role of galanin in energy balance is questionable at present.

Orexins are two newly discovered neuropeptides localized in neurons within and around the lateral and posterior hypothalamus (143); they bind and activate two G protein-coupled receptors. Orexin expression increases with starvation, and central administration of these proteins induces hyperphagia (143). The role of these proteins as mediators of food intake regulation requires further investigation.

Cocaine- and Amphetamine-Regulated Transcript (CART) is a new anorectic peptide closely associated with leptin and NPY in animal models (96). Food-deprived rats show a marked decrease in CART mRNA in the arcuate nucleus, and in animals showing disruptions in leptin signaling, CART is almost completely absent. Intracerebral injections of CART completely block the feeding response induced by NPY, suggesting that CART may be an endogenous inhibitor of food intake in normal animals.

Studies in animal models have increased our knowledge of the role of certain genes in energy balance. It is very likely that many of these pathways are redundant and impact on both energy intake and energy expenditure via stimulation/inhibition of the autonomic nervous system. New neuropeptides and pathways are likely to be discovered in the near future, which will enhance our understanding of the regulation of energy balance and will also show the complexity and redundancy of this regulation.

Additional Mediators of Energy Balance

Uncoupling Proteins (UCP) are mitochondrial proteins that generate heat and burn calories by short-circuiting ATP production from substrate oxidation by returning hydrogen ions to the inner mitochondrial membrane (95). UCP1 is found primarily in brown adipose tissue and is therefore unlikely to play a role in human obesity. UCP2 is widely distributed in adult human tissues and is upregulated, or increased, in white adipose tissue in response to fat feeding (60), whereas UCP3 is largely found in skeletal muscle (182). We have recently reported a negative correlation between UCP3 mRNA levels and BMI and a positive correlation between UCP3 mRNA levels and RMR in Pima Indians (150), suggesting a role for this newly discovered protein in energy balance in humans.

Glucagon-Like Peptide-1 (GLP-1) is a peptide derived from enteroendocrine cells in the intestine. GLP-1 acts on the pancreas to stimulate insulin secretion and thus appears to have a significant role in carbohydrate metabolism (76). Specific GLP-1 receptors have been found in the paraventricular nucleus of the hypothalamus; therefore, GLP-1 is also thought to act centrally to inhibit food intake (175) most likely due to taste aversion (75).

Cholecystokinin (CCK) is released from the intestine after consumption of a meal and appears to affect satiety through stimulation of the vagus nerve. Large doses of exogenous CCK lower meal size, but lead to a compensatory increase in meal frequency, resulting in no overall change in caloric intake (185). It is possible that CCK has a central satiety effect that may play an inhibitory role in the development of obesity (101).

Genetic Studies

Most of these described predictors of weight gain are genetically determined metabolic traits (33). We recently performed a genomic scan to search for linkage of these traits to obesity in Pima Indians and found significant linkage for percent body fat, 24-hour energy expenditure, and 24-hour respiratory quotient to different regions of the genome (115). Using different analytical techniques, we found evidence of linkage (logarithm of odds [LOD] > 2.0) at chromosomes 11q21-q22 and 18q21 for percent body fat; 11q23-q24 for metabolic rate; and 1p31-p21 and 20q11.2 for 24-hour respiratory quotient. Possible candidate genes include the leptin receptor at 1p31 and the agouti signaling protein at 20q11.2. The combination of genome-wide scans in different populations and new techniques designed to identify differential expression of genes in tissues from obese and lean subjects will allow detection of the genes underlying the susceptibility of individuals to obesity.

Behavioral Influences on the Development of Obesity

The development of obesity can be influenced by numerous behavioral and environmental factors ranging from family socioeconomic status to perceptions of body image; from birth order to child care provisions; and from mother's education level to the number of hours spent sleeping at night. Parental obesity is reported to more than double the risk of adult obesity among both obese and nonobese children under 10 years of age (189), and some of this influence is likely to be due to environmental as well as genetic factors. Garn et al. (67) found that low-income boys and girls showed a greater long-term increase in body fatness over an 18-year period than children from higher income families. Lissau-Lund-Sørensen and Sørensen (103) reported that the quality of the dwelling in which 9- to 10-year-old children were reared was the single most important risk factor for overweight in young adulthood (assessed 10 years later): being reared in a poor area increased the risk of becoming overweight more than three times compared to being reared in a relatively more prosperous area, even after controlling for parental education and occupation.

Obesity in children has also been associated with family size (86), birth order (124), social class position (44, 136), marital status (160), level of social support (47, 68) and parental support (102), family functioning (190), and parental education level and occupation (103), among other factors. These factors can profoundly influence both dietary and activity habits, ultimately affecting the development of obesity (see table 5.2).

Table 5.2   Influence of the Family Environment on the Prevalence of Obesity in Children

Variable	Effect	Variable	Effect
Family size		Level of parental support	
Single child	+	Strong	−
> 1 child	−		
		Family functioning	
Birth order		High	−
Youngest of many	+		
		Parental education	
Social class		Advanced	−
Upper class	−		
		Parental occupational status	
Marital status		High	−
Single parent household	+		
Level of social support			
Strong	−		

+ effect denotes a positive correlation between the prevalence of obesity and the familial variable while a − effect denotes a negative correlation.

Energy Intake

Familial influences on food choices have been extensively studied. In the Netherlands, family resemblance in fat intake has been found between parents and children from 1-3 years of age up to 21 years of age (55). This may be due to common meal consumption (118). However, Klesges et al. (94) found that the threat of parental knowledge of their food intake behavior resulted in young children choosing more nutritious foods and fewer calories than when they were told there would be no monitoring, suggesting that parental influences go beyond meal sharing. Preferences for fatty foods and total fat intake have been found to correlate with parental BMI in 3- to 5-year-old children (56) and specifically maternal obesity in 4- to 7-year-old children (114). Mela (109) suggested that learning plays a major role in the formation of most sensory preferences, and these early preferences for dietary fat may represent learned behaviors.

Eating behavior may have a considerable influence on the development of obesity. Binge eating (197), overeating, and impulsive eating (148) occur with greater frequency among the obese (198); these behaviors have been linked to situational antecedents such as eating in the car; snacking while bored, depressed, tired, or irritable; and overeating in response to meal skipping (148). Obese children eat faster and do not slow down their rate of eating toward the end of the meal as much as normal weight children (13). Indeed, in infants, a vigorous feeding style characterized by rapid suckling, at higher pressure, with a shorter interval between bursts of suckling is associated with higher caloric intake and greater adiposity at 1-2 years of age (2). And it appears that during the first few years of life, controls over food intake become more complex as children learn to eat in response to the presence of palatable food, the social setting, and their emotional states (15). Eating behavior can clearly affect the total number of calories consumed, and thus the development of obesity.

The reported effects of alcohol consumption on total calorie intake are conflicting. Colditz et al. (32) found no effect of alcohol on body weight gain, whereas de Castro and Orozco (42) found that calories from ingested alcohol have an additive effect on overall intake. Due to a transitory reduction in lipid oxidation, alcohol does not inhibit the intake of other substrates (165), but displaces fat oxidation up to 74% of the energy content of alcohol (112). In the Quebec Family Study, alcohol intake was associated with a high daily caloric intake (173) and fat gain, particularly in the trunk area (172).

An important psychological variable influencing the development of obesity is body image perception. In many cultures, a large body size is considered a symbol of wealth and prestige, although that appears to be changing. In Samoa, for example, where large body size is prevalent, people do not have as strongly negative a view of obesity as is common in the United States (22). And the Ojibway-Cree people of northwestern Ontario, Canada who express dissatisfaction with their own large body image, still prefer a larger body size than Caucasian populations (69).

Energy Expenditure

Behavioral and environmental influences on energy expenditure generally relate to the physical activity component of energy expenditure. It is well known that activity habits are established at a young age; studies on intervention strategies in adults show that those people who can most readily be persuaded to take up more activity are those who, as children, were physically active at school and during their leisure (88). Because physical activity will be extensively discussed in the following chapters, only brief mention will be made here.

In children, major behavioral influences obviously emanate from the parents. An early study measuring RMR and energy intake in 3- to 4-year-old children found that children with at least one obese parent consumed less energy, but also expended less energy, especially in physical activity, than those with no obese parents (73). Roberts et al. (135) used doubly labeled water to measure total energy expenditure (TEE) in 3-month-old infants of lean and obese mothers and found that TEE was 20% lower (mostly due to reduced activity) in those who were overweight at one year of age. In this study, 50% of the children of overweight mothers were themselves overweight, whereas none of the infants of lean mothers were overweight. In contrast, Davies et al. (39) found that no aspect of infant energy expenditure, in children of similar age as those in the Roberts et al. study (135), was related to parental BMI. In 4- to 6-year-old children, Goran et al. (71) found no relationship between parental BMI and TEE, RMR, or activity energy expenditure in the offspring. Despite inconsistent results in this area, it is notable that Epstein et al. (52) found that offspring of obese parents rated the enjoyment of physical activity lower than offspring of lean parents.

It may be that physical activity in the young is more strongly related to the activity, rather than the obesity, level of the parents. Gottlieb and Chen (72) reported that parental exercise, father's occupation, and ethnicity were significantly related to the overall frequency of exercise in 7th and 8th grade students. Sallis et al. (146) studied 3- to 5-year-old children and found that the effects of parental role modeling on child physical activity levels extended to free-play settings beyond the confines of the home environment. Klesges et al. (93) determined that parental participation can increase children's activity levels, especially in children having one or both parents overweight, and that activity patterns track among siblings. Clearly, activity patterns of children are strongly influenced by the familial environment.

A sedentary lifestyle, characterized not only by a lack of vigorous exercise, but by increased physical inactivity (29), represents a significant risk factor for obesity development, especially for children. Several reports have found that today's young children expend only 30-40% more calories in physical activity energy expenditure over the RMR (62, 70, 144). This is in contrast to the World Health Organization (196) recommendation that the food intake requirement for children 4 to 6 years of age be 70-100% above the RMR. Several studies have linked television and video viewing to body fatness in both children (6, 23, 48) and adults (57, 134). We have recently shown that the risk of obesity in young adulthood is associated with increased time spent watching television in childhood (145). As computer usage becomes more popular and computer ownership more accessible; as more children spend more time in child care facilities where sedentary activities are often used to "babysit" children; as parents keep children indoors due to fear of the danger and traffic of city streets; and, as fiscal constraints force educational facilities to sacrifice physical education programs, physical inactivity will increase in our population and, very likely, the prevalence of obesity as well.

Critical Periods for the Development of Obesity

The nutritional programming hypothesis put forth by Lucas (104) suggests that there are sensitive periods when nutritional manipulations may program obesity development, perhaps by changes in gene expression, preferential clonal selection of adapted cells in specific tissues, and/or differential proliferation of tissue cell types. Identifying key periods in life for the development of obesity may allow us to develop strategies for prevention and treatment of this disease.

Childhood

Dietz (46) has identified three periods during childhood that are critical for the development of obesity: the intrauterine period, the period between the ages of 4 and 6 years when adiposity rebound can occur (138), and adolescence.

The Dutch famine data from World War II indicate that maternal starvation during the last trimester of pregnancy protects against obesity in the offspring, possibly by influencing adipose tissue cellularity (125). In contrast, starvation during the first two trimesters promotes obesity in the offspring, perhaps due to an impairment of food intake regulation (125). These results seem to indicate that limiting maternal weight gain during the third trimester of pregnancy may be beneficial to the offspring.

In youth, adiposity rebound is defined as the onset of the second period of rapid growth in body fat, beginning at about 5-6 years of age. Rolland-Cachera et al. (137) found that an early age of adiposity rebound (< 5.5 years) is associated with greater adiposity in early adulthood. These results were confirmed by Siervogel et al. (157) in the Fels Longitudinal Growth Study. Dietz (45) offered three mechanisms that may explain why adiposity rebound contributes to adult obesity: This period may be critical for learning dietary and exercise behaviors; early adiposity rebound may reflect early maturation; and, children with early adiposity rebound may be those who, in utero, were exposed to gestational diabetes.

In adolescence, rapid maturation is associated with greater obesity in adulthood (179). In the Harvard Growth Study (113), adolescent weight in males was associated with increased mortality from coronary heart disease, ischemic heart disease, and colorectal cancer, independently of the effect adolescent weight had on adult weight. It is possible that this increased risk is related to the central deposition of body fat that takes place in males during adolescence (45).

Adulthood

In adult women, pregnancy often represents a risk period for obesity development. Excessive maternal weight is associated with increased rates of labor problems, Cesarean section, and fetal macrosomia, among other problems (30). More importantly, obesity is associated with parity; permanent weight gains of 1.5-5.0 kg for each pregnancy have been reported (79, 140).

Smoking cessation also results in weight gain, perhaps due to the effect of nicotine on energy expenditure and appetite (120). Flegal et al. (59) estimate that the 10-year weight gain associated with smoking cessation was 4.4 kg in men and 5.0 kg in women.

Although weight gain during the time of menopause is a commonly reported symptom, the evidence linking weight gain to menopause is conflicting. Some studies have not confirmed the postmenopausal increase in weight (78, 193), but rather have found a change in body-fat-mass and body-fat distribution (170, 183) at this time. Moreover, hormone replacement therapy appears to modulate these changes and prevents the central distribution of body fat commonly reported after menopause (77). As more of the population ages, it is likely that this area of research will attract greater attention.

Future Directions

The rising tide of obesity worldwide presents us with two main challenges: to treat the individuals who are currently obese and to prevent obesity in those who are still lean. Although the strategies for treatment and prevention are very different, both are doomed if our understanding of the disease is inadequate; from this understanding flow the paradigms upon which clinicians and public health practitioners base their programs and actions. Given the limited success of treatment programs and the total failure of public health programs, it is important to reexamine the paradigms upon which these programs are based.

The move from the static energy balance equation (equation 5.2) to the human physiological model of dynamic nutrient balance equations represents a paradigm shift away from assuming that the origins of obesity lie *only* in metabolic "defects," psychological "abnormalities," or genetic "mutations" within the individual to assuming that the differences between individuals represent the normal spectrum of physiological or genetic variance. The corollary of accepting that the "pathology" does not lie only within the individual in many cases, is that it probably lies within the environment. Indeed, the increasing prevalence of obesity has emerged in parallel with an increase in dietary fat content and with a decrease in physical activity (83, 122). Therefore, more detailed research on effective ways of modifying the behaviors associated with these environmental risk factors is necessary if we are going to adequately address this emerging epidemic. A new paradigm may emerge from such research indicating that, in many cases, obesity may result from "normal physiological variability within a pathoenvironment."

Historically, epidemics are controlled only when the environmental factors behind them are included in the resolution paradigms and the ensuing public health actions. Diverse examples of this are infectious diseases (draining the swamps for malaria, infection control policies in hospitals), road crashes (median barriers, speed policing), smoking (taxation, smoke-free legislation), and heart disease (saturated fat content of manufactured food, payments to dairy and meat producers based on protein, not fat, content). The obesity epidemic needs a similar environmental approach to complement the current educational, behavioral, genetic, pharmacological, and surgical approaches.

A greater understanding of the physiology of obesity will come from a variety of sources. Foremost among these is the current research isolating genes related to obesity. The discovery of new neuropeptides involved in the control of energy intake and energy expenditure is a good example of what molecular and genetic research approaches can teach us. Discoveries of new pathophysiological pathways underlying the susceptibility to obesity, in combination with the identification of the environmental determinants of obesity, will allow for the development of coherent public health strategies designed to both treat and prevent obesity at the population level.

References

1. Acheson, K.J., Y. Schutz, T. Bessard, E. Ravussin, E. Jéquier, and J.P. Flatt. 1984. Nutritional influences on lipogenesis and thermogenesis after a carbohydrate meal. *American Journal of Physiology* 246:E62-E70.
2. Agras, W.S., H.C. Kraemer, R.I. Berkowitz, A.F. Korner, and L.D. Hammer. 1987. Does a vigorous feeding style influence early development of adiposity? *Journal of Pediatrics* 110:799-804.
3. Akabayashi, A., J.I. Koenig, Y. Watanabe, J.T. Alexander, and S.F. Leibowitz. 1994. Galanin-containing neurons in the paraventricular nucleus: a neurochemical marker for fat ingestion and body weight gain. *Proceedings of the National Academy of Sciences, U.S.A.* 91:10375-10379.
4. Albu, J., M. Shur, M. Curi, L. Murphy, S.B. Heymsfield, and F.X. Pi-Sunyer. 1997. Resting metabolic rate in obese, premenopausal black women. *American Journal of Clinical Nutrition* 66:531-538.
5. Alpert, S. 1990. Growth, thermogenesis, and hyperphagia. *American Journal of Clinical Nutrition* 52:784-792.
6. Anderson, R.E., C.J. Crespo, S.J. Bartlett, L.J. Cheskin, and M. Pratt. 1998. Relationship of physical activity and television watching with body weight and level of fatness among children. *Journal of the American Medical Association* 279:938-942.
7. Anonymous. 1988. Exercise and energy balance. *Lancet* 1:392-394.
8. Ashford, J., M. Sawyer, and R.G. Whitehead. 1986. High levels of energy expenditure in obese women. *British Medical Journal* 292:983-987.
9. Astrup, A., B. Buemann, N.J. Christensen, and S. Toubro. 1994. Failure to increase lipid oxidation in response to increasing dietary fat content in formerly obese women. *American Journal of Physiology* 266:E592-E599.
10. Astrup, A., B. Buemann, P. Western, S. Toubro, A. Raben, and N.J. Christensen. 1994. Obesity as an adaptation to a high-fat diet: evidence from a crosssectional study. *American Journal of Clinical Nutrition* 59:350-355.
11. Astrup, A., B. Buemann, S. Toubro, C. Ranneries, and A. Raben. 1996. Low resting metabolic rate in subjects predisposed to obesity: a role for thyroid status. *American Journal of Clinical Nutrition* 63:879-883.
12. Bandini, L.G., D.A. Schoeller, H. Cyr, and W.H. Dietz. 1990. Validity of reported energy intake in obese and non-obese adolescents. *American Journal of Clinical Nutrition* 52:421-425.

13. Barkeling, B., S. Ekman, and S. Rössner. 1992. Eating behaviour in obese and normal weight 11-year-old children. *International Journal of Obesity and Related Metabolic Disorders* 16:355-360.
14. Bell, E.A., V.H. Castellanos, C.L. Pelkman, M.L. Thorwart, and B.J. Rolls. 1998. Energy density of foods affects energy intake in normal-weight women. *American Journal of Clinical Nutrition* 67:412-420.
15. Birch, L.L. 1998. Psychological influences on the childhood diet. *Journal of Nutrition* 128(Suppl.):407S-410S.
16. Bittencourt, J.C., F. Presse, C. Arias, C. Peto., J. Vaughan, J.L. Nahon, W. Vale, and P.E. Sawchenko. 1992. The melanin-concentrating hormone system of the rat brain: an immuno- and hybridization histochemical characterization. *Journal of Comparative Neurology* 319:218-245.
17. Bogardus, C., S. Lillioja, E. Ravussin, W. Abbott, J. Zawadzki, A. Young, W.C. Knowler, R. Jacobowitz, and P.P. Moll. 1986. Familial dependence of the resting metabolic rate. *New England Journal of Medicine* 315:96-100.
18. Boothby, W.M., J. Berkson, and H.L. Dunn. 1936. Studies of the energy of metabolism of normal individuals: a standard for basal metabolism with a nomogram for clinical application. *American Journal of Physiology* 116:468-484.
19. Bouchard, C., A. Tremblay, A. Nadeau, J.P. Després, G. Thériault, M.R. Boulay, G. Lortie, C. Leblanc, and G. Fournier. 1989. Genetic effect in resting and exercise metabolic rates. *Metabolism* 38:364-370.
20. Bowen, D.J., and N.E. Grunberg. 1990. Variations in food preference and consumption across the menstrual cycle. *Physiology and Behavior* 47:287-291.
21. Bray, G.A. 1993. The nutrient balance hypothesis: peptides, sympathetic activity, and food intake. *Annals of the New York Academy of Sciences* 676:223-241.
22. Brewis, A.A., S.T. McGarvey, J. Jones, and B.A. Swinburn. 1998. Perceptions of body size in Pacific Islanders. *International Journal of Obesity and Related Metabolic Disorders* 22:185-189.
23. Buchowski, M.S., and M. Sun. 1996. Energy expenditure, television viewing and obesity. *International Journal of Obesity and Related Metabolic Disorders* 20:236-244.
24. Buffenstein, R., S.D. Poppitt, R.M. McDevitt, and A.M. Prentice. 1995. Food intake and the menstrual cycle: a retrospective analysis, with implications for appetite research. *Physiology and Behavior* 58:1067-1077.
25. Calloway, D.H., and E. Zannie. 1980. Energy requirements and energy expenditure of elderly men. *American Journal of Clinical Nutrition* 33:2088-2092.
26. Caro, J.F., J.W. Kolaczynski, M.R. Nyce, J.P. Ohannesian, I. Opentanova, W.H. Goldman, R.B. Lynn, P.-L. Zhang, M.K. Sinha, and R.V. Considine. 1996. Decreased cerebrospinal-fluid/serum leptin ratio in obesity: a possible mechanism for leptin resistance. *Lancet* 348:159-161.

27. Caro, J.F., M.K. Sinha, J.W. Kolaczynski, P.L. Zhang, and R.V. Considine. 1996. Leptin: the tale of an obesity gene. *Diabetes* 45:1455-1462.
28. Carpenter, W.H., T. Fonong, M.J. Toth, P.A. Ades, J. Calles-Escandon, J.D. Walston, and E.T. Poehlman. 1998. Total daily energy expenditure in free-living older African-Americans and Caucasians. *American Journal of Physiology* 274:E96-E101.
29. Ching, P.L.Y.H., W.C. Willett, E.B. Rimm, G.A. Colditz, S.L. Gortmaker, and M.J. Stampfer. 1996. Activity level and risk of overweight in male health professionals. *American Journal of Public Health* 86:25-30.
30. Cnattingius, S., R. Bergstrom, L. Lipworth, and M.S. Kramer. 1998. Prepregnancy weight and the risk of adverse pregnancy outcomes. *New England Journal of Medicine* 338:147-152.
31. Cohn, S.H., D. Vartzky, S. Yasumura, A. Sawitsky, I. Zanzi, A. Vaswani, and K.J. Ellis. 1980. Compartmental body composition based on total-body nitrogen, potassium, and calcium. *American Journal of Physiology* 239:E524-E530.
32. Colditz, G.A., E. Giovannucci, E.B. Rimm, M.J. Stampfer, B. Rosner, F.E. Speizer, E. Gordis, and W.C. Willett. 1991. Alcohol intake in relation to diet and obesity in women and men. *American Journal of Clinical Nutrition* 54:49-55.
33. Comuzzie, A.G., and D.B. Allison. 1998. The search for human obesity genes. *Science* 280:1374-1377.
34. Comuzzie, A.G., J.E. Hixson, L. Almasy, B.D. Mitchell, M.C. Mahaney, T.D. Dyer, M.P. Stern, J.W. MacCluer, and J. Blangero. 1997. A major quantitative trait locus determining serum leptin levels and fat mass is located on human chromosome 2. *Nature Genetics* 15:273-276.
35. Considine, R.V., M.K. Sinha, M.L. Heiman, A. Kriauciunas, T.W. Stephens, M.R. Nyce, J.P. Ohannesian, C.C. Marco, L.J. McKee, T.L. Bauer, and J.F. Caro. 1996. Serum immunoreactive-leptin concentrations in normal-weight and obese humans. *New England Journal of Medicine* 334:292-295.
36. Cooling, J., and J.E. Blundell. 1998. Are high-fat and low-fat consumers distinct phenotypes? Differences in the subjective and behavioural response to energy and nutrient challenges. *European Journal of Clinical Nutrition* 52:193-201.
37. Cunningham, J.J. 1991. Body composition as a determinant of energy expenditure: a synthetic review and a proposed general prediction equation. *American Journal of Clinical Nutrition* 54:963-969.
38. D'Alessio, D.A., E.C. Kavle, M.A. Mozzoli, J. Smalley, M. Polansky, Z.V. Kendrick, L.R. Owen, M.C. Bushman, G. Boden, and O.E. Owen. 1988. Thermic effect of food in lean and obese men. *Journal of Clinical Investigation* 81:1781-1789.
39. Davies, P.S.W., J.C.K. Wells, C.A. Fieldhouse, J.M.E. Day, and A. Lucas. 1995. Parental body composition and infant energy expenditure. *American Journal of Clinical Nutrition* 61:1026-1029.

40. de Castro, J.M. 1993. Genetic influences on daily intake and meal patterns of humans. *Physiology & Behavior* 53:777-782.
41. de Castro, J.M. 1993. Independence of genetic influences on body size, daily intake and meal patterns of humans. *Physiology & Behavior* 54:633-639.
42. de Castro, J., and S. Orozco. 1990. Moderate alcohol intake and spontaneous eating patterns of humans: evidence of unregulated supplementation. *American Journal of Clinical Nutrition* 52:246-253.
43. de Jonge, L., and G.A. Bray. 1997. The thermic effect of food and obesity: a critical review. *Obesity Research* 5:622-631.
44. De Spiegelaere M., M. Dramaix, and P. Hennart. 1998. The influence of socioeconomic status on the incidence and evolution of obesity during early adolescence. *International Journal of Obesity and Related Metabolic Disorders* 22:268-274.
45. Dietz, W.H. 1998. Childhood weight affects adult morbidity and mortality. *Journal of Nutrition* 128(Suppl.):411S-414S.
46. Dietz, W.H. 1994. Critical periods in childhood for the development of obesity. *American Journal of Clinical Nutrition* 59:955-959.
47. Dietz, W.H. 1983. Childhood obesity: Susceptibility, cause, and management. *The Journal of Pediatrics* 103:676-686.
48. Dietz, W.H., and S.L. Gortmaker. 1985. Do we fatten our children at the television set? Obesity and television viewing in children and adolescents. *Pediatrics* 75:807-812.
49. Doucet, E., and A. Tremblay. 1997. Food intake, energy balance and body weight control. *European Journal of Clinical Nutrition* 51:846-855.
50. Eckel, R.H. 1992. Insulin resistance: an adaptation for weight maintenance. *Lancet* 340:1452-1453.
51. Epel, S.E., R. Lapidus, K.B. Horgen, S. Kingston, J. Ickovics, and K. Brownell. 1998. Who's eating and when? The predictive power of stress reactivity in a laboratory study of stress-induced eating. *Annals of Behavioral Medicine* 20(Abstract):S032.
52. Epstein, L.H., A. Valoski, R.R. Wing, K.A. Perkins, M. Fernstrom, B. Marks, and J. McCurley. 1989. Perception of eating and exercise in children as a function of child and parent weight status. *Appetite* 12:105-118.
53. Erickson, J.C., J.E. Clegg, and R.D. Palmiter. 1996. Sensitivity to leptin and susceptibility to seizures of mice lacking neuropeptide Y. *Nature* 381:415-421.
54. Ferraro, R., S. Lillioja, A.M. Fontvieille, R. Rising, C. Bogardus, and E. Ravussin. 1992. Lower sedentary metabolic rate in women compared with men. *Journal of Clinical Investigation* 90:780-784.
55. Feunekes, G.U., A. Stafleu, C. de Graaf, and W.A. van Stavern. 1997. Family resemblance in fat intake in the Netherlands. *European Journal of Clinical Nutrition* 51:793-799.
56. Fisher, J.O., and L.L. Birch. 1995. Fat preferences and fat consumption of 3- to 5-year-old children are related to parental adiposity. *Journal of the American Dietetic Association* 95:759-764.

57. Fitzgerald, S.J., A.M. Kriska, M.A. Pereira, and M.P. de Courten. 1997. Associations among physical activity, television watching, and obesity in adult Pima Indians. *Medicine and Science in Sports and Exercise* 29:910-915.

58. Flegal, K.M., M.D. Carroll, R.J. Kuczmarski, and C.L. Johnson. 1998. Overweight and obesity in the United States: prevalence and trends, 1960-1994. *International Journal of Obesity and Related Metabolic Disorders* 22:39-47.

59. Flegal, K.M., R.P. Troiano, E.R. Pamuk, R.J. Kuczmarski, and S.M. Campbell. 1995. The influence of smoking cessation on the prevalence of overweight in the United States. *New England Journal of Medicine* 333:1165-1170.

60. Fleury, C., M. Neverova, S. Collins, S. Raimbault, O. Champigny, C. Levi-Meyrueis, F. Bouillaud, M.F. Saldin, R.S. Surwit, D. Ricquier, and C.H. Warden. 1997. Uncoupling protein-2: a novel gene linked to obesity and hyperinsulinemia. *Nature Genetics* 15:269-272.

61. Flier, J.S., and E. Maratos-Flier. 1998. Obesity and the hypothalamus: novel peptides for new pathways. *Cell* 92:437-440.

62. Fontvieille, A.M., I.T. Harper, R.T. Ferraro, M. Spraul, and E. Ravussin. 1993. Daily energy expenditure by five-year-old children, measured by doubly labeled water. *Journal of Pediatrics* 123:200-207.

63. Foster, G.D., T.A. Wadden, and R.A. Vogt. 1997. Resting energy expenditure in obese African-American and Caucasian women. *Obesity Research* 5:1-8.

64. Frankenfield, D.C., E.R. Muth, and W.A. Rowe. 1998. The Harris-Benedict studies of human basal metabolism: history and limitations. *Journal of the American Dietetic Association* 98:439-445.

65. Froidevaux, F., Y. Schutz, L. Christin, and E. Jéquier. 1993. Energy expenditure in obese women before and during weight loss, after refeeding and in the weight-relapse period. *American Journal of Clinical Nutrition* 57:35-42.

66. Gantz, I., Y. Konda, T. Tashiro, Y. Shimoto, H. Miwa, G. Munzert, S.J. Watson, J. DelValle, and T. Yamada. 1993. Molecular cloning of a novel melanocortin receptor. *Journal of Biological Chemistry* 268:8246-8250.

67. Garn, S.M., P.J. Hopkins, and A.S. Ryan. 1984. Differential fatness gain of low income boys and girls. *American Journal of Clinical Nutrition* 34:1465-1468.

68. Gerald, L.B., A. Anderson, G.D. Johnson, C. Hoff, and R.F. Trimm. 1994. Social class, social support and obesity risk in children. *Child Care, Health and Development* 20:145-163.

69. Gittelsohn, J., S.B. Harris, A.L. Thorne-Lyman, A.J.G. Hanley, A. Barnie, and B. Zinman. 1996. Body image concepts differ by age and sex in an Ojibway-Cree community in Canada. *Journal of Nutrition* 126:2990-3000.

70. Goran, M.I., W.H. Carpenter, and E.T. Poehlman. 1993. Total energy expenditure in 4- to 6-yr-old children. *American Journal of Physiology* 264:E706-711.

71. Goran, M.I., W.H. Carpenter, A. McGloin, R. Johnson, J.M. Hardin, and R.L. Weinsier. 1995. Energy expenditure in children of lean and obese parents. *American Journal of Physiology* 268:E917-E924.
72. Gottlieb, N.H., and M.-S. Chen. 1985. Sociocultural correlates of childhood sporting activities: their implications for heart health. *Social Science and Medicine* 21:533-539.
73. Griffiths, M., and P.R. Payne. 1976. Energy expenditure in small children of obese and non-obese parents. *Nature* 260:698-700.
74. Griffiths, M., P.R. Payne, A.J. Stunkard, J.P.W. Rivers, and M. Cox. 1990. Metabolic rate and physical development in children at risk of obesity. *Lancet* 336:76-77.
75. Gunn, I., D. O'Shea, and S.R. Bloom. 1997. Control of appetite—the role of glucagon-like peptide-1 (7-36) amide. *Journal of Endocrinology* 155:197-200.
76. Gutniak, M., C. Orskov, J.J. Holst, B. Ahren, and S. Efendic. 1992. Antidiabetogenic effect of glucagon-like peptide-1 (7-36) amide in normal subjects and patients with diabetes mellitus. *New England Journal of Medicine* 326:1316-1322.
77. Haarbo, J., U. Marslew, A. Gotfredsen, and C. Christiansen. 1991. Post-menopausal hormone replacement therapy prevents central distribution of body fat after menopause. *Metabolism* 40:1323-1326.
78. Hamman, R.F., P.H. Bennett, and M. Miller. 1975. The effect of menopause on serum cholesterol in American (Pima) Indian women. *American Journal of Epidemiology* 102:164-169.
79. Harris, H.E., G.T. Ellison, and M. Holliday. 1997. Is there an independent association between parity and maternal weight gain? *Annals of Human Biology* 24:507-519.
80. Harris, J.A., and F.G. Benedict. 1919. *A biometric study of basal metabolism in man.* Publication No. 279. Washington, D.C.: The Carnegie Institute.
81. Heitmann, B.L., L. Lissner, T.I.A. Sørensen, and C. Bengtsson. 1995. Dietary fat intake and weight gain in women genetically predisposed for obesity. *American Journal of Clinical Nutrition* 61:1213-1217.
82. Heymsfield, S.B., P.C. Darby, L.S. Muhlheim, D. Gallagher, C. Wolper, and D.B. Allison. 1995. The calorie: myth, measurement, and reality. *American Journal of Clinical Nutrition* 62(Suppl.):1034S-41S.
83. Hill, A.J., P.J. Rogers, and J.E. Blundell. 1995. Techniques for the experimental measurement of human eating behavior and food intake: a practical guide. *International Journal of Obesity and Related Metabolic Disorders* 19:361-375.
84. Hill, J.O., and J.C. Peters. 1998. Environmental contributions to the obesity epidemic. *Science* 280:1371-1374.
85. Hoag, S., J.A. Marshall, R.H. Jones, and R.F. Hamman. 1995. High fasting insulin levels associated with lower rates of weight gain in persons with normal glucose tolerance: the San Luis Valley Diabetes Study. *International Journal of Obesity and Related Metabolic Disorders* 19:175-180.

86. Jacoby, A., D.G. Altman, J. Cook, W.W. Holland, and A. Elliot. 1975. Influence of some social and environmental factors on the nutrient intake and nutritional status of schoolchildren. *British Journal of Preventive Medicine* 29:116-120.

87. Jakicic, J.M., and R.R. Wing. 1998. Differences in resting energy expenditure in African-American vs. Caucasian overweight females. *International Journal of Obesity and Related Metabolic Disorders* 22:236-242.

88. James, W.P.T. 1995. A public health approach to the problem of obesity. *International Journal of Obesity and Related Metabolic Disorders* 19(Suppl. 3):S37-S43.

89. Jéquier E., and Y. Schutz. 1983. Long-term measurements of energy expenditure in humans using a respiratory chamber. *American Journal of Clinical Nutrition* 38:989-998.

90. Kaiyala, K.J., S.C. Woods, and M.W. Schwartz. 1995. New model for the regulation of energy balance and adiposity by the central nervous system. *American Journal of Clinical Nutrition* 62(Suppl.):1123S-1134S.

91. Kaplan, A.S., B.S. Zemel, and V.A. Stallings. 1996. Differences in resting energy expenditure in prepubertal black children and white children. *Journal of Pediatrics* 129:643-647.

92. Keys, A., H.L. Taylor, and F. Grande. 1987. Basal metabolism and age of adult man. *Metabolism* 22:5979-5987.

93. Klesges, R.C., L.H. Eck, C.L. Hanson, C.K. Haddock, and L.M. Klesges. 1990. Effects of obesity, social interactions, and physical environment on physical activity in preschoolers. *Health Psychology* 9:435-449.

94. Klesges, R.C., R.J. Stein, L.H. Eck, T.R. Isbell, and L.M. Klesges. 1991. Parental influence on food selection in young children and its relationships to childhood obesity. *American Journal of Clinical Nutrition* 53:859-864.

95. Klingenberg, M. 1992. Mechanism and evolution of the uncoupling protein of brown adipose tissue. *Trends in Biochemical Science* 15:108-112.

96. Kristensen, P., M.E. Judge, L. Thim, U. Ribel, K.N. Christjansen, B.B. Wulff, J.T. Clausen, P.B. Jensen, O.D. Madsen, N. Vrang, P.J. Larsen, and S. Hastrup. 1998. Hypothalamic CART is a new anorectic peptide regulated by leptin. *Nature* 393:72-76.

97. Krude, H., H. Biebermann, W. Luck, R. Horn, G. Brabant, and A. Gruters. 1998. Severe early-onset obesity, adrenal insufficiency and red hair pigmentation caused by POMC mutations in humans. *Nature Genetics* 19:155-157.

98. Kuczmarski, R.J., K.M. Flegal, S.M. Campbell, and C.L. Johnson. 1994. The increasing prevalence of overweight among US adults. The National Health and Nutrition Examination Surveys, 1960-1991. *Journal of the American Medical Association* 272:205-211.

99. Larson, D.E., R.T. Ferraro, D.S. Robertson, and E. Ravussin. 1995. Energy metabolism in weight-stable postobese individuals. *American Journal of Clinical Nutrition* 62:735-739.

100. Leibel, R.L. 1990. Is obesity due to a heritable difference in "set-point" for adiposity? *Western Journal of Medicine* 153:429-431.

101. Lieverse, R.J., J.B.M.J. Jansen, A.A.M. Masclee, and C.B.H.W. Lamers. 1995. Satiety effects of a physiological dose of cholecystokinin in humans. *Gut* 36:176-179.

102. Lissau, I., and T.I.A. Sørensen. 1994. Parental neglect during childhood and increased risk of obesity in young adulthood. *Lancet* 343:324-327.

103. Lissau-Lund-Sørensen, I., and T.I.A. Sørensen. 1992. Prospective study of the influence of social factors in childhood on risk of overweight in young adulthood. *International Journal of Obesity and Related Metabolic Disorders* 16:169-175.

104. Lucas, A. 1998. Programming by early nutrition: an experimental approach. *Journal of Nutrition* 128(Suppl.):401S-406S.

105. Ludwig, D.S., K.G. Mountjoy, J.B. Tatro, J.A. Gillette, R.C. Frederich, J.S. Flier, and E. Maratos-Flier. 1998. Melanin-concentrating hormone: a functional melanocortin antagonist in the hypothalamus. *American Journal of Physiology* 274:E627-E633.

106. Maffei, M., J. Halaas, E. Ravussin, R.E. Pratley, G.H. Lee, Y. Zhang, H. Fei, S. Kim, R. Lallone, S. Ranganathan, P.A. Kern, and J.M. Friedman. 1995. Leptin levels in human and rodent: measurement of plasma leptin and *ob* RNA in obese and weight-reduced subjects. *Nature Medicine* 1:1155-1161.

107. Martini, M.C., J.W. Lampe, J.L. Slavin, and M.S. Kurzer. 1994. Effect of the menstrual cycle on energy and nutrient intake. *American Journal of Clinical Nutrition* 60:898-899.

108. McNeil, G., A.C. Bruce, A. Ralph, and W.P.T. James. 1988. Inter-individual differences in fasting nutrient oxidation and the influence of diet composition. *International Journal of Obesity and Related Metabolic Disorders* 12:445-463.

109. Mela, D.J. 1995. Understanding fat preference and consumption: application of behavioural sciences to a nutritional problem. *Proceedings of the Nutrition Society* 54:453-464.

110. Miller, M.W., D.M. Duhl, H. Vrieling, S.P. Cordes, M.M. Ollmann, B.M. Winkes, and G.S. Barsh. 1993. Cloning of the mouse agouti gene predicts a secreted protein ubiquitously expressed in mice carrying the lethal yellow mutation. *Genes and Development* 7:454-467.

111. Miltenberger, R.J., R.L. Mynatt, J.E. Wilkenson, and R.P. Woychik. 1997. The role of the *agouti* gene in the yellow obese syndrome. *Journal of Nutrition* 127(Suppl. 9S):1902S-1907S.

112. Murgatroyd, P.R., M.L.H.M. Van de Ven, G.R. Goldberg, and A.M. Prentice. 1996. Alcohol and the regulation of energy balance: overnight effects on diet-induced thermogenesis and fuel storage. *British Journal of Nutrition* 75:33-45.

113. Must, A., P.F. Jacques, G.E. Dallal, C.J. Bajema, and W.H. Dietz. 1992. Long-term morbidity and mortality of overweight adolescents. A follow-up

study of the Harvard Growth Study of 1922 to 1935. *New England Journal of Medicine* 327:1350-1355.

114. Nguyen, V.T., D.E. Larson, R.K. Johnson, and M.I. Goran. 1996. Fat intake and adiposity in children of lean and obese parents. *American Journal of Clinical Nutrition* 63:507-513.

115. Norman, R.A., P.A. Tataranni, R. Pratley, D.B. Thompson, R.L. Hanson, M. Prochazka, L. Baier, M.G. Ehm, H. Sakul, T. Foroud, W.T. Garvey, D. Burns, W.C. Knowler, P.H. Bennett, C. Bogardus, and E. Ravussin. 1998. Autosomal genomic scan for loci linked to obesity and energy metabolism in Pima Indians. *American Journal of Human Genetics* 62:659-668.

116. Odeleye, O.E., M. de Courten, D.J. Pettitt, and E. Ravussin. 1997. Fasting hyperinsulinemia is a predictor of increased body weight gain and obesity in Pima Indian children. *Diabetes* 46:1341-1345.

117. Ogden, C.L., R.P. Troiano, R.R. Briefel, R.J. Kuczmarski, K.M. Flegal, and C.L. Johnson. 1998. Prevalence of overweight among preschool children in the United States, 1971 through 1994. *Pediatrics* 99:E1-E7.

118. Oliveria S.A., R.C. Ellison, L.L. Moore, M.W. Gillman, E.J. Garrahie, and M.R. Singer. 1992. Parent-child relationships in nutrient intake: the Framingham Children's Study. *American Journal of Clinical Nutrition* 56:593-598.

119. Ollmann, M.M., B.D. Wilson, Y.K. Yang, J.A. Kerns, Y. Chen, I. Gantz, and G.S. Barsh. 1997. Antagonism of central melanocortin receptors in vitro and in vivo by agouti-related protein. *Science* 278:135-138.

120. Perkins, K.A. 1993. Weight gain following smoking cessation. *Journal of Consulting and Clinical Psychology* 61:768-777.

121. Prentice, A.M. 1998. Manipulation of dietary fat and energy density and subsequent effects on substrate flux and food intake. *American Journal of Clinical Nutrition* 67(Suppl.):535S-541S.

122. Prentice, A.M., and S.A. Jebb. 1995. Obesity in Britain: gluttony or sloth? *British Medical Journal* 311:437-439.

123. Raben, A., J.J. Holst, N.J. Christensen, and A. Astrup. 1996. Determinants of postprandial appetite sensations: macronutrient intake and glucose metabolism. *International Journal of Obesity and Related Metabolic Disorders* 20:161-169.

124. Ravelli, G.P., and L. Belmont. 1979. Obesity in nineteen-year-old men: family size and birth order association. *American Journal of Epidemiology* 109:66-70.

125. Ravelli, G.P., Z.A. Stein, and M.W. Susser. 1976. Obesity in young men after famine exposure in utero and early infancy. *New England Journal of Medicine* 295:349-353.

126. Ravussin, E., S. Lillioja, W.C. Knowler, L. Christin, D. Freymond, W.G.H. Abbott, V. Boyce, B.V. Howard, and C. Bogardus. 1988. Reduced rate of energy expenditure as a risk factor for body weight gain. *New England Journal of Medicine* 318:467-472.

127. Ravussin, E., S. Lillioja, T.E. Anderson, L. Christin, and C. Bogardus. 1986. Determinants of 24-hour energy expenditure in man: methods and results using a respiratory chamber. *Journal of Clinical Investigation* 78:1568-1578.

128. Ravussin, E., R.E. Pratley, M. Maffei, H. Wang, J.M. Friedman, P.H. Bennett, and C. Bogardus. 1997. Relatively low plasma leptin concentrations precede weight gain in Pima Indians. *Nature Medicine* 2:238-240.

129. Ravussin, E., and B.A. Swinburn. 1996. Energy expenditure and obesity. *Diabetes Reviews* 4:403-422.

130. Ravussin, E., and B.A. Swinburn. 1993. Metabolic predictors of obesity: cross-sectional versus longitudinal data. *International Journal of Obesity and Related Metabolic Disorders* 17(Suppl. 3):S28-S31.

131. Ravussin, E., and B. Swinburn. 1992. Pathophysiology of obesity. *Lancet* 340:404-408.

132. Ravussin, E., and P.A. Tataranni. Dietary fat and human obesity. 1997. *Journal of the American Dietetic Association* 97(Suppl. 1):S42-S46.

133. Ravussin, E., F. Zurlo, R. Ferraro, and C. Bogardus. 1990. Energy expenditure in man: determinants and risk factors for body weight gain. In *Progress in Obesity Research*, eds. Y. Oomura, S. Tarui, S. Inoui, T. Shimazu, 175-182. London: John Libbey & Company Ltd.

134. Rissanen, A.M., M. Heliövaaram, P. Knekt, A. Reunanen, and A. Aromaa. 1991. Determinants of weight gain and overweight in adult Finns. *European Journal of Clinical Nutrition* 45:419-430.

135. Roberts, S.B., J. Savage, W.A. Coward, B. Chew, and A. Lucas. 1988. Energy expenditure and intake in infants born to lean and overweight mothers. *New England Journal of Medicine* 318:461-466.

136. Rolland-Cachera, M.F., and F. Bellisle. 1986. No correlation between adiposity and food intake: why are working class children fatter? *American Journal of Clinical Nutrition* 44:779-787.

137. Rolland-Cachera, M.F., M. Deheeger, M. Guilloud-Bataille, P. Avons, E. Patois, and M. Sempe. 1987. Tracking the development of adiposity from one month of age to adulthood. *Annals of Human Biology* 14:219-229.

138. Rolland-Cachera, M.F., M. Deheeger, F. Bellisle, M. Sempe, M. Guilloud-Bataille, and E. Patois. 1984. Adiposity rebound in children: a simple indicator for predicting obesity. *American Journal of Clinical Nutrition* 39:129-135.

139. Rosenbaum, M., M. Nicolson, J. Hirsch, E. Murphy, F. Chu, and R.L. Leibel. 1997. Effects of weight change on plasma leptin concentrations and energy expenditure. *Journal of Clinical Endocrinology and Metabolism* 82:3647-3654.

140. Rössner, S., and A. Ohlin. 1995. Pregnancy as a risk factor for obesity: lessons from the Stockholm Pregnancy and Weight Development Study. *Obesity Research* 3(Suppl. 2):267S-275S.

141. Roza, A.M., and H.M. Shizgal. 1984. The Harris-Benedict equation reevaluated: resting energy requirements and the body cell mass. *American Journal of Clinical Nutrition* 40:168-182.

142. Saad, M.F., S.A. Alger, F. Zurlo, J.B. Young, C. Bogardus, and E. Ravussin. 1991. Ethnic differences in sympathetic nervous system-mediated energy expenditure. *American Journal of Physiology* 261:E789-E794.

143. Sakurai, T., A. Amemiya, M. Ishii, I. Matsuzaki, R.M. Chemelli, H. Tanaka, S.C. Williams, J.A. Richardson, G.P. Kozlowski, S. Wilson, J.R.S. Arch, R.E. Buckingham, A.C. Haynes, S.A. Carr, R.S. Annan, D.E. McNulty, W.-S. Liu, J.A. Terrett, N.A. Elshourbagy, D.J. Bergsma, and M. Yanagisawa. 1998. Orexins and orexin receptors: a family of hypothalamic neuropeptides and G protein-coupled receptors that regulate feeding behavior. *Cell* 92:573-585.

144. Salbe, A.D., A.M. Fontvieille, I.T. Harper, and E. Ravussin. 1997. Low levels of physical activity in 5-year-old children. *Journal of Pediatrics* 131:423-429.

145. Salbe, A.D., C. Weyer, A.M. Fontvieille, and E. Ravussin. 1998. Low levels of physical activity and time spent viewing television at 9 years of age predict weight gain 8 years later in Pima Indian children. *International Journal of Obesity and Related Metabolic Disorders* 22(Suppl. 4)(Abstract):S10.

146. Sallis, J.F., T.L. Patterson, T.L. McKenzie, and P.R. Nader. 1988. Family variables and physical activity in preschool children. *Journal of Developmental and Behavioral Pediatrics* 9:57-61.

147. Saltzman, E., and S.B. Roberts. 1996. Effects of energy imbalance on energy expenditure and respiratory quotient in young and older men: a summary of data from two metabolic studies. *Aging* (Milano) 8:370-378.

148. Schlundt, D.G., J.O Hill, T. Sbrocco, J. Pope-Cordle, and T. Kasser. 1990. Obesity: a bioenergetic or biobehavioral problem. *International Journal of Obesity and Related Metabolic Disorders* 14:815-828.

149. Schofield, W.N. 1985. Predicting basal metabolic rate, new standards and review of previous work. *Human Nutrition: Clinical Nutrition* 39C(Suppl. 1):5-14.

150. Schrauwen, P., J. Xia, C. Bogardus, R.E. Pratley, and E. Ravussin. 1999. Skeletal muscle uncoupling protein 3 expression is a determinant of energy expenditure in Pima Indians. *Diabetes* 48:146-149.

151. Schutz Y., T. Bessard, and E. Jéquier. 1984. Thermogenesis measured over a whole day in obese and non-obese women. *American Journal of Clinical Nutrition* 40:542-552.

152. Schwartz, R.S., L.F. Jaeger, and R.C. Veith. 1988. Effect of clonidine on the thermic effect of feeding in humans. *American Journal of Physiology* 254:R90-R94.

153. Schwartz, M.W., E. Peskind, M. Raskind, E.J. Boyko, and D. Porte, Jr. 1996. Cerebrospinal fluid leptin levels: relationship to plasma levels and to adiposity in humans. *Nature Medicine* 2:589-593.

154. Schwartz, M.W., R.J. Seeley, L.A. Campfield, P. Burn, and D.G. Baskin. 1996. Identification of targets of leptin action in rat hypothalamus. *Journal of Clinical Investigation* 98:1101-1106.

155. Seidell, J.C., D.C. Muller, J.D. Sorkin, and R. Andres. 1992. Fasting respiratory exchange ratio and resting metabolic rate as predictors of weight gain: the Baltimore Longitudinal Study on Aging. *International Journal of Obesity and Related Metabolic Disorders* 16:667-674.

156. Shock, N.W., and M.J. Yiengst. 1955. Age changes in basal respiratory measurements and metabolism in males. *Journal of Gerontology* 10:31-40.

157. Siervogel, R.M. A.F. Roche, S. Guo, D. Mukherjee, and W.C. Chumlea. 1991. Patterns of change in weight/stature^2 from 2 to 18 years: findings from long-term serial data for children in the Fels Longitudinal Growth Study. *International Journal of Obesity and Related Metabolic Disorders* 15:479-485.

158. Smith, B.K., D.A. York, and G.A. Bray. 1994. Chronic cerebroventricular galanin does not induce sustained hyperphagia or obesity. *Peptides* 15:1267-1272

159. Snitker, S., P.A. Tataranni, and E. Ravussin. 1998. Respiratory quotient is inversely associated with muscle sympathetic nerve activity. *Journal of Clinical Endocrinology and Metabolism* 83:3977-3979.

160. Sobal, J., and A.J. Stunkard. 1989. Socioeconomic status and obesity: A review of the literature. *Psychological Bulletin* 105:260-275.

161. Solomon, S.J., M.S. Kurzer, and D.H. Calloway. 1982. Menstrual cycle and basal metabolic rate in women. *American Journal of Clinical Nutrition* 36:611-616.

162. Spraul, M., E. Ravussin, A.M. Fontvieille, R. Rising, D.E. Larson, and E.A. Anderson. 1993. Reduced sympathetic nervous activity. A potential mechanism predisposing to body weight gain. *Journal of Clinical Investigation* 92:1730-1735.

163. Stephens, T.W., M. Basinski, P.K. Bristow, J.M. Bue-Valleskey, S.G. Burgett, L. Craft, J. Hale, J. Hoffmann, H.M. Hsiung, A. Kriauciunas, W. MacKelllar, P.R. Rosteck, Jr., B. Schoner, D. Smith, F.C. Tinsley, X-Y. Zhang, and M. Helman. 1995. The role of neuropeptide Y in the antiobesity action of the *obese* gene product. *Nature* 377:530-532.

164. Stubbs, R.J., P. Ritz, W.A. Coward, and A.M. Prentice. 1995. Covert manipulation of the ratio of dietary fat to carbohydrate and energy density: effect on food intake and energy balance in free-living men eating ad libitum. *American Journal of Clinical Nutrition* 62:330-337.

165. Suter, P., Y. Schutz, and E. Jéquier. 1992. The effect of ethanol on fat storage in healthy subjects. *New England Journal of Medicine* 326:983-987.

166. Swinburn, B.A., B.L. Nyomba, M.F. Saad, F. Zurlo, I. Raz, W.C. Knowler, S. Lillioja, C. Bogardus, and E. Ravussin. 1991. Insulin resistance associated with lower rates of weight gain in Pima Indians. *Journal of Clinical Investigation* 88:168-173.

167. Tataranni, P.A., and E. Ravussin. 1995. Variability in metabolic rate: biological sites of regulation. *International Journal of Obesity and Related Metabolic Disorders* 19:S102-106.

168. Tataranni, P.A., D.E. Larson, S. Snitker, and E. Ravussin. 1995. Thermic effect of food in humans: methods and results from use of a respiratory chamber. *American Journal of Clinical Nutrition* 61:1013-1019.

169. Tataranni, P.A., J.B. Young, C. Bogardus, and E. Ravussin. 1997. A low sympathoadrenal activity is associated with body weight gain and development of central adiposity in Pima Indian men. *Obesity Research* 5:341-347.

170. Tchernof, A., and E.T. Poehlman. 1998. Effects of the menopause transition on body fatness and body fat distribution. *Obesity Research* 6:246-254.

171. Toubro, S., T.I.A. Sørensen, C. Hindsberger, N.J. Christensen, and A. Astrup. 1998. Twenty-four-hour respiratory quotient: the role of diet and familial resemblance. *Journal of Clinical Endocrinology and Metabolism* 83:2758-2764.

172. Tremblay, A., B. Buemann, G. Thériault, and C. Bouchard. 1995. Body fatness in active individuals reporting low lipid and alcohol intake. *European Journal of Clinical Nutrition* 49:824-831.

173. Tremblay, A., E. Wouters, M. Wenker, S. St-Pierre, C. Bouchard, and J.-P. Després. 1995. Alcohol and a high-fat diet: a combination favoring overfeeding. *American Journal of Clinical Nutrition* 62:639-644.

174. Troiano, R.P., K.M. Flegal, R.J. Kuczmarski, S.M. Campbell, and C.L. Johnson. 1995. Overweight prevalence and trends for children and adolescents. *Archives of Pediatric and Adolescent Medicine* 149:1085-1091.

175. Turton, M.D., D. O'Shea, I. Gunn, S.A. Beak, C.M. Edwards, K. Meeran, S.J. Choi, G.M. Taylor, M.M. Heath, P.D. Lambert, J.P. Wilding, D.M. Smith, M.A. Ghatei, J. Herbert, and S.R. Bloom. 1996. A role for glucagon-like peptide-1 in the central regulation of feeding. *Nature* 379:69-72.

176. Tzankoff, S.P. and A.H. Norris. 1977. Effect of muscle mass decrease on age-related BMR changes. *Journal of Applied Physiology* 43:1001-1006.

177. Valdez, R., B.D. Mitchell, S.M. Haffner, H.P. Hazuda, P.A. Morales, A. Monterrosa, and M.P. Stern. 1994. Predictors of weight change in a bi-ethnic population. The San Antonio Heart Study. *International Journal of Obesity and Related Metabolic Disorders* 18:85-91.

178. Valtuena, S., J. Salas-Salvado, and P.G. Lorda. 1997. The respiratory quotient as a prognostic factor in weight-loss rebound. *International Journal of Obesity and Related Metabolic Disorders* 21:811-817.

179. van Lenthe, F.J., H.C.G. Kemper, and W. van Mechelen. 1996. Rapid maturation in adolescence results in great obesity in adulthood: The Amsterdam Growth and Health Study. *American Journal of Clinical Nutrition* 64:18-24.

180. Vaughn, L., F. Zurlo, and E. Ravussin. 1991. Aging and energy expenditure. *American Journal of Clinical Nutrition* 53:821-825.

181. Vettor, R., N. Zarjevski, I. Cusin, F. Rohner-Jeanrenaud, and B. Jeanrenaud. 1994. Induction and reversibility of an obesity syndrome by

intracerebroventricular neuropeptide Y administration to normal rats. *Diabetologia* 37:1202-1208.

182. Vidal-Puig, A., G. Solanes, D. Grujic, J.S. Flier, and B.B. Lowell. 1997. UCP3: an uncoupling protein homologue expressed preferentially and abundantly in skeletal muscle and brown adipose tissue. *Biochemica Biophysica Research Communications* 235:79-82.

183. Wang, Q., C. Hassager, P. Ravn, S. Wang, and C. Christiansen. 1994. Total and regional body-composition changes in early postmenopausal women: age-related or menopause-related? *American Journal of Clinical Nutrition* 60:843-848.

184. Webb, P. 1986. 24-hour energy expenditure and the menstrual cycle. *American Journal of Clinical Nutrition* 44:614-619.

185. West, D.B., D. Fey, and S.C. Woods. 1984. Cholecystokinin persistently suppresses meal size but not food intake in free-feeding rats. *American Journal of Physiology* 246:R776-R787.

186. Westerterp, K.R., W.P.H.G. Verboeket-van de Venne, M.S. Westerterp-Plantenga, E.J.M. Velthuis-te Wierik, C. de Graaf, and J.A. Westrate. 1996. Dietary fat and body fat: an intervention study. *International Journal of Obesity and Related Metabolic Disorders* 20:1022-1026.

187. Westerterp-Plantenga, M.S., W.J. Pasman, M.J.W. Yedema, and N.E.G. Wijckmans-Duijsens. 1996. Energy intake adaptation of food intake to extreme densities of food by obese and non-obese women. *European Journal of Clinical Nutrition* 50:401-407.

188. Weyer, C., S. Snitker, R. Rising, E. Ravussin, and C. Bogardus. 1999. Determinants of energy expenditure and fuel utilization in man: effects of body composition, age, sex, ethnicity, and glucose tolerance in 916 subjects. *International Journal of Obesity and Related Metabolic Disorders* 203:715-722.

189. Whitaker, R.C., J.A. Wright, M.S. Pepe, K.D. Seidel, and W.H. Dietz. 1997. Predicting obesity in young adulthood from childhood and parental obesity. *New England Journal of Medicine* 337:869-873.

190. Wilkins, S.C., O.W. Kendrick, K.R. Stitt, N. Stinett, and V.A. Hammarlund. 1998. Family functioning is related to overweight in children. *Journal of the American Dietetic Association* 98:572-574.

191. Willett, W.C. 1998. Is dietary fat a major determinant of body fat? *American Journal of Clinical Nutrition* 67(Suppl.):556S-562S.

192. Willett, W.C., and M. J. Stampfer. 1986. Total energy intake: implications for epidemiologic analyses. *American Journal of Epidemiology* 124(1):17-27.

193. Wing, R.R., K.A. Matthews, L.H. Kuller, E.N. Meilahn, and P.L. Plantinga. 1991. Weight gain at the time of menopause. *Annals of Internal Medicine* 151:97-102.

194. Wing, R.R., M.K. Sinha, R.V. Considine, W. Lang, and J.F. Caro. 1996. Relationship between weight loss maintenance and changes in serum leptin levels. *Hormone and Metabolism Research* 28:698-703.

195. World Health Organization. 1998. *Obesity: Preventing and managing the global epidemic*. Report of a WHO consultation on obesity, Geneva, 3-5 June 1997. Geneva: World Health Organization.

196. World Health Organization. 1985. *Energy and protein requirements*. Report of a joint FAO/WHO/UNU expert consultation. Technical report series 724. Geneva: World Health Organization.

197. Yanovski, S.Z. 1993. Binge eating disorders: current knowledge and future directions. *Obesity Research* 1:306-318.

198. Yanovski, S.Z., M. Leet, J.A. Yanovski, M. Flood, P.W. Gold, H.R. Kissileff, and B.T. Walsh. 1992. Food selection and intake of obese women with binge eating disorder. *American Journal of Clinical Nutrition* 56:975-980.

199. Yanovski, S.Z., J.C. Reynolds, A.J. Boyle, and J.A. Yanovski. 1997. Resting metabolic rate in African-American and Caucasian girls. *Obesity Research* 5:321-325.

200. Yost, T., D.R. Jensen, and R.H. Eckel. 1995. Weight regain following sustained weight reduction is predicted by relative insulin sensitivity. *Obesity Research* 3:583-587.

201. Zurlo, F., R. Ferraro, A.M. Fontvieille, R. Rising, C. Bogardus, and E. Ravussin. 1992. Spontaneous physical activity and obesity: cross-sectional and longitudinal studies in Pima Indians. *American Journal of Physiology* 263:E296-E300.

202. Zurlo, F., S. Lillioja, A. Esposito-Del Puente, B.L. Nyomba, I. Raz, M.F. Saad, B.A. Swinburn, W.C. Knowler, C. Bogardus, and E. Ravussin. 1990. Low ratio of fat to carbohydrate oxidation as a predictor of weight gain: study of 24-h RQ. *American Journal of Physiology* 259:E650-E657.

CHAPTER 6

Assessment of Human Energy Expenditure

Christopher L. Melby, Richard C. Ho, and James O. Hill

Department of Food Science and Human Nutrition, Colorado State University, Ft. Collins, CO, U.S.A., and Center for Human Nutrition, University of Colorado Health Sciences Center, Denver, CO, U.S.A.

The laws of thermodynamics dictate that an energy surplus is at the root of all obesity; if energy intake (E_{IN}) exceeds energy expenditure (E_{OUT}), energy storage will occur. Accurate assessment of both E_{IN} and E_{OUT} is important in understanding energy balance and body-weight regulation. In this chapter, we focus on the energy expenditure side of the energy balance equation. We will briefly discuss the principles of calorimetry, and then describe the methods used to assess the components of energy expenditure, giving special attention to the variety of methods used to assess the energy expended in physical activity.

Calorimetry

The source of energy for metabolic work in the human body is the phosphoanhydride bond of adenosine triphosphate (ATP) (41). All of the energy released from hydrolysis of ATP molecules that is used for metabolic work is degraded to heat and must be released from the body if internal temperature is to remain constant. The oxidation of macronutrients provides the energy necessary to resynthesize the ATP molecules (i.e., regenerate ATP from ADP and P_i), but less than 50% of the potential energy available from the oxidation of the macronutrients is actually conserved in the phosphoanhydride bonds of ATP. This results in significant heat loss accompanying nutrient oxidation (41). As the newly synthesized ATP molecules undergo hydrolysis to provide energy for metabolic work, the energy released will be transformed into heat. Energy expenditure can be accurately assessed by measuring heat released or, indirectly, by measuring the amount of oxygen consumed in the oxidation of macronutrients necessary for ATP synthesis (38).

Direct Calorimetry

Direct calorimeters measure the actual heat released by the body (133). If body heat storage remains unchanged from the time the person enters the calorimeter until leaving, the amount of heat released equals the amount of heat produced and provides a measure of energy expenditure. Room calorimeters have been built that are capable of measuring heat production over an extended period of time, typically at least 24 hours. These calorimeters have a slow response time because of their size and the time lag between when heat is produced by the body and when heat is measured by the calorimeter. This problem is compounded during exercise, when increased heat storage elevates body-core temperature. This stored heat is released slowly over time and is eventually detected by the calorimeter. Therefore, when the measurement period involves exercise, a relatively long period is needed for accurate assessment of energy expenditure by direct calorimetry.

Substrate oxidation rates cannot be determined by direct calorimetry. Although direct calorimetry is considered the gold standard for measuring energy expenditure, few laboratories throughout the world routinely use direct calorimetry to determine energy expenditure in human subjects.

Indirect Calorimetry

Indirect calorimetry, in contrast with direct calorimetry, is technically less difficult, requires less costly equipment, and allows calculation of both energy expenditure and rates of substrate oxidation. The most common types of indirect calorimetry are the respiratory gas exchange, doubly labeled water, and labeled bicarbonate methods, all of which are described below.

Respiratory Gas Exchange

Indirect calorimetry by respiratory gas exchange is based on determination of the rates of oxygen consumption ($\dot{V}O_2$) and carbon dioxide production ($\dot{V}CO_2$). These quantities are derived by analyzing the concentrations of oxygen and carbon dioxide in inspired and expired air. The respiratory quotient (RQ) is calculated as $\dot{V}O_2/\dot{V}CO_2$, is a unitless value, and can be used to estimate substrate oxidation rates. Energy expenditure is then estimated by calculating the energy equivalent of each liter of oxygen consumed. Energy expenditure can be determined most accurately if substrate oxidation is also calculated (40), but it can be estimated based on a typical pattern of substrate oxidation in subjects that consume a mixed diet (134).

Indirect calorimetry can be used to calculate substrate oxidation rates if urinary nitrogen is also assessed during the measurement period. Urinary nitrogen levels determine protein oxidation. When the oxygen consumed and the carbon dioxide produced by protein oxidation are subtracted from the total measured amounts of oxygen consumed and carbon dioxide produced, the nonprotein respiratory quotient (NPRQ) can be calculated and used to determine the relative amounts of carbohydrate and fat oxidized in a given time period (40, 61, 134). The NPRQ for glucose is 1.0 and for fatty acids is near 0.70. Thus relatively more oxygen is needed to metabolize fatty acids to CO_2 and H_2O than is needed to metabolize carbohydrates.

Care must be taken when using indirect calorimetry to estimate substrate oxidation during periods of exercise, because RER can be influenced by bicarbonate ion kinetics, which can be altered by exercise. During periods of bicarbonate ion depletion and repletion, the RQ determined from respiratory gas exchange measurements at the mouth will not accurately represent the true cellular RQ; thus, the RQ cannot be reliably used to determine the energy equivalent of oxygen consumption and substrate oxidation rates at any given time when bicarbonate ion concentrations are changing. During high-intensity exercise, respiratory $\dot{V}CO_2$ values are often higher than cellular CO_2 production, thus elevating the measured RQ above the true cellular RQ (e.g., during measurement of max $\dot{V}O_2$, the RQ value should be well above 1.0 because of the additional amounts of CO_2 exhaled beyond what is produced in cellular respiration).

Conversely, for a period of time following high-intensity exercise, the measured RQ may be much lower than the true cellular RQ due to replenishment of the body's bicarbonate ion concentrations. For example, during the first hour following an intense bout of weightlifting, the RQ will often drop as low as 0.60 (84). While this value is outside the usual range, it is not due to measurement error. The $\dot{V}CO_2$ is reduced as some of the metabolically produced CO_2, which would normally be exhaled, is used to re-equilibrate the bicarbonate pool. Estimates of fuel utilization will be inaccurate if made over the short period when measured RQ does not reflect the true cellular RQ. However, if the period of indirect calorimetry encompasses these transient periods of depletion and repletion of bicarbonate ions, the average RQ during the entire period can be used to estimate fuel utilization for the entire period.

Doubly Labeled Water
The doubly labeled water (DLW) technique developed by Lifson (72) is another method of indirect calorimetry. Unlike respiratory gas exchange measurements, DLW is useful for assessing energy expended during free-living conditions outside the laboratory environment (109). This method is based on the subjects consuming a dose of two stable isotopes of water as ^2H_2O and $H_2^{18}O$. These isotopes equilibrate with total body water and then are eliminated over time. The ^2H isotope is eliminated from the body only in water, while ^{18}O is eliminated in both water and carbon dioxide. The differential rate of disappearance between the two isotopes provides an estimate of carbon dioxide production, a value which in turn can be converted to energy expenditure based on the estimated RQ.

In practice, the RQ is either estimated from the composition of the usual diet (i.e., food quotient) or is assumed to be 0.85, a value typical of a daily RQ in a person eating a mixed diet (15). The doubly labeled water technique allows assessment of energy expenditure over relatively long periods of time (10-14 days) in free-living subjects. There is generally good agreement between energy expenditure measured by the doubly labeled water and measurement by respirometry (113). The technique has been estimated to have an accuracy of 1-3% (113). The major disadvantages of the doubly labeled water technique are its relatively high cost, need for specialized equipment, and inability to measure substrate oxidation rates.

Labeled Bicarbonate Method

Another method used to estimate CO_2 production and energy expenditure is the labeled bicarbonate method (38). A dose of $NaH^{14}CO_3$ or $NaH^{13}CO_3$ is administered, which over time is diluted by the body's own CO_2 production from cellular macronutrient oxidation. The magnitude of the isotopic dilution during the measurement period can then be used to determine the rate of CO_2 production, which in a manner similar to the doubly labeled water method, allows for the calculation of energy expenditure. The recovery of the labeled carbons is accomplished by sampling breath, blood, or saliva. Recently, Elia et al. (37) used 24-hour urine samples to recover the label and measure the specific activity of urea. They found that this method provided estimates of energy expenditure that were very similar to those obtained from respiratory gas exchange in the calorimeter over a range of 1.35-1.75 times resting metabolic rate.

Components of Energy Expenditure

For measurement purposes, total daily energy expenditure can be divided into three separate components: resting metabolic rate (RMR), the thermic effect of food (TEF), and the energy expended in physical activity (EE_{ACT}) (55).

Resting Metabolic Rate

Resting metabolic rate (RMR), the amount of energy needed by the human body to maintain normal physiologic processes during rest in a post-absorptive state (10-12 hours after the last meal at which time macronutrients are no longer being absorbed into tissues for assimilation), comprises approximately 70% of daily energy expenditure in humans (121). It is usually measured in the early morning hours after an overnight fast, prior to any physical activity. Resting metabolic rate is most conveniently measured by indirect calorimetry using a ventilated canopy or a mouthpiece to measure respiratory gas exchange. Typically, RMR is measured 15-60 minutes following a period of rest and equilibration of the equipment. Bullough and Melby (26) and others (128) have found RMR values to be similar whether the subject slept at home and reported to the laboratory the morning of the test (outpatient protocol), or slept overnight in the laboratory (inpatient protocol).

However, Berke et al. (14) found RMR to be higher when elderly subjects had RMR measured using an outpatient protocol.

A variety of factors are known to influence RMR, with fat-free mass (FFM) recognized as the major determinant (5). Of the various tissue components that make up the fat-free mass, internal organs with high metabolic activity, including liver, brain, heart, lungs, and kidneys, contribute most to the RMR of the fat-free body mass (35). Resting metabolic rate is lower in women than men and decreases with advancing age (5). However, much of the variance in the relationships between RMR, gender, and age can be explained by the covariation of gender and age with fat-free weight (5). Various thermogenic hormones (thyroid hormones, epinephrine, etc.) also are known to influence the rate of resting energy expenditure (125). RMR appears to be influenced by genetic factors, with a higher degree of concordance in RMR between monozygotic than dizygotic twins (18). While RMR should be measured at a time in which subjects are free of psychological stress, Schmidt and colleagues (108) reported a positive correlation in young men between RMR values and values on the trait anxiety scale.

There are conflicting data regarding how physical activity affects RMR. Some cross-sectional studies, in which RMR and physical activity/fitness are measured within the same timeframe, show an elevated RMR in endurance-trained as compared to sedentary individuals (27, 92, 93). Arciero et al. (5) found peak $\dot{V}O_2$ to be a significant predictor of RMR in 500 healthy men and women. Other studies have not found differences in RMR between endurance-trained and sedentary subjects (23, 24, 76). Sharp et al. (121) found that the level of aerobic fitness was not significantly related to 24-hour energy expenditure or to any component of 24-hour energy expenditure in subjects studied in a whole-room calorimeter.

Some of the discrepancy in studies may be related to the measurement of RMR in relation to the time of the last exercise bout(s). Studies that found a higher RMR in exercise-trained subjects tended to measure RMR within 24 hours of the last exercise bout, whereas those that found no difference tended to measure RMR 48-56 hours after the last exercise bout. This suggests that an elevation of RMR in trained athletes may reflect the acute perturbations of strenuous exercise, rather than an adaptation to training. However, it appears that exercise (endurance or resistance) must be strenuous and of long duration in order for RMR to be elevated when measured one or two days after the exercise (25, 82, 83).

It is also possible that energy intake in relation to energy expenditure can affect RMR. Poehlman et al. (93) suggested that the combination of high energy expenditure and high energy intake (high energy flux or turnover) could elevate RMR in endurance-trained athletes, even when they were in energy balance (93). A study from our laboratory provides some support for this hypothesis (25). Resting metabolic rate in exercise-trained individuals was influenced by the total flux of energy through the body (total energy expenditure at steady state). Also, in trained versus untrained subjects, RMR was elevated under acute conditions of high energy expenditure and high energy intake, but this elevation was attenuated as the time interval increased from the last exercise bout to the measurement of RMR. These data suggest that RMR may be chronically elevated in individuals who engage in daily, high-intensity, prolonged exercise and whose energy intake is sufficient to maintain energy balance, due to an effect of acute exercise rather than to an adaptation to chronic exercise.

Thermic Effect of Food

The thermic effect of food (TEF) accounts for about 8-10% of daily energy expenditure (and hence energy intake if expenditure and intake are in balance) (56). There appear to be both obligatory and facultative components to TEF. The former involves the energy costs of digestion, absorption, and assimilation of macronutrients; while the latter involves an elevation of energy expenditure above that attributed to the obligatory component, perhaps due to increased sympathetic nervous system activity stimulated by increasing plasma insulin concentrations.

The thermic effect of food is best measured by indirect calorimetry (97). Following an RMR measurement, as described above, the subject is provided with a meal of known energy and macronutrient composition. Energy expenditure is measured for a period of time after the meal while the subject continues to lie or sit at rest. TEF is usually calculated as the area under the response curve generated by plotting energy expenditure vs. time and is expressed as a percentage of total ingested energy. Reed and Hill (97) reviewed how TEF varies with characteristics of the subjects and of the ingested meal. Depending on the size of the test meal, TEF may last for up to 6 hours (97).

There is no consistent picture of how physical activity affects TEF. Some investigators report an increase in TEF with increased physical fitness level (54), others report the opposite (45, 71, 93), and some find no effect (88). Witt et al. (136) suggested that TEF may be influenced by the time interval between the last exercise bout and the TEF measurement. Several studies in which TEF was measured 12-24 hours after strenuous exercise (71, 93) found TEF to be lower in trained individuals; this was not always the case in studies in which TEF was measured 36 hours or more following exercise (32, 54).

Other studies have focused on the combined effect of exercise and food intake on thermogenesis. Segal and Gutin (115) suggested that there is a synergistic effect of acute exercise and food intake on metabolic rate such that the combined effects of exercise and energy intake on thermogenesis are more than additive. However, Gilbert and Misner (44) found that meal consumption 30 minutes prior to exercise did not increase TEM during exercise in either trained or untrained subjects. With so much conflicting data, it is difficult to arrive at a single conclusion regarding exercise and TEF. However, in all likelihood the effect of exercise on TEF is fairly small, with the benefits of exercise on weight control resulting far more from the increased energy expenditure during exercise than from its impact on this specific component of 24-hour energy expenditure.

Energy Expenditure of Physical Activity

The energy expended during physical activity (EE_{ACT}) may represent between 15 and 50% of total energy expenditure, depending on the amount of physical activity performed and the body mass of the subject (74). Because this component of energy expenditure is the most variable, both within and between subjects (95), its potential role in body-weight regulation and in the etiology of obesity deserves close examination.

The amount of energy expended during physical activity is determined by the amount of activity performed and the efficiency with which it is performed. The total amount of physical activity performed over a day is the sum of planned exercise, activities associated with daily living (e.g., walking, stair climbing, etc.), and unproductive muscular activities such as fidgeting and shivering, sometimes referred to as nonexercise physical activity thermogenesis. In comparing energy expenditure among individuals, EE_{ACT} is more variable than the other components of energy expenditure. For example, Ravussin et al. (95) measured physical activity by radar in a whole-room calorimeter and found that the between-subject variation in energy expended in physical activity ranged from 830-4180 kJ/day. These large differences in physical activity were seen in subjects confined to a small room with no access to exercise equipment. Differences in amount of physical activity performed may be even greater in a free-living situation.

Note that energy expenditure is related to body size. For example, the energy cost of walking a mile will be greater for an individual who weighs 100 kg than a person who weighs only 60 kg. However, a small lean person who is moderately active may expend a similar number of kJ per day as a tall obese person who is sedentary.

Methods of Assessing Physical Activity

Accurate assessment of physical activity in free-living subjects presents a formidable challenge for researchers. It is important to realize that amount of physical activity performed is not synonymous with energy expended in physical activity. For example, a short bout of strenuous exercise may produce the same energy expenditure as a much longer bout of less intense exercise. Some of the techniques we will review can measure only the amount (and/or patterns) of physical activity; others measure only the energy expended in physical activity; and some measure both. Table 6.1 provides summary information regarding the various methods used to assess physical activity energy expenditure.

The amount of physical activity performed by free-living subjects has been expressed in time spent in physical activity (hours, minutes), in units of movement (counts), or even as numerical scores derived from responses to a questionnaire. Energy expended in physical activity has been expressed as total energy (kJ), work performed (watts), and metabolic equivalents (METs; one MET is equal to the rate

Table 6.1 Characteristics of Physical Activity Assessment Techniques

Assessment method	Group		Study		Subject		Limitations	Validity	Reliability
	Age	Size	Cost	Time	Time	Effort			
Direct calorimetry	Infant-Elderly	Small	$$$$	Hours-days	Hours-days	Movement	Cost, lab setting	High	High
Indirect calorimetry									
Respiratory gas exchange	Adolescent-Elderly	Moderate	$$	Minutes-hours	Minutes-hours	Artificial activity	Lab setting	High	High
Doubly labeled water	Infant-Elderly	Small	$$$$	2-3 weeks	Minutes/day	Urine collection	Cost, only ADEE	High	High
Bicarbonate-urea	Infant-Elderly	Small	$$$	Days-weeks	Minutes/day	Urine collection	Cost, invasive	High	?
Dietary intake	Adolescent-Elderly	Large	$$	Days	After meals	Record intake	Only TDEE	Moderate	Moderate
PA index/job classification	Employed only	Large	$	Days	Minutes	ID work activity	Only occupational PA	Low	Moderate
Surveys									
PA diary	Adolescent-Elderly	Large	$$	One day	Throughout day	Detail activity	Time	Moderate	Moderate
Recall questionnaire	Adolescent-Elderly	Large	$	1-7 days	Less than 1 hour	Recall activity	Subject memory	Moderate	Moderate
Quantitative history	Adolescent-Elderly	Large	$	Weeks-year	Less than 1 hour	Recall activity	Subject memory	Moderate	Moderate
General survey	Adolescent-Elderly	Large	$	Days-year	Minutes	ID activity	Not specific	Moderate	Moderate
Activity monitors									
HR monitoring	Infant-Elderly	Moderate	$$$	Minutes-days	1 hour +	Calibration test	Only at moderate levels	Moderate	High
Pedometers	Child-Elderly	Large	$	1-3 days	Very low	Wear on belt	Only walking/running	Moderate	Moderate
Uniaxial accelerometers	Infant-Elderly	Moderate	$$$	1-3 days	Very low	Wear on belt	Unidimensional	Moderate	Moderate
Triaxial accelerometers	Infant-Elderly	Moderate	$$$	1-3 days	Very low	Wear on belt	Cost, static activities	Moderate	Moderate
Kinematic analysis	Child-Elderly	Small	$$$	Minutes-hours	Minutes-hours	Detailed analysis	Limited movements	Moderate	Moderate

Adapted from LaPorte, Montoye, and Caspersen 1985(70)

of resting oxygen consumption, which for nonobese adults is approximately 3.5 ml·kg^{-1}·min^{-1} of oxygen, or 1 kcal·kg body weight^{-1}·h^{-1}). The use of METs provides a way of expressing the energy cost of activities as multiples of RMR. Recently, an extensive compendium of MET values for many types of activities was published by Ainsworth et al. (2).

The physical activity level (PAL) is a means of providing information about relative differences in usual level of physical activity. The PAL is calculated as total daily energy expenditure (TEE; usually determined using doubly labeled water) divided by RMR. For sedentary subjects the PAL is around 1.5 (33). This value can increase to around 3.5-4.5 under extreme exercise conditions.

LaPorte, Montoye, and Caspersen (70) found that over 30 different methods for assessing physical activity have been used in research studies, each with specific strengths and weaknesses. The major techniques for assessing physical activity levels and energy expended in physical activity are reviewed in the following sections.

Doubly Labeled Water

The doubly labeled water technique has been considered as a near gold standard for assessing the energy expended in physical activity under free-living conditions. This technique allows EE$_{ACT}$ to be estimated if RMR and TEF are known. The EE$_{ACT}$ is determined by subtracting 24-hour RMR and TEF from measured total energy expenditure. Often, this involves extrapolating 24-hour RMR from a single measurement made during a 30- to 60-minute time period and assuming TEF to be about 10% of total energy expenditure (47). Although total EE can be measured over a two- to three-week period, this method provides no information regarding amount, duration, intensity, or frequency of physical activity. In large epidemiological studies, use of the technique is not feasible due to cost, labor, time constraints, and the need to characterize physical activity.

Physical Activity Index

The Physical Activity Index (PAI) is frequently used to analyze data in large-scale epidemiologic studies. It provides an estimate of energy expenditure during a typical 24-hour period. The number of hours of sleep is reported; and the amount of time spent in light, moderate, and heavy activity are multiplied by the corresponding intensities and associated energy costs (6, 127). Job classification has been used as a physical activity index by ranking occupations according to activity levels (87). However, this type of physical activity index does not account for nonoccupational and leisure-time physical activity. In post-industrialized societies, as jobs and occupations become more mechanized and involve less physical work, the importance of accurately measuring nonoccupational physical activity becomes even more important.

Physical Activity Questionnaires

The use of questionnaires has generally been considered the most practical method of assessing EE_{ACT} in large population studies. Currently, there are over 40 different questionnaires that have been developed to assess physical activity. Depending on the specific questionnaire, subjects may be asked to provide information regarding type, time, duration, intensity, and frequency of physical activity. Activities are commonly grouped into light, moderate, and high intensities, which usually correspond to specific levels of energy expenditure or MET values. Some questionnaires provide numerical endpoints that group or rank individuals according to levels of physical activity. Commonly used questionnaires estimate occupational physical activity (3), leisure-time physical activity (68, 89, 127, 138), or both (28, 51, 63, 98, 103, 120). Population-specific questionnaires test males and/or females, in various age categories. Table 6.2 provides information about various commonly used physical activity questionnaires.

Table 6.2   Physical activity questionnaires

Questionnaire	Type of administration	Type of activity	Time frame	Measurement scale
Diary				
Edholm et al. (34)	SAQ	Detailed	2 weeks	kcal
LaPorte et al. (69)	SAQ	Detailed	12 hours	kcal
Recall				
Seven Day Recall (16)	Interview	Habitual	1 week	METs
Harvard Alumni (89)	SAQ	Leisure	1 week	kcal/week
Five Cities Project (103)	Interview	Habitual	1 week	kcal/day
Quantitative history				
Tecumseh (98)	Interview	Leisure	Past 12 months	METs
Stanford Usual Activity (103)	Interview	Habitual	3 months	Score
Cardia Physical Activity (122)	Interview	Habitual	Past 12 months	Weighted score
Minnesota LTPA (42)	Interview	Leisure	Past 12 months	METs
General				
HIP (120)	SAQ or interview	Habitual	1 week	28-point score
Lipid Research Clinics (51)	Interview	Habitual	1 week	Classification
Framingham (63)	Interview	Habitual	1 day	Daily index
Baecke (8)	Interview	Habitual	Past 12 months	METs or score

SAQ = Self-Administered Questionnaire.

Design
Questionnaires can be characterized on the basis of four components: mode of administration (personal or telephone interview, self-administration, mail survey),

time frame of report (from a few minutes to up to a year), specific characteristics of physical activity assessed (type, frequency, duration, intensity), and a calculated summary output (activity category, METS, kJ, rank, score). The types of physical activity measured range from the broad (occupational or leisure time) to the specific (gardening, climbing stairs). Physical activity may be recorded on a daily, weekly, or yearly schedule. Measures of physical activity intensity may range from choices such as "active" or "inactive" to specific levels (low, moderate, heavy). Some questionnaires have as few as 2 questions and others as many as 100 questions. Physical activity questionnaires fall into one of the following categories: diary surveys, recall surveys, quantitative histories surveys, and general surveys (70).

Diary surveys are typically self-administered and require subjects to record their daily activities for a short period of time (e.g., 24 hours). If the net energy costs of specific daily activities have been determined for the subject by indirect calorimetry, or if one uses estimated energy costs of physical activity from previously published data, diary surveys can provide estimates of energy expenditure (1, 34). This type of survey is most useful for ranking subjects or providing a general description of their physical activity patterns. Activity diaries are time consuming for the subjects, and data analysis can be burdensome for the investigator. In general, EE_{ACT} estimated from activity diaries correlated only modestly with EE_{ACT} determined by accelerometers (85).

Physical activity recall surveys retrospectively assess, by personal or telephone interviews, information about physical activity during the past 1-7 days. Reported time and intensity spent engaged in specific physical activities are converted to kJ based on previously published intensity values (70). Bouchard et al. (17) showed a high correlation (r = 0.96) between repeat administrations of a 3-day physical activity recall. Repeat administrations of the 7-day Stanford PA recall produced reliability correlations of 0.65, 0.08, 0.31, and 0.61 for light, moderate, hard, and very hard activity groups, respectively (103). The Physical Activity Scale for the Elderly (PASE) is designed specifically for older populations and has shown moderate reliability (r = 0.75) (131).

Quantitative history surveys are similar to the recall methods, but involve a longer period of reporting, with individuals typically reporting specific physical activity patterns over a year. Scoring is generally translated into energy expenditure values (kJ or METs). This method has been shown to accurately discriminate between the most and least active groups. The Minnesota Leisure Time Physical Activity (LTPA) Questionnaire, as the name suggests, quantifies only leisure time activity. The Minnesota LTPA Questionnaire has demonstrated good reliability (r = 0.69) in epidemiological research (42, 59). The Baecke questionnaire (8) has shown good reliability as well (r = 0.88). Several studies found a significant relationship between estimated energy expenditure from the Baecke questionnaire and physical activity levels as measured by an accelerometer (49, 77), while others have not (85). Because physical activity recalls and histories are retrospective assessments, they do not suffer from the problem of instrument-induced alterations of activity patterns. However, these recall and history assessments may be subject to more inaccuracies due to the time delay between actually performing the activity and recalling it.

General physical activity surveys provide less detail than the previously described instruments. They are typically less complicated to administer and score, providing a brief estimation of occupational and leisure time physical activity. The time frame for measuring activity may vary from a day, to a week, to a year. Individual activity patterns are usually scored and then individuals ranked or classified accordingly. Numerous questionnaires of this type have been used in large well-known population studies (8, 51, 63, 104, 120).

Major Limitations of Physical Activity Questionnaires
The major limitations to the use of physical activity questionnaires include the subjective nature of the instrument and the dependence on the accurate recall of the subject. Many instruments do not provide accurate conversion of subjective measures of physical activity intensity to accurate determinations of energy expenditure. Many of these questionnaires are limited to specific levels of literacy, language, and cultures, and have limited utility outside of their use in the specific population groups for which they were originally developed. Questionnaires also lack the sensitivity to detect small but significant differences in physical activity between groups, or changes in physical activity over time within the same group (130).

Validity of Physical Activity Questionnaires
Self-reports of physical activity have been compared with other methods of assessing EE_{ACT} in order to estimate their validity. The best comparison is with measurement of EE_{ACT} obtained by indirect calorimetry or doubly labeled water. Many studies have reported a relatively poor correlation between EE_{ACT} measured by physical activity questionnaires and EE_{ACT} measured by doubly labeled water or indirect calorimetry (1, 57, 94, 114). Better results have been obtained when comparisons have been made with heart rate monitoring and accelerometers (62, 85), but these findings are questionable considering the inconsistency of the latter measures.

Heart Rate Monitoring

Heart rate (HR) is related to oxygen consumption and, under some circumstances, can be useful in estimating energy expenditure (116). Commonly, HR monitors consist of an analog component for conditioning and processing of an ECG signal. Many varieties have a digital component for HR recording at various time intervals. Typically, electrodes are placed on the chest and the device is worn on the belt.

The relationship between heart rate and oxygen consumption varies substantially from person to person, and the most accurate way of using heart rate monitoring to estimate energy expenditure involves "calibrating" this relationship in each subject before the study is initiated. This involves measuring heart rate and oxygen consumption by a standard protocol (treadmill, cycle ergometer) over a range of work intensities. The within-person correlation between oxygen consumption and heart rate on a treadmill or cycle ergometer at moderate-high intensity commonly exceeds 0.95 (29). Regression equations are then developed for each individual based on their respective HR-$\dot{V}O_2$ curve (111, 124).

Major Limitations of Heart Rate Monitoring

A major limitation to the heart rate monitoring technique is that the relationship between heart rate and oxygen consumption is much weaker during low-level physical activity than during more intense activities. Thus, the accuracy of this technique tends to increase with subjects who spend time in more intense physical activity. The technique would be less precise in largely sedentary individuals (1, 80, 105). Some investigators have shown that because of the inaccurate HR-$\dot{V}O_2$ relationship at low HRs (57, 74), HRs below 120 bpm should not be considered valid (100). Instead, it has been suggested that when utilizing HR monitoring, a threshold HR should be determined for each individual (the mean of the highest HR for standing activity and the lowest HR of given activities). Heart rates above the given threshold are used to calculate energy expenditure depending on each individual's regression line. Below the threshold, EE is predicted based on the individual's resting $\dot{V}O_2$ (101).

Additionally, variables such as level of fitness (64), environmental conditions [ambient temperature (116), humidity, altitude], food intake, emotional status (13), body position, regional muscle involvement (79, 129), static vs. dynamic exercise, and continuous vs. intermittent exercise can cause changes in heart rate independent of oxygen consumption.

Validity of Heart Rate Monitoring

The accuracy of heart rate monitoring has been compared with other techniques for assessing the energy expended in physical activity with mixed results. Schulz et al. (111) and Livingstone et al. (73) compared estimated EE by heart rate monitoring with estimated EE using doubly labeled water. They used different HR-$\dot{V}O_2$ regression equations and found that none deviated significantly from the doubly labeled water technique in estimating energy expended in physical activity. Others (31) found that HR monitoring overestimated EE by 3-16% compared to the values obtained from a room calorimeter. Several studies have reported that HR monitoring to estimate energy expenditure (as determined by dietary records and body-mass changes over time) was no more accurate than activity diaries (1, 62).

Pedometers

Pedometers measure vertical oscillations and record a total count of the movements. Devices may be worn on the ankle, wrist, or belt. The PEDOBOY pedometer uses a spring attached to a pendulum that is displaced when the foot strikes the ground. It records vertical accelerations by triggering the movement of cogwheels on the pedometer dial (117).

Pedometers are designed specifically to evaluate walking, but have historically been considered relatively inaccurate and unreliable at counting steps and distance walked (43, 130). The newer electronic pedometers have better accuracy (12). Pedometers may estimate the number of steps only, or they may additionally estimate variables such as step rate, distance traveled, and energy expenditure. If distance is estimated, the pedometer must be individually calibrated to stride length. It appears that pedometers are useful in estimating low to moderate occupational activities (jobs requiring a combination of sitting, standing, and walking), yet fail to discriminate between moderate and heavy work (117). Additionally, pedometers are unable to detect static work such as upper-body movements like lifting, and typically only measure movement in one direction.

The large-scale integrated activity monitor (LSI) is purely a movement counter. It contains a mercury switch that closes upon exposure to a 3% incline or decline. Consecutive closures of the switch trigger an internal counter to record the number of movements (16 closures = 1 count). Washburn and LaPorte (132) found that the LSI was more accurate at recording counts during fast walking when worn at the hip versus the back. The LSI has been shown to be insensitive to very low and very high levels of activity (67).

Major Limitations of Pedometers

There are many limitations to using pedometers to measure energy expenditure. They are unable to measure movement while cycling (unless worn on the ankle), record intensity of activities, detect static work, or discriminate between uphill and downhill terrain. They also do not accurately record very slow walking speeds (66, 106, 107). Variations in stride length during fast walking or running result in inaccurate distance calculations. Also, when distance is measured, any actual movement that does not result in horizontal travel will still register as distance covered. Thus, vertical movements that are accounted for by accelerometers are not measured by most pedometers. The stride rates recorded by pedometers increase curvilinearly with $\dot{V}O_2$ (66). Because pedometers translate vertical oscillations to stride distance, accurate readings are less applicable with individuals exhibiting gait abnormalities. Due to problems regarding the mercury switch, the LSI is apparently insensitive to very low and very high physical activity intensity (67).

Validity of Pedometers

Pedometers have yet to be tested exhaustively for validity and reliability, although it is recognized that the newer electronic pedometers are much more accurate than the older mechanical versions. It has been suggested that habitual physical activity can be estimated relatively accurately with one week of measurement recordings (48, 101). Bassett et al. (12) reported that an electronic pedometer was accurate to within 0.3% of the actual number of steps walked at 4.8 km/h, and recorded 100.7 and 100.6% of steps taken on the left and right sides, respectively. Avons et al. (7) reported that actometers (activity monitors) worn on the ankle, wrist, and waist together accounted for 91% of the change in an individual's EE during various activities in a calorimeter. Pedometers have the highest correlation when walking is the major mode of energy expenditure (65). Variations of pedometer location (ankle, waist) also contributed to the variability of measurement. Pedometers worn on the waist produce more accurate values than those worn on the ankle (39). When worn on the ankle or knee, the LSI was shown to produce poor estimates of $\dot{V}O_2$ during cycling (58). However, because of the simplicity and convenience of pedometers, they could prove to be valuable tools in some population studies.

Uniaxial Accelerometers

Uniaxial accelerometers utilize a time-sampling mechanism that allows chrono-logical measurements of frequency, intensity, and duration of movement. These devices are designed to be worn on the belt, and they utilize a mercury switch transducer to quantify physical movement. Data can be collected over intervals ranging from minutes to weeks.

Accelerometers have been used in the assessment of EE_{ACT} for nearly two decades (137). Previous studies have shown a significant relationship between vertical accelerations produced by body movement and oxygen consumption (86). This relationship can be used to estimate EE_{ACT} for free-living individuals. The most common unidimensional motion sensor is the Caltrac (Hemokinetics, Madison, WI). The Caltrac is preprogrammed to estimate resting metabolic rate based on a formula using age, height, weight, and gender. This microcomputer then measures additional energy expenditure by translating vertical acceleration measurements into units of energy (86). The Caltrac (78 g) contains a cantilevered piezoceramic beam that produces a voltage in response to vertical accelerations and decelerations. This voltage is translated to estimate energy expenditure for given activities (118). It measures activity intensity and quantity. The device records activity counts or estimates energy expenditure based on movement, age, gender, weight, and height. Each activity count corresponds approximately to 0.101 kcal/kg (102). The follow-ing regression equation described by Montoye et al. (86) is used to estimate oxygen consumption, which is then used in the calculation of EE:

$$\dot{V}O_2 \text{ (mL/kg/min)} = 8.2 + 0.08 \text{ (counts/minute)}$$

Major Limitations of Uniaxial Accelerometers
Although uniaxial accelerometers detect changes in speed, they cannot detect changes in grade (incline) (81). Electronic accelerometers are not adaptable to swimming and cannot detect movement distal to the device (cycling, rowing, skiing). Furthermore, accelerometers do not detect EE due to static movements (lifting, carrying) or isometric contractions (pushing, pulling). Additionally, the Caltrac does not detect differences between rest, sleep, and sedentary activities, and therefore resting EE values are reportedly 9% higher than values found in standard tables (53). Although subject to many of the same limitations, the Computer Science and Applications (CSA) accelerometer overcomes some of the shortcomings of the Caltrac technology. The CSA is lighter than the Caltrac (70 g vs. 78 g) and may be worn on the belt, ankle,

or waist, thus detecting movements in addition to vertical accelerations. The CSA may also be used to measure and record heart rates while simultaneously measuring and recording activity counts. Furthermore, the CSA possesses a higher activity count resolution and the ability to store data for up to six weeks, when it then may be downloaded to computer. However, unlike the Caltrac, the CSA cannot be programmed to estimate EE. Rather, regression equations must be used to convert body mass and activity counts into calories or METs.

Validity of Uniaxial Accelerometers

Studies have reported good reproducibility from the Caltrac (r = 0.94) and have found that estimates of EE are more accurate than with either the LSI or the Calcount (86, 102, 119). Montoye et al. (86) reported a correlation coefficient of 0.79 between Caltrac estimates and $\dot{V}O_2$ measurements. Bray and Morrow (22) reported a reliability correlation for the Caltrac of r > 0.99. Based on activity counts per minute, Washburn and LaPorte (132) found that the overall reliability and validity of the Caltrac ($r^2 = 0.79$ and $r^2 = 0.73$, respectively) were much better than the LSI ($r^2 = 0.39$ and $r^2 = 0.21$, respectively). Similar results have also been reported by Montoye et al. (86), who also showed that the Caltrac was sensitive to changes in walking speed typically encountered in daily life. The LSI was significantly less sensitive compared to the Caltrac, and was influenced by body position.

When Caltrac estimates (MET minutes/day) were compared to 4-week versions of the Minnesota Leisure Time Physical Activity Questionnaire and data from the Survey of Activity, Fitness, and Exercise, Richardson et al. (99) reported only modest associations over the course of one year. These devices are most accurate when multiple units are worn simultaneously (wrist, ankle, waist) and body mass is used with movement counts to estimate energy expenditure (r = 0.95) (81). The correlation is less impressive when a single unit is used with body mass to predict EE (r = 0.85). Janz (60) found that the CSA accurately measured children's physical activity levels using heart rate as a criterion. Using a model that considers body mass as well as hip, ankle, and waist activity counts, Melanson and Freedson (81) reported that the CSA accurately predicted EE compared to actual EE determined by indirect calorimetry.

Numerous laboratory studies have reported good validity and reliability with the Caltrac and CSA activity monitors when the exercise protocols involved treadmill walking or jogging (9, 10, 11, 81, 86, 90, 102). One study showed that the average daily EE was 4.1% lower for the Caltrac compared to the doubly labeled water technique. This difference was not statistically significant (50). Conversely, field studies have reported less encouraging results, particularly with activities other than walking (102, 132, 135).

Triaxial Accelerometers

Uniaxial accelerometers relate unidimensional accelerations (vertical plane) to EE_{ACT}. Although vertical accelerations comprise much of EE_{ACT}, accelerations in the two horizontal axes contribute to EE_{ACT} as well. Thus, a device containing three separate uniaxial accelerometers was developed to measure physical activity in three dimensions.

Design

Triaxial accelerometers (Tritrac-R3D) have the same time sampling as their uniaxial counterparts (Caltrac, CSA), but assess activity in the mediolateral (x), anteroposterior (y), and vertical (z) dimensions, which include vector magnitudes. A free-moving electrical transducer detects movement variations in these three dimensions, providing a linear transformation of physical movement into an electrical signal for analog processing. The movement counts reflect velocity over time, which can then be programmed to compute kJ/unit time. The estimation of EE_{ACT} is calculated internally using an unpublished regression equation. Data is collected in 1-15 minute intervals, with a maximum of 14 days of collection. While the Tritrac-R3D is worn on the belt, other models such as the actigraph are worn on the wrist.

Actigraphy has been traditionally used to assess physical activity via wrist movements. Actigraphs generally give no information regarding movement type or intensity, and have commonly been used with individuals suffering from conditions such as Parkinson's disease, Alzheimer's disease, and sleep disorders. The actigraph allows for device sensitivity selection (to the nearest 0.01 g or 0.05 g) and measures movement changes in frequency and intensity. Newly developed actigraphs contain an accelerometer component and are more sensitive to a greater range of free-living physical activities (96). The actigraph monitor utilizes a slightly different motion detection mechanism than the Tritrac-R3D (91).

Major Limitations of Triaxial Accelerometers

These devices produce outputs based on acceleration due to body movement, gravitational acceleration, external vibrations, and accelerations caused by excessive movements of the sensor (19, 91). Accelerometers are insensitive to movements that involve changes in resistance (weightlifting, cycling) and incline (hills, stairs). A single device is also limited to detecting local body movements; if worn on the leg, it will not resolve arm movements. However, the use of multiple accelerometers increases the accuracy of the EE prediction. The accuracy and sensitivity of accelerometers in the vertical plane are accommodated by limitations in the horizontal plane. Finally, accelerometers do not record static movements.

Most motion sensors are limited by the inability to accurately characterize the variety of movements (resistance, intensity) that occur in a free-living environment. While certain models attenuate artificial movements (20), others are susceptible to external vibration/movement artifacts (operating machinery, riding in a vehicle). Many of the older piezoceramic accelerometers seem to consistently lose sensitivity with age, while the more recently developed lightweight piezoresistive models maintain extended stability with age (ICSensors type 3031-010).

Validity of Triaxial Accelerometers

Meijer et al. (80) reported that the triaxial accelerometer more accurately predicted energy expenditure during low-level activities than heart rate monitoring. This same study found that the triaxial accelerometer was highly associated with energy intake ($r = 0.99$, $p = 0.025$) during one week in free-living adults.

Eston et al. (39) compared estimates of energy expenditure using a Digiwalker pedometer, a WAM uniaxial accelerometer, and a Tritrac-R3D triaxial accelerometer. They found that energy expenditure was best estimated using a triaxial accelerometer. Eston et al. (39) also reported that the triaxial accelerometer had a higher correlation with $\dot{V}O_2$ than did HR, and thus concluded that these devices should not be validated against HR. In nonregulated children's activities, the triaxial accelerometer is more accurate than heart rate monitoring.

A high correlation ($r = 0.92$) was found between estimated EE measured by an accelerometer and a 24-hour respiratory chamber protocol (112). Other studies have found significant correlations in EE_{ACT} by triaxial accelerometers, indirect calorimetry, and PA questionnaires (21, 30, 78). Sugimoto et al. (126) reported that the actigraph accelerometer accurately measured physical activity when compared to physical activity records, although physical activity was overestimated while individuals engaged in static activities. When using the wrist-worn actigraph, activities such as playing video games typically produce activity readings similar to walking (91). However, all of these studies concluded that the accelerometer consistently underestimated EE in free-living conditions. Chen and Sun (30) found that the Tritrac significantly underestimated EE throughout all walking activities. Similar to HR monitoring, accelerometers are less accurate at low intensities (EE < 4 METS). Advanced regression equations may be utilized to improve the accuracy of devices such as the Tritrac (30).

Many validation studies have been undertaken in laboratory settings, under controlled and limited activities. Therefore, reports of validity and reliability using any of these units have yet to be established in free-living situations. Due to the possibility that HR estimates of EE are unreliable and inaccurate, validation of accelerometers against HR monitoring methods seems suspect. The studies that validate accelerometers against calorimetry and the doubly labeled water are more appropriate.

Summary

There are a variety of methods for assessment of total energy expenditure, and its components—resting metabolic rate, the thermic effect of feeding, and energy expenditure of physical activity. Indirect calorimetry using respiratory gas exchange measurement provides a useful, convenient, and relatively inexpensive means of accurately measuring both RMR and TEF. Exercise energy expenditure performed in a laboratory can also be readily and accurately measured by indirect calorimetry. However, apart from the use of isotopic dilution techniques, indirect calorimetry cannot be readily used to measure free-living EE_{ACT}, and these dilution techniques (i.e., use of doubly labeled water and the labeled bicarbonate-urea method) are only feasible for studying small numbers of subjects in their free-living environments. Hence, other less valid methods are typically used for studies using larger numbers of subjects. The major limitation of all of these techniques is the lack of adequate validation. A common criterion by which these methods can be compared has yet to be accepted. Although some of these instruments have been validated using the doubly labeled water technique, others have been validated only against respiratory gas exchange in a laboratory environment or by comparisons to other questionnaires.

We have reviewed the strengths and weaknesses of physical activity questionnaires, HR monitoring, pedometers, and accelerometers. The variety of energy expenditure assessment techniques provides the investigator with an array of methods to choose from based on the population to be studied, the sample size, and the conditions under which physical activity energy expenditure is to be measured. None of these techniques is perfect for providing accurate assessment of physical activity patterns and the energy expended in physical activity in free-living subjects. Because most of the current techniques are not multidimensional, combining various methods where possible may provide the most accurate means of assessing physical activity.

There is particular concern about the ability of these techniques to assess mild- to moderate-intensity physical activity, considering the current focus on recommending that the public increase participation in the low- and moderate-intensity activities to promote overall health. A better understanding of the relationship among physical activity levels, health, and chronic disease risk is needed; therefore, the development of more accurate, inexpensive means of measuring physical activity outside of the controlled laboratory environment should be a high research priority.

Acknowledgments

Supported by NIH grants DK48520 and DK42549 and The Colorado Agricultural Experiment Station, Project #616.

References

1. Acheson, K.J., I.T. Campbell, O.G. Edholm, D.S. Miller, and M.J. Stock. 1980. The measurement of daily energy expenditure—an evaluation of some techniques. *American Journal of Clinical Nutrition* 33: 1155-1164.
2. Ainsworth, B.E., W.L. Haskell, A.S. Leon, D.R. Jacobs, Jr., H.J. Montoye, J.F. Sallis, and R.S. Paffenbarger, Jr. 1993. Compendium of physical activities. Classification of energy costs of human physical activities. *Medicine and Science in Sports and Exercise* 25: 71-80.

3. Ainsworth, B.E., D.R. Jacobs, A.S. Leon, M.T. Richardson, and H.J. Montoye. 1993. Assessment of the accuracy of physical activity questionnaire occupational data. *Journal of Occupational Medicine* 3510: 1017-1027.

4. Aminian, K., V. Robert, E. Jequier, and Y. Schutz. 1995. Incline, speed, and distance assessment during unconstrained walking. *Medicine and Science in Sports and Exercise* 272: 226-234.

5. Arciero, P., M.I. Goran, and E.T. Poehlman. 1993. Resting metabolic rate is lower in women than in men. *Journal of Applied Physiology* 75:2514-2520.

6. Astrand, P.O., and K. Rodahl. 1977. *Textbook of Work Physiology*. New York: McGraw-Hill, 465.

7. Avons, P., P. Garthwaite, H.L. Davies, P.R. Murgatroyd, and W.P.T. James. 1988. Approaches to estimating physical activity in the community: calorimetric validation of actometers and heart rate monitoring. *European Journal of Clinical Nutrition* 42: 185-196.

8. Baecke, J.A., J. Burema, and J.E. Frijters. 1982. A short questionnaire for the measurement of habitual physical activity in epidemiological studies. *American Journal of Clinical Nutrition* 36: 936-942.

9. Ballor, D.L., L.M. Burke, D.V. Knudson, J.R. Olson, and H.J. Montoye. 1989. Comparison of three methods of estimating energy expenditure: Caltrac, heart rate, and video analysis. *Research Quarterly for Exercise and Sport* 604: 362-368.

10. Balogun, J.A., N.T. Farina, E. Fay, K. Rossmann, and L. Pozyc. 1986. Energy cost determination using a portable accelerometer. *Physical Therapy* 667: 1102-1109.

11. Balogun, J.A., D.A. Martin, and M.A. Clendenin. 1989. Calorimetric validation of the Caltrac accelerometer during level walking. *Physical Therapy* 69: 501-509.

12. Bassett, D.R., B.E. Ainsworth, S.R. Leggett, C.A. Mathien, J.A. Main, D.C. Hunter, and G.E. Duncan. 1996. Accuracy of five electronic pedometers for measuring distance walked. *Medicine and Science in Sports and Exercise* 288: 1071-1077.

13. Bateman, S.C., R. Goldsmith, K.F. Jackson, H.P. Smith, and V.S. Mattocks. 1970. Heart rate of training captains engaged in different activities. *Aerospace Medicine* 41: 425-429.

14. Berke, E.M., A.W. Gardner, M.I. Goran, and E.T. Poehlman. 1992. Resting metabolic rate and the influence of the pretesting environment. *American Journal of Clinical Nutrition* 55: 626-629.

15. Black, A.E., A.M. Prentice, and W.A. Coward. 1986. Use of food quotients to predict respiratory quotients for the doubly-labelled water method of measuring energy expenditure. *Human Nutrition: Clinical Nutrition* 40C: 381-391.

16. Blair, S.N., W.L. Haskell, P. Ho, R.S. Paffenbarger, K.M. Vranizan, J.W. Farquhar, and P.D. Wood. 1985. Assessment of habitual physical activity by a seven-day recall in a community survey and controlled experiments. *American Journal of Epidemiology* 1225: 794-804.

17. Bouchard, C., A. Tremblay, C. LeBlanc, G. Lortie, R. Savard, and G. Theriault. 1983. A method to assess energy expenditure in children and adults. *American Journal of Clinical Nutrition* 37: 461-467.

18. Bouchard, C., A. Tremblay, A. Nadeau, J.P. Despres, G. Theriault, M.R. Boulay, G. Lortie, C. Leblanc, and G. Fournier. 1989. Genetic effect in resting and exercise metabolic rates. *Metabolism* 38: 364-370.

19. Bouten, C.V.C., K.T.M. Koekkoek, M. Verduin, R. Kodde, and J.D. Janssen. 1997. A triaxial accelerometer and portable data processing unit for the assessment of daily physical activity. *IEEE Transactions on Biomedical Engineering* 443: 136-147.

20. Bouten, C.V.C., W. Verboeket-Van de Venne, K.R. Westerterp, M. Verduin, and J.D. Janssen. 1996. Daily physical activity assessment: comparison between movement registration and doubly labeled water. *Journal of Applied Physiology* 812: 1019-1026.

21. Bouten, C.V.C., K.R. Westerterp, M. Verduin, and J.D. Janssen. 1994. Assessment of energy expenditure for physical activity using a triaxial accelerometer. *Medicine and Science in Sports and Exercise* 2612: 1516-1523.

22. Bray, M.S., and J.R. Morrow. 1993. Accuracy and reliability of the Caltrac accelerometer in a field setting. *Research Quarterly in Exercise and Sport* (Suppl. A-69).

23. Broeder, C.E., K.A. Burrhus, L.S. Svanevik, and J.H. Wilmore. 1992a. The effects of aerobic fitness on resting metabolic rate. *American Journal of Clinical Nutrition* 55: 795-801.

24. Broeder, C.E., K.A. Burrhus, L.S. Svanevik, and J.H. Wilmore. 1992b. The effects of either high-intensity resistance or endurance training on resting metabolic rate. *American Journal of Clinical Nutrition* 55: 802-810.

25. Bullough, R.C., M.A. Harris, C.G. Gillette, and C.L. Melby. 1995. Interaction of acute changes in exercise energy expenditure and energy intake on resting metabolic rate. *American Journal of Clinical Nutrition* 61: 473-481.

26. Bullough, R.C., and C.L. Melby. 1993. Effect of inpatient versus outpatient measurement protocol on resting metabolic rate and respiratory exchange ratio. *Annals of Nutrition and Metabolism* 37: 24-32.

27. Burke C.M., R.C. Bullough, and C.L. Melby. 1993. Resting metabolic rate and postprandial thermogenesis by level of aerobic fitness in young women. *European Journal of Clinical Nutrition* 47: 575-85.

28. Cassel, J., S. Heyden, A.G. Bartel, B.H. Kaplan, H.A. Tyroler, J.C. Cornoni, and C.G. Hames. 1971. Occupation and physical activity and coronary heart disease. *Archives of Internal Medicine* 28: 920-928.

29. Ceesay, S.M., A.M. Prentice, K.C. Day, P.R. Murgatroyd, G.R. Goldberg, and W. Scott. 1989. The use of heart rate monitoring in the estimation of energy expenditure: a validation study using indirect whole-body calorimetry. *British Journal of Nutrition* 61: 175-186.

30. Chen, K.Y., and M. Sun. 1997. Improving energy expenditure estimation by using a triaxial accelerometer. *Journal of Applied Physiology* 836: 2112-2122.
31. Dauncey, M.J., and W.P.T. James. 1979. Assessment of the heart rate method for determining energy expenditure in man, using a whole body calorimeter. *British Journal of Nutrition* 42: 1-13.
32. Davis, J.R., A.R. Tagliafero, R. Ketzer, T. Gerardo, J. Nichols, and J. Wheeler. 1983. Variations in dietary induced thermogenesis and body fatness with aerobic capacity. *European Journal of Physiology* 50: 319-329.
33. Durnin, J.V.G.A. 1967. *Energy, Work, and Leisure*. London, U.K.: Heinemann Educational Books.
34. Edholm, O.G., and J.G. Fletcher. 1955. The energy expenditure and food intake of individual men. *British Journal of Nutrition* 9: 286-300.
35. Elia, M. 1992. Organ and tissue contribution to metabolic rate. In *Energy Metabolism: Tissue Determinants and Cellular Corollaries*, ed. J.M. Kinney and H.N. Tucker. New York: Raven Press, 61-79.
36. Elia, M., N.J. Fuller, and P.R. Murgatroyd. 1992. Measurment of bicarbonate turnover in humans: applicability to estimation of energy expenditure. *American Journal of Physiology* 26: E75-E88.
37. Elia, M., M.G. Jones, G. Jennings, S.D. Poppitt, N.J. Fuller, P.R. Murgatroyd, and S.A. Jebb. 1995. Estimating energy expenditure from specific activity of urine urea during lengthy subcutaneous $NaH^{14}CO_3$ infusion. *American Journal of Physiology* 32: E172-E182.
38. Elia, M., and G. Livesay. 1992. Energy expenditure and fuel selection in biological systems: the theory and practice of calculations based on indirect calorimetry and tracer methods. In *Metabolic Control of Eating, Energy Expenditure and the Bioenergetics of Obesity,* ed. A.P. Simopoulos. New York: Karger.
39. Eston, R.G., A.V. Rowlands, and D.K. Ingledew. 1998. Validity of heart rate, pedometry, and accelerometry for predicting the energy cost of children's activities. *Journal of Applied Physiology* 841: 362-371.
40. Ferrannini, E. 1988. The theoretical bases of indirect calorimetry: a review. *Metabolism* 37: 287-301.
41. Flatt, J.P. 1992. Energy Cost of ATP Synthesis. In *Energy Metabolism: Tissue Determinants and Cellular Corollaries,* eds. J.M. Kinney and H.N. Tucker. New York: Raven Press, Ltd.
42. Folsom, A.R., D.R. Jacobs, C.J. Caspersen, O. Gomez-Marin, and J. Knudsen. 1986. Test-retest reliability of the Minnesota Leisure Time Physical Activity Questionnaire. *Journal of Chronic Disease* 397: 505-511.
43. Gayle, R., H.J. Montoye, and J. Philpot. 1977. Accuracy of pedometers for measuring distance walked. *Research Quarterly for Exercise and Sport* 48: 632-636.
44. Gilbert, J.A., and J.E. Misner. 1993. Failure to find increased TEM at rest and during exercise in aerobically trained and resistance trained subjects. *International Journal of Sports Nutrition* 3: 55-66.

45. Gilbert, J.A., J.E. Misner, R.A. Boileau, L. Ji, and M.H. Slaughter. 1991. Lower thermic effect of a meal post-exercise in aerobically trained and resistance-trained subjects. *Medicine and Science in Sports and Exercise* 23: 825-830.

46. Godin, G., and R.J. Shephard. 1985. A simple method to assess exercise behavior in the community. *Canadian Journal of Applied Sport Science* 10: 141-146.

47. Goran, M.I., and E.T. Poehlman. 1992. Endurance training does not enhance total energy expenditure in healthy elderly persons. *American Journal of Physiology* 263: E950-E957.

48. Gretebeck, R.J., and H.J. Montoye. 1992. Variability of some objective measures of physical activity. *Medicine and Science in Sports and Exercise* 2410: 1167-1172.

49. Gretebeck, R., and H.J. Montoye. 1990. A comparison of six physical activity questionnaires with Caltrac accelerometer recordings. Abstract. *Medicine and Science in Sports and Exercise* 22: S79.

50. Gretebeck, R., H.J. Montoye, and W. Porter. 1991. Comparison of the doubly labeled water method for measuring energy expenditure with Caltrac accelerometer. Abstract. *Medicine and Science in Sports and Exercise* 23: S60.

51. Haskell, W.L., H.L. Taylor, P.D. Wood, H. Schrott, and G. Heiss. 1980. Strenuous physical activity, treadmill exercise test performance and plasma high-density lipoprotein cholesterol. *Circulation* 62 (Suppl. IV): 53-61.

52. Haskell, W.L., M.C. Yee, A. Evans, and P.J. Irby. 1993. Simultaneous measurement of heart rate and body motion to quantitate physical activity. *Medicine and Science in Sports and Exercise* 251: 109-115.

53. Hemokinetics, Inc. 1987. Caltrac technical application note. 2923 Osmudsen Road, Madison, WI.

54. Hill, J.O., S.B. Heymsfield, C.B. McManus, and M. DiGirolamo. 1984. Meal size and thermic response to food in male subjects as a function of maximum aerobic capacity. *Metabolism* 33: 743-749.

55. Hill J.O., M.J. Pagliassotti, and J.C. Peters. 1994. Nongenetic determinants of obesity and fat topography. In *Genetic Determinants of Obesity*, ed. C. Bouchard, 35-48. Boca Raton: CRC Press, Inc.

56. Horton, E.S. 1983. Introduction: an overview of the assessment and regulation of energy balance in humans. *American Journal of Clinical Nutrition* 38: 972-977.

57. Hoyt, R.W., T.E. Jones, C.J. Baker-Fulco, D.A. Schoeller, R.B. Schoene, R.S. Schwartz, E.W. Askew, and A. Cymerman. 1994. Doubly labeled water measurement of human energy expenditure during exercise at high altitudes. *American Journal of Physiology* 266: R966-971.

58. Hunter, G.R., H.J. Montoye, J.G. Webster, R. Demment, L.L. Ji, and A. Ng. 1989. The validation of a portable accelerometer for estimating energy expenditure in bicycle riding. *Journal of Sports Medicine and Physical Fitness* 29(3): 218-222.

59. Jacobs, D.R., B.E. Ainsworth, T.J. Hartman, and A.S. Leon. 1993. A simultaneous evaluation of 10 commonly used physical activity questionnaires. *Medicine and Science in Sports and Exercise* 251: 81-91.

60. Janz, K.F. 1994. Validation of the CSA accelerometer for assessing children's physical activity. *Medicine and Science in Sports and Exercise* 26(3): 369-375.

61. Jequier, E., K.J. Acheson, and Y. Schutz. 1987. Assessment of energy expenditure and fuel utilization in man. *Annual Review of Nutrition* 7: 187-208.

62. Kalkwarf, H.J., J.D. Haas, A.Z. Belko, R.C. Roach, and D.A. Roe. 1989. Accuracy of heart-rate monitoring and activity diaries for estimating energy expenditure. *American Journal of Clinical Nutrition* 49: 37-43.

63. Kannel, W.B., and P. Sorlie. 1979. Some health benefits of physical activity: the Framingham study. *Archives of Internal Medicine* 139: 857-861.

64. Kappagoda, C., R. Linden, and J. Newell. 1979. Effects of the Canadian Air Force Training Program on submaximal exercise test. *Quarterly Journal of Experimental Physiology* 64: 185-204.

65. Kashiwazaki, H., T. Inaoka, T. Suzuki, and Y. Kondo. 1986. Correlations of pedometer readings with energy expenditure in workers during free-living daily activities. *European Journal of Applied Physiology* 54: 585-590.

66. Kemper, H.C.G., and R. Verschuur. 1977. Validity and reliability of pedometers in habitual activity research. *European Journal of Applied Physiology* 37: 71-82.

67. Klesges, R.C., L.M. Klesges, A.M. Swenson, and A.M. Pheley. 1985. A validation of two motion sensors in the prediction of child and adult physical levels. *American Journal of Epidemiology* 122: 400-410.

68. Lamb, K.L., and D.A. Brodie. 1990. The assessment of physical activity by leisure-time physical activity questionnaires. *Sports Medicine* 103: 159-180.

69. LaPorte, R.E., L.H. Kuller, D.J. Kupfer, R.J. McPartland, G. Matthews, and C. Caspersen. 1979. An objective measure of physical activity for epidemiologic research. *American Journal of Epidemiology* 109: 158-168.

70. LaPorte, R.E., H.J. Montoye, and C.J. Caspersen. 1985. Assessment of physical activity in epidemiologic research: problems and prospects. *Public Health Reports* 100: 131-146.

71. LeBlanc, J., P. Diamond, J. Cote, and A. Labrie. 1984. Hormonal factors in reduced postprandial heat production of exercise-trained subjects. *Journal of Applied Physiology* 57: 772-776.

72. Lifson, N. 1966. Theory of use of the turnover rates of body water for measuring energy and material balance. *Journal of Theoretical Biology* 12: 46-74.

73. Livingstone, M.B., W.A. Coward, A.M. Prentice, P.S. Davies, J.J. Strain, P.G. McKenna, C.A. Mahoney, J.A. White, C.M. Stewart, and M.J. Kerr. 1992. Daily energy expenditure in free-living children: comparison of heart-rate monitoring with the doubly labeled water ($^2H_2{}^{18}O$) method. *American Journal of Clinical Nutrition* 56: 343-352.

74. Livingstone, M.B., A.M. Prentice, W.A. Coward, S.M. Ceesay, J.J. Strain, P.G. McKenna, G.B. Nevin, M.E. Barker, and R.J. Hickey. 1990. Simultaneous measurement of free-living energy expenditure by the doubly labeled water method and heart-rate monitoring. *American Journal of Clinical Nutrition* 52: 59-65.

75. Luke, A., K.C. Make, N. Barkey, R. Cooper, and D. McGee. 1997. Simultaneous monitoring of heart rate and motion to assess energy expenditure. *Medicine and Science in Sports and Exercise* 291: 144-148.

76. Lundholm, K., G. Holm, L. Lindmark, B. Larsson, L. Sjostrom, and P. Bjorntorp. 1986. Thermogenic effect of food in physically well-trained elderly men. *European Journal of Applied Physiology* 55: 486-492.

77. Mahoney, M., and P. Freedson. 1990. Assessment of physical activity from Caltrac and Baecke questionnaire techniques. Abstract. *Medicine and Science in Sports and Exercise* 22: S80.

78. Matthews, C.E., and P.S. Freedson. 1995. Field trial of a three dimensional activity monitor: comparison with self-report. *Medicine and Science in Sports and Exercise* 277: 1071-1078.

79. McCloskey, D., and K. Streatfield. 1975. Muscular reflex stimuli to the cardiovascular system during isometric contraction of muscle groups of different mass. *Journal of Physiology* 230: 431-441.

80. Meijer, G.A., K.R. Westerterp, H. Koper, and F.T. Hoor. 1989. Assessment of energy expenditure by recording heart rate and body acceleration. *Medicine and Science in Sports and Exercise* 213: 343-347.

81. Melanson, E.L., and P.S. Freedson. 1995. Validity of the Computer Science and Applications, Inc. CSA activity monitor. *Medicine and Science in Sports and Exercise* 276: 934-940.

82. Melby, C.L., W.D. Schmidt, and D. Corrigan. 1990. Resting metabolic rate in weight-cycling collegiate wrestlers compared with physically active, noncycling control subjects. *American Journal of Clinical Nutrition* 52: 409-14.

83. Melby, C.L., C. Scholl, G. Edwards, and R. Bullough. 1993. Effect of acute resistance exercise on postexercise energy expenditure and resting metabolic rate. *Journal of Applied Physiology* 75: 1847-1853.

84. Melby, C.L., T. Tincknell, and W.D. Schmidt. 1992. Energy expenditure following a bout of non-steady state resistance exercise. *Journal of Sports Medicine and Physical Fitness* 32: 128-135.

85. Miller, D.J., P.S. Freedson, and G.M. Kline. 1994. Comparison of activity levels using the Caltrac accelerometer and five questionnaires. *Medicine and Science in Sports and Exercise* 263: 376-382.

86. Montoye, H.J., R.A. Washburn, R.S. Servais, A. Ertl, J.G. Webster, and F.J. Nagle. 1983. Estimation of energy expenditure by a portable accelerometer. *Medicine and Science in Sports and Exercise* 15: 403-407.

87. Morris, J.N., A. Kagan, D.C. Pattison, and M.J. Gardner. 1966. Incidence and prediction of ischemic heart disease in London businessmen. *Lancet* 2: 553-559.

88. Owen, O.E., E. Kavle, and R.S. Owens. 1986. A reappraisal of caloric requirements in healthy women. *American Journal of Clinical Nutrition* 44: 1-19.

89. Paffenbarger, R.S., A.L. Wing, and R.T. Hyde. 1978. Physical activity as an index of heart attack risk in college alumni. *American Journal of Epidemiology* 108: 161-175.

90. Pambianco, G., R.T. Wing, and R. Robertson. 1990. Accuracy and reliability of the Caltrac accelerometer for estimating energy expenditure. *Medicine and Science in Sports and Exercise* 22: 858-862.

91. Patterson, S.M., D.S. Krantz, L.C. Montgomery, P.A. Deuster, S.M. Hedges, and L.E. Nebel. 1993. Automated physical activity monitoring: validation and comparison with physiological and self-report measures. *Psychophysiology* 30: 296-305.

92. Poehlman, E.T., T.L. McAuliffe, D.R. Van Houten, and E. Danforth, Jr. 1990. Influence of age and endurance training on metabolic rate and hormones in healthy men. *American Journal of Physiology* 259: E66-E72.

93. Poehlman, E.T., C.L. Melby, S.F. Badylak, and J. Calles. 1989. Aerobic fitness and resting energy expenditure in young adult males. *Metabolism* 38: 85-90.

94. Racette, S., D.A. Schoeller, and R.F. Kushner. 1995. Comparison of heart rate and physical activity recall with doubly labeled water in obese women. *Medicine and Science in Sports and Exercise* 27: 126-133.

95. Ravussin, E., S. Lillioja, T.E. Anderson, L. Christin, and C. Bogardus. 1986. Determinants of 24-hour energy expenditure in man: Methods and results using a respiratory chamber. *Journal of Clinical Investigation* 78:1568-1578.

96. Redmond, D.P., and F.W. Hegge. 1985. Observations on the design and specification of a wrist-worn human activity monitoring system. *Behavior, Research Method, Instruments, & Computers* 17: 659-669.

97. Reed, G.W., and J.O. Hill. 1996. Measuring the thermic effect of food. *American Journal of Clinical Nutrition* 63:164-169.

98. Reiff, G.G., R.D. Montoye, R.D. Remington, J.A. Napier, H.L. Metzner, and F.H. Epstein. 1967. Assessment of physical activity by questionnaire and interview. *Journal of Sports Medicine and Physical Fitness* 7: 135-142.

99. Richardson, M.T., A.S. Leon, D.R. Jacobs, B.E. Ainsworth, and R. Serfass. 1995. Ability of the Caltrac accelerometer to assess daily physical activity levels. *Journal of Cardiopulmonary Rehabilitation* 15: 107-113.

100. Riddoch, C.J., and C.A.G. Boreham. 1995. The health-related physical activity of children. *Sports Medicine* 192: 86-102.

101. Rowlands, A.V., R.G. Eston, and D.K. Ingledew. 1997. Measurement of physical activity in children with particular reference to the use of heart rate and pedometry. *Sports Medicine* 244: 258-272.

102. Sallis, J.F., M.J. Buono, J.J. Roby, D. Carlson, and J.A. Nelson. 1990. The Caltrac accelerometer as a physical activity monitor for school-aged children. *Medicine and Science in Sports and Exercise* 225: 698-703.

103. Sallis, J.F., W.L. Haskell, P.D. Wood, S.P. Fortmann, T. Rogers, S.N. Blair, and R.S. Paffenbarger, Jr. 1985. Physical activity assessment methodology in the five-city project. *American Journal of Epidemiology* 121: 91-106.

104. Salonen, J.T., W.L. Puska, and J. Tuomilehto. 1982. Physical activity and risk of myocardial infarction, cerebral stroke and death. *American Journal of Epidemiology* 115: 526-537.

105. Saris, W.H.M. 1986. Habitual activity in children: methodology and finding in health and disease. *Medicine and Science in Sports and Exercise* 18: 253-263.

106. Saris, W.H.M., and R.A. Binkhorst. 1977a. The use of pedometer and actometer in studying daily physical activity in man. Part I: reliability of pedometer and actometer. *European Journal of Applied Physiology* 37: 219-228.

107. Saris, W.H.M., and R.A. Binkhorst. 1977b. The use of pedometer and actometer in studying daily physical activity in man. Part II: validity of pedometer and actometer measuring the daily physical activity. *European Journal of Applied Physiology* 37: 229-235.

108. Schmidt, W.D., P.J. O'Connor, J.B. Cochrane, and M. Cantwell. 1996. Resting metabolic rate is influenced by anxiety in college men. *Journal of Applied Physiology* 80: 638-642.

109. Schoeller, D.A., and C.R. Fjeld. 1991. Human energy metabolism: what have we learned from the doubly labeled water method? *Annual Review of Nutrition* 11: 355-371.

110. Schoeller, D.A., and P. Webb. 1984. Five-day comparison of the doubly labeled water method with respiratory gas exchange. *American Journal of Clinical Nutrition* 40: 153-158.

111. Schulz, S., K.R. Westerterp, and K. Bruck. 1989. Comparison of energy expenditure by the doubly labeled water technique with energy intake, heart rate, and activity recording in man. *American Journal of Clinical Nutrition* 49: 1146-1154.

112. Schutz, Y., F. Froidevauz, and E. Jequier. 1988. Estimation of 24 h energy expenditure by a portable accelerometer. *Proceedings of the Nutrition Society* 47: 23A.

113. Seal, J.L., J.M. Conway, and J.J. Canary. 1993. Seven-day validation of doubly labeled water method using indirect calorimetry. *Journal of Applied Physiology* 74: 402-409.

114. Seale, J.L., W.V. Rumpler, J.M. Conway, and C.W. Miles. 1990. Comparison of doubly labeled water, intake-balance, and direct- and indirect calorimetry methods for measuring energy expenditure in adult man. *American Journal of Clinical Nutrition* 52: 66-71.

115. Segal, K.R., and B. Gutin 1983. Thermic effects of food and exercise in lean and obese women. *Metabolism* 32: 581-589.

116. Sengupta, A., D. Saka, S. Muklonadhyay, and P. Gosivamn. 1979. Relationship between pulse rate and energy expenditure during graded work at different temperatures. *Ergonomics* 22: 1207-1215.

117. Sequeira, M.M., M. Rickenbach, V. Wietlisbach, D. Tullen, and Y. Schutz. 1995. Physical activity assessment using a pedometer and its comparison with a questionnaire in a large population survey. *American Journal of Epidemiology* 1429: 989-999.

118. Servais, S.B., J.G. Webster, and H.J. Montoye. 1984. Estimating human energy expenditure using an accelerometer device. *Journal of Clinical Engineering* 9: 159-170.

119. Servais, S.B., J.G. Webster, and H.J. Montoye. 1982. Estimating human energy expenditure using an accelerometer device. *IEEE Frontiers of Engineering in Health Care* 8: 309-312.

120. Shapiro, S., E. Weinblatt, C.W. Frank, and R.V. Sager. 1965. The HIP study of incidence and prognosis of coronary heart disease. *Journal of Chronic Diseases* 18: 527-558.

121. Sharp, T.A., G.W. Reed, M. Sun, N.N. Abumrad, and J.O. Hill. 1992. Relationship between aerobic fitness level and daily energy expenditure in weight stable humans. *American Journal of Physiology* 263: E121-128.

122. Sidney, S.D., D.R. Jacobs, W.L. Haskell, et al. 1991. Comparison of two methods of assessing physical activity in the CARDIA study. *American Journal of Epidemiology* 133: 1231-1245.

123. Sims E.A., and E. Danforth. 1987. Expenditure and storage of energy in man. *Journal of Clinical Investigation* 79: 1019-1025.

124. Spurr, G.B., A.M. Prentice, P.R. Murgatroyd, G.R. Goldberg, J.C. Reina, and N.T. Christman. 1988. Energy expenditure from minute-by-minute heart rate recording: comparison with indirect calorimetry. *American Journal of Clinical Nutrition* 48: 552-559.

125. Staten, M.A., D.E Matthews, P.E. Cryer, and M. Bier. 1987. Physiological increments in epinephrine stimulate metabolic rate in humans. *American Journal of Physiology* 253: E322-E330.

126. Sugimoto, A., Y. Hara, T.W. Findley, and K. Yoncmoto. 1997. A useful method for measuring daily physical activity by a three-direction monitor. *Scandinavian Journal of Rehabilitation Medicine* 29(1): 37-42.

127. Taylor, H.L., D.R. Jacobs, Jr., B. Schucker, J. Knudsen, A.S. Leon, and G. Debacker. 1978. A questionnaire for the assessment of leisure-time physical activity. *Journal of Chronic Diseases* 31: 741-755.

128. Turley, K.R., P.J. McBride, and J.H. Wilmore. 1993. Resting metabolic rate measured after subjects spent the night at home versus at a clinic. *American Journal of Clinical Nutrition* 58: 141-144.

129. Vokac, Z., H. Bell, H. Bautz-Holter, and K. Rodahl. 1975. Oxygen uptake/heart rate relationship in leg and arm exercise, sitting and standing. *Journal of Applied Physiology* 39: 54-59.

130. Washburn, R.A., M.K. Chin, and H.J. Montoye. 1980. Accuracy of pedometer in walking and running. *Research Quarterly for Exercise and Sport* 51: 695-702.

131. Washburn, R.A., K. Smith, A. Jette, and C. Janney. 1993. The Physical Activity Scale for the Elderly PASE: development and evaluation. *Journal of Clinical Epidemiology* 46: 153-162.
132. Washburn, R.A., and R.E. LaPorte. 1988. Assessment of walking behavior: effect of speed and monitor position on two objective physical activity monitors. *Research Quarterly for Exercise and Sport* 591: 83-85.
133. Webb, P. 1991. The measurement of energy expenditure. *Journal of Nutrition* 121:1897-1901.
134. Weir, J.B. 1949. New methods for calculating metabolic rate with special reference to protein. *Journal of Physiology London* 109:1-9.
135. Williams, E., R. Klesges, C. Hanson, and L. Eck. 1989. A prospective study of the reliability and convergent validity of three physical activity measures in a field research trial. *Journal of Clinical Epidemiology* 42: 1161-1170.
136. Witt, K.A., J.T. Snook, T.M. O'Dorisio, D. Aivony, and W.B. Malarkey. 1993. Exercise training and dietary carbohydrate: effects on selected hormones and the thermic effect of feeding. *International Journal of Sport Nutrition* 3: 272-289.
137. Wong, T.C., J.B. Webster, H.J. Montoye, and R.A. Washburn. 1981. Portable accelerometer device for measuring human energy expenditure. *IEEE Transactions on Biomedical Engineering* 286: 467-471.
138. Yasin, S., M.R. Alderson, J.W. Marr, C. Pattison, and J.N. Morris. 1967. Assessment of habitual activity apart from occupation. *British Journal of Preventive Social Medicine* 21: 163-169.

CHAPTER 7

The Assessment of Energy and Nutrient Intake in Humans

Klaas R. Westerterp, PhD
Department of Human Biology, Maastricht University, Maastricht, The Netherlands

The increase in obesity in our society is combined with the often permanent availability of highly palatable foods. However, dietary surveys show a decline of both energy intake and fat intake. The question is whether this is a true decline or a result of the increasing awareness of the negative consequences of a high energy and fat intake, leading to a discrepancy between real intake and measured intake. Thus, the accurate assessment of energy and nutrient intake is of crucial importance for national health campaigns.

Energy Intake

Methodology

The measurement of habitual food consumption in humans is one of the hardest tasks in energy balance studies. The two basic problems are the accurate determination of a subject's customary food intake and the conversion of this information to nutrient and energy intake. Regarding the first problem, any technique used to measure food intake should not be so intense as to interfere with the subject's dietary habits and thus alter the parameter being measured. The second problem concerns the necessary duration of the measurement; i.e., a period of time sufficient to obtain a true reflection of habitual food intake. The standard methods of determining food intake are as follows:

1. Indirect determination from information on group consumption, inspection of family budgets, larder inventories, figures for agricultural production, etc.
2. Estimation by recollection of food consumed over the last day, week, month, or even longer
3. Measurement and recording of food intake as eaten

This chapter focuses on methods of measuring food intake at the individual level, i.e., dietary recall (method 2) and dietary record (method 3). The dietary record method requires subjects to record the types and amounts of all foods consumed over a given time period. The foods are weighed or recorded in household measures like cups and spoons. This information is converted to weight or volume by measuring the actual devices used or adopting standard values from reference tables. The dietary recall method uses the subjects' report of intake over the previous 24-hour period (24-hrecall) or the report of customary intake over the previous week up to the past year (diet history). Here, the same methods are used to quantify the reported intake from information on portion size. Food models, volume models, and photographs can help individuals remember the amounts consumed.

Alternatives to the dietary record and dietary recall methods are the double-portion technique and the technique of supplying subjects with their daily food and doing the measurement of food quantity and quality in the laboratory. With the double-portion technique, subjects have to collect from every food item consumed an equivalent amount for later analysis. One usually analyzes a mixed sample of the collected foods per 24-hour interval. When food is supplied, subjects have to be carefully instructed to consume only the food provided and return all the leftovers.

The information on the quality of food consumption, i.e., macronutrient ratio, will be closer to "real life" in the dietary recall than in the situation where food is supplied. On the other hand, the quantitative information is superior when food is supplied. Generally, one wants subjects to be in energy balance, i.e., food intake should cover the subject's energy needs. This is very difficult, if not impossible, to check for. Sources of error in dietary reporting are errors due to poor memory, inaccurate estimation of food amounts, and wishful thinking. Wishful thinking results in overweight people systematically underreporting their intake (see figure 7.5) and underweight people, such as anorectics, probably overreporting their intake. Finally, if respondents are required to write down and even weigh or measure what they eat, they may alter their usual dietary habits either to make recording easier or to hide their habits.

The length of time that food intake should be measured to determine habitual intake is debatable. If people live by a regular activity pattern (five days of work and two days of leisure on the weekend), it would seem reasonable to suppose that their social and dietary habits are determined by this pattern of activity. Measurement of food intake should include samples of both workday and weekend intake, and it would of course be preferable to measure intake for the whole week. The length of the observation period is primarily determined by the level of day-to-day variability and the level of accuracy desired.

Basiotis et al. (2) evaluated the number of days necessary to measure food intake with confidence. In their study, 29 healthy individuals (16 females and 13 males) between the ages of 20 and 55 kept daily food intake records for 365 consecutive days as they consumed their customary diets. They were students and scientists, trained to record the types and portions of their food. Daily intake of energy; the macronutrients carbohydrate, protein, and fat; and many micronutrients were calculated with standard reference tables. The average intake of nutrients over the year was assumed to reflect the individual's "usual" intake, and standard deviation the daily variability. From this information, the number of days of dietary records needed for estimated individual and group intake to be within 10% of usual intake was calculated. The results indicated that the number of days of food intake records needed to predict the usual nutrient intake of an individual varied substantially among individuals for the same nutrient and within individuals for different nutrients. For individual estimates of food intake, food energy required the fewest days (on average 31) and micronutrients the most days (average of 433 days for vitamin A). The amount of time needed to estimate mean nutrient intake for the group was considerably shorter, ranging from 3 days for food energy to 41 days for vitamin A. For larger groups even fewer days would be needed. To achieve a level of precision of 10% in a single individual, at least four weeks are required. Thus, measuring food intake at the level of the individual is a tedious job and likely to interfere with the subject's dietary habits.

Acheson et al. (1) studied food intake for 12 males who spent one year on an Antarctic base. From their data they evaluated three methods of determining individual food intake: dietary record, dietary recall, and the double portion technique with bomb calorimetry. The period of investigation of intake for these individuals varied between 6 and 12 months. The tedious job of weighing and recording food intake and collecting duplicate food samples requires a high degree of motivation for the subjects; in the confined area of a resident base camp, encouragement by the investigator makes it is easier to accomplish these tasks. During the study the food intake of each subject was determined for at least one week of each month. During the week of study the subjects weighed and recorded all food consumed. Once during this week the subjects were asked to record everything they ate during the previous 24 hours. The dietary record method and the use of food composition tables underestimated energy intake by a mean of 7% in comparison with analysis of duplicate meals by bomb calorimetry. Errors of greater than 20% occurred in energy intake determined with dietary recall. In 62 of 68 occasions, recall underestimated actual food consumption. The authors suggested that this discrepancy might have been smaller if the subjects had had a more skilled and persistent interviewer. Errors of greater than 20% are unacceptable in an energy balance study.

The cross-check procedure is an attempt to improve the recall procedure. In this procedure, a trained interviewer asks the subject about the food eaten in the recent past and cross-checks this information against data on food purchases. With a skilled interviewer and a cooperative subject, one assumes the standard deviation of the estimate to be 10%, but the error is probably greater. An interviewer can find out what people *think* they eat, but this is often far from reality. Garrow (8) concluded that however intensively the eating habits of people may be studied, it is impossible to predict their energy intake over a period of a week with an accuracy of better than 10%.

Food Provisioning

The most obvious way to check self-reports of food intake is "supervised" feeding according to the reported intake. De Vries et al. (6) compared self-reported energy intake, calculated from 3-day food records, with actual intakes needed to maintain body weight during controlled trials that lasted from 6 to 9 weeks. In 269 free-living healthy and mainly normal-weight young adults, reported energy intake was 1.2 ± 1.6 MJ/d lower than required for body weight. The bias was smaller in men (-8 ± 13%) than in women (-12 ± 14%). There was no relationship between the underestimation of energy requirements with the body-mass index of the subjects, probably because nearly all the subjects were nonobese (mean ± SD body mass index, 22.1 ± 2.4 kg/m²).

Velthuis-te Wierik et al. (21) examined food provisioning in slightly heavier subjects (body mass index 24.6 ± 2.2 kg/m²). In this study, energy intake was restricted to approximately 80% of habitual intake. Habitual intake was estimated with a 7-day dietary record and checked with doubly labeled water over a two-week run-in period for eight subjects. Figure 7.1 shows the weight change of the subjects plotted as a function of the difference between energy intake and energy expenditure over the run-in period. The regression line (change in body mass (kg) = 0.034 [energy intake–energy expenditure (MJ)] + 0.056) denotes an energy equivalent per kg body mass change of 34 MJ (8000 kcal); this is within the range of 30-35 MJ/kg for obese subjects losing weight with a mix of 70-80% fat mass and 20-30% fat-free mass on an energy restricted diet (23). The difference between energy intake and

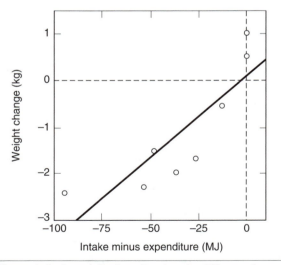

Figure 7.1   Change in body weight plotted against the difference between energy intake and energy expenditure. Energy intake was determined from food provisioning, based on a dietary record of habitual intake over 7 days. Energy expenditure was measured by the doubly labeled water method in eight men over 14 days. The line was calculated by means of linear regression (adapted from reference 21).

energy expenditure over the run-in period implied a mean underestimation of energy requirements of $16 \pm 13\%$. The result was that the subjects were restricted to 80% of habitual energy intake, as aimed for, plus the underestimation of habitual intake resulting in a weight reduction of 7.4 ± 1.7 kg over 10 weeks. The overall weight reduction was similar to the weight reduction on very low calorie diets, although subjects received a diet that still contained 9.2 MJ/d. This illustrates the success a moderately energy-restricted diet in daily life can have when there is close control of energy intake by food provisioning.

An example of underreporting of energy intake in relation to body mass index is a study in the elderly on the estimation of energy intake to feed subjects at energy balance (17). Subjects included 17 men and 11 women, age 70 ± 5 y with body mass index 25.2 ± 3.1 kg/m^2. Energy intake was measured with a dietary record or a dietary questionnaire, and subjects were fed for 3 weeks according to this intake. Energy expenditure was measured simultaneously with doubly labeled water to verify the energy requirements. Body weight decreased significantly during the intervention period and was related to the discrepancy between energy intake and energy expenditure. The underestimation of habitual intake was closely related to the body mass index, subjects with a higher body mass showing a larger underestimation.

Reference Methods

A number of reference methods are available to verify the results of dietary assessment. Four methods that are commonly employed involve the analysis of the following data:

- Urine nitrogen
- Total energy expenditure
- Resting metabolic rate and physical activity
- Total water loss

The principles, underlying assumptions, and some examples of each method will be discussed in the following sections.

Urine Nitrogen Analysis

Analysis of urine nitrogen is one of the first developed independent measures of dietary intake in free-living individuals. When sufficient time is allowed, urine nitrogen concentration provides an objective measure of habitual protein intake. Potential sources of error are incomplete collection of urine and disturbances of protein balance. Bingham et al. (3, 4) determined the value of 24-hour urine nitrogen excretion as a way of validating dietary methods. Daily nitrogen intake and excretion was measured in eight subjects who consumed their usual diets for 28 days. Completeness of urine collections was verified by using p amino benzoic acid (PABA). Protein balance was indicated by the constancy of fat-free mass during the observation interval. Eight 24-hour urine collections were sufficient to estimate dietary nitrogen intake to within 81 ± 5% (SD).

An early example of the use of daily urinary nitrogen excretion as a control of the accuracy of recorded intake in obese subjects is a study by Warnold et al. (22). They observed obese patients on an energy-restricted diet. Patients recorded a protein intake of 46 g/d; the expected protein intake, calculated on the basis of nitrogen losses, was 87 g/d. This indicated that the patients did not accurately record their intake.

Drawbacks of this method are that (1) subjects have to consume PABA with their meals and (2) collect all of their urine for several days. Using the urine nitrogen method to validate dietary nitrogen intake as a basis for conclusions about the validity of total dietary energy intake implies that reporting validity is the same for all macronutrients, protein, carbohydrate, fat, and alcohol. However, this assumption is not necessarily valid, especially for obese individuals, as we will discuss in a later section.

Total Energy Expenditure

In subjects who are stable with respect to weight and body composition, energy intake equals energy expenditure. As with nitrogen balance, energy balance does not imply precise balance every day, but rather that over a period of several weeks energy expenditure should equal habitual energy intake. Thus, total energy expenditure provides an objective measure of habitual energy intake. Total energy expenditure can be measured in free-living subjects over a period of one to three weeks using the doubly labeled water method. The method is noninvasive and does not interfere with the behavior of a subject. At the beginning of an experiment, the subject collects a urine sample at night. Then, the subject consumes a dose of doubly labeled water (approximately 100-150 ml water with 10 atom percent ^{18}O and 5 atom percent ^{2}H is sufficient for an adult). On the morning after administration of a dose and weekly thereafter, the subject collects a urine sample again.

A recent example of the perfect relationship between total energy expenditure and energy intake is a study in elite endurance athletes (19). Four females and four males of the Swedish national cross-country ski team were studied during a preseason training camp. Energy intake was calculated from dietary records of weighed food items. Females recorded food intake over the first 5 days and males for 4 days of a 7-day observation period of total energy expenditure with doubly labeled water. Body mass did not change significantly (+0.2 ± 0.5 kg) during the observation period. Energy intake closely matched energy expenditure (figure 7.2). The discrepancy between energy intake and energy expenditure as a percentage of energy expenditure was -0.6 ± 3.3 % (mean ± SD). The setting of the observation was probably optimal for an energy balance study. Dietitians assisted with the recording of food intake during the joint meals, and subjects were highly motivated to maintain energy balance for optimal performance at a level of energy turnover up to 4.5 times basal metabolic rate. High levels of energy turnover on their own are no guarantee for a close match between recorded intake and measured expenditure.

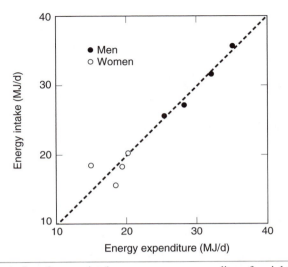

Figure 7.2   A plot of energy intake versus energy expenditure for eight cross-country skiers over one week in a training camp. Energy intake was measured with a weighed dietary record; energy expenditure was measured simultaneously with doubly labeled water. Data from (19).

In order to evaluate energy requirements at a high level of energy turnover, four participants of the Tour de France cycle race noted their daily food consumption during the 22-day tournament in specially designed diaries (25). A trained nutritionist gave instructions on filling in the diaries, made weekly checks of the cyclists' diaries, and cross-checked the diaries with information about the food supply during the race, including caloric beverages. Energy expenditure was measured simultaneously over three subsequent 7-day intervals with doubly labeled water. All four subjects managed to cover their energy expenditure with energy intake, in view of their unchanged body energy reserves. However, recorded intake was lower than measured expenditure (figure 7.3). The discrepancy showed a systematic increment from (mean ± SD) 13 ± 8% in the first week to 21 ± 10% in the second week, and to 35 ± 4% in the third week. Subjects probably cannot and will not accurately recall what they have eaten at the end of the day or whenever intake is recorded in the diary. Additionally, the longer the observation lasts, the larger the discrepancy between recorded intake and measured expenditure, i.e., the larger the underreporting of energy intake.

An effect comparable with that of the cyclists was measured in subjects who were preparing to run a half marathon (24). Measurement of energy intake, with a 7-day dietary record, was performed before the start of training (0 weeks), and 8, 20, and 40 weeks after the start of training. At week 0 the difference between energy intake and simultaneously measured energy expenditure (with doubly labeled water) was -4 ± 16%. However, reported energy intake was unchanged at week 40, whereas energy expenditure was increased 21 ± 10% (figure 7.4). Subjects therefore had

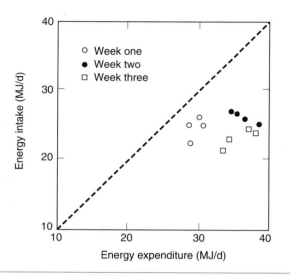

Figure 7.3   Energy intake plotted against energy expenditure in four cycle racers over three subsequent weeks of the Tour de France. Energy intake was measured with a dietary record; energy expenditure was measured simultaneously with doubly labeled water. Data from (25).

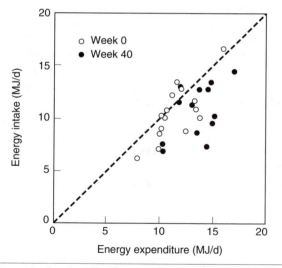

Figure 7.4   Energy intake plotted against energy expenditure in subjects at the start (week 0) and at the end (week 40) of a training period to run a half marathon. Energy intake was measured with a dietary record; energy expenditure was measured simultaneously with doubly labeled water. Data from (24).

significantly underreported energy intake. The difference between energy intake and measured energy expenditure at week 40 was -19 ± 17%, which is approximately equivalent to the increase in energy requirements. In this example, the increase in underreporting with time was coincident with an increase in energy expenditure. Thus, the subjects might not have been aware of the increase in intake, i.e., they were recording unchanged portion sizes while actual portion sizes increased, or they forgot to record extra snacks.

Discrepancies between recorded intake and measured expenditure are typical for obese subjects, as shown in figure 7.5. Obese people often suggest that their intake is normal or even lower than normal, and they consequently suggest a low level of energy expenditure as the reason for overweight. Prentice et al. (18) showed that energy expenditure in obese people is significantly higher than in lean controls. They observed a mean difference between reported energy intake and measured energy expenditure of 33%, whereas the corresponding difference in the lean group was only 2%, which is in line with results presented in figure 7.5. The reason for underreporting intake in the obese subjects might be their wish to eat less and their consequent underestimation of real intake. Lean people are not expected to have any reason for underreporting.

The doubly labeled water method is the standard for validating records of energy intake in free-living subjects. Unfortunately, the availability and costs of ^{18}O-labeled water and the need of high-tech facilities for sample analysis restrict widespread application. An alternative is to estimate energy requirements from estimated resting/basal metabolic rate and estimated physical activity, the two main components of daily energy expenditure.

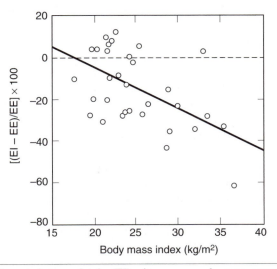

Figure 7.5 Reported energy intake (EI) minus measured energy expenditure (EE), expressed as a percentage of EE and plotted against the body mass index with the calculated linear regression line. Data from (15, 24).

Resting Metabolic Rate and Physical Activity

Goldberg et al. (9) suggested a method of validating recorded energy intake. They expressed energy intake as a multiple of resting metabolic rate (RMR), which is estimated from equations based on sex, age, height, and age (this multiple is also known as the physical activity level, PAL). They adopted a cutoff limit of PAL to screen for underreporting, assuming that recorded intakes below $1.35 \times$ RMR were unlikely to represent habitual intake.

PAL ranges from a theoretical value of 1.1 in someone with zero activity (RMR plus 10% to account for diet-induced energy expenditure) to values greater than 2 in highly active subjects. It is likely that recorded intakes below $1.35 \times$ RMR are inaccurate, unless the individual stayed in bed all day. On the other hand, the discrepancy between recorded intake and habitual intake might be as large for someone with a recorded intake of $1.2 \times$ RMR as for someone with a recorded intake of $1.8 \times$ RMR, depending on the physical activity level of the two subjects. The ratio of recorded intake to RMR allows one to exclude "unlikely" results, but unfortunately does not imply that the remaining results are valid.

Devices are now available to monitor physical activity. Bouten et al. (5) validated a triaxial accelerometer as an instrument for objectively quantifying physical-activity-related energy expenditure. Combined with measured resting metabolic rate, it was possible to explain 85% of the variation in total energy expenditure measured with doubly labeled water. The triaxial accelerometer for movement registration (tracmor) was attached to the low back of subjects using an elastic belt around the waist, with measurement directions along the anteroposterior, mediolateral, and longitudinal axes of the trunk. The tracmor has now been miniaturized to a 30-gram instrument measuring $7 \times 2 \times 0.8$ cm, which can be worn invisibly and without hindrance. Estimated or measured resting metabolic rate in combination with tracmor-assessed physical activity allows large-scale validation of energy intake measurements.

Total Water Loss

Underreporting of habitual intake may be due to underrecording and undereating. Goris and Westerterp (10) evaluated each of these potential errors by comparing reported food intake and water intake with energy expenditure and water loss (11). When subjects record food intake, they simultaneously record water intake. Water balance is preserved and is therefore an independent indicator for underrecording. To minimize errors in measuring portion size and other errors in recording intake, 22 female dietitians (age 35 ± 9 y and body mass index 21.9 ± 2.3 kg/m^2) were the subjects. Energy intake and water intake were measured over a one-week interval with a weighed dietary record. Energy expenditure was estimated from resting metabolic rate and measured with a ventilated hood at the beginning and end of the week. Physical activity was measured on all seven days with a triaxial accelerometer for movement registration. Water loss was estimated with deuterium-labeled water (^2H_2O). Energy and water balance were checked by measuring empty body weight (precision ± 0.1 kg) one week before the start, at the start, and at the end of the observation interval of seven days. Mean energy intake and water intake were, respectively, 8.5 ± 1.0 MJ/d and 2.4 ± 0.4 l/d. Energy intake as a multiple of resting metabolic rate was 1.4 ± 0.2 (range 1.1 to 1.7). The change in body weight from the nonrecording week to the recording week was 0.10 ± 0.59 to -0.54 ± 0.80 kg (paired t-test; p = 0.02). In a multiple regression analysis, recorded energy intake was explained by resting metabolic rate (p = 0.02), physical activity (p = 0.04), and body-weight change (p = 0.002) (explained variation 65%, p = 0.0003). Recorded water intake plus calculated metabolic water closely matched measured water loss (r = 0.92; p = 0.0001). The close match between recorded water intake and measured water loss indicated a high recording precision. Moreover, the body-weight change over the recording week indicated undereating with respect to energy balance. This method (i.e., measurement of water turnover and change in body weight) was applied in obese subjects as well, showing 12% underrecording and 26% undereating (10). Undereating might be a goal in itself for obese subjects, i.e., record food intake to lose weight. Unfortunately, it is unlikely that the undereating will last long enough to be significant for long-term weight loss.

Macronutrient Intake

Validation of dietary reporting at macronutrient level is of course more complex than validation of total energy intake. The methodology is a combination of dietary reporting with the reference methods described above, i.e., urine nitrogen output and total energy expenditure, or urine nitrogen output and resting metabolic rate and physical activity. Presently, data are available only from the latter combination. Heitmann and Lissner (12) studied macronutrient composition as part of a study on determinants of mortality from cardiovascular disease in obese subjects. Danish citizens aged 35, 45, 55, and 65 years were randomly selected from a population sample of 4581 subjects. A total of 323 participants were interviewed about their diets in the previous month and their physical activity during leisure and work; a PABA-validated 24-hour urine sample was collected from each participant. Degree of obesity was positively associated with underreporting of total energy and protein. Correction of this underreporting for energy resulted in an overreporting of protein relative to the other macronutrients. The results suggest a differential reporting pattern for different foods. Obese people prefer to omit reporting fatty foods and foods rich in carbohydrates, i.e., snack foods. This was confirmed by a recent study of Goris et al. (10) where the reported energy percentage from fat was underreported as measured with doubly labeled water.

Energy Density of Food Intake and Overweight

Several studies have shown that the diet of overweight subjects has a higher energy density that that of nonoverweight individuals. Jiang and Hunt (13) reported the analyses of the diets of 11 adult men who had collected a double portion of their food over 7 days. The energy density of the diet, including drinks, ranged from 1.8 kJ/g in normal-weight subjects to 3.9 kJ/g in overweight subjects. This difference is most likely explained by a higher fat content in the diets of overweight subjects. Dreon et al. (7) related diet composition, as measured with a 7-day food record, to body composition, as measured with hydrostatic weighing, in 155 middle-aged men. Subjects with a higher percentage of body fat consumed a diet with relatively more fat and less carbohydrate. Tremblay et al. (20) and Miller et al. (16) made similar observations. In all four studies mentioned in this paragraph (7, 13, 16, 20), there was no correlation between energy intake and indexes for overweight or obesity. We will return to this later.

Obesity is caused by a relatively high energy intake and a sedentary lifestyle. Obese individuals often deny part of their energy intake (figure 7.5). Westerterp-Plantenga et al. (28) explained the difference in diet composition between obese and nonobese subjects as a cultural and physiological adaptation. They analyzed energy intake in 68 women (34 obese and 34 nonobese) matched for age (20-50 years) and selected on the basis of completing food intake diaries accurately, i.e., underreporting < 10% of their estimated energy intake. The food types that were consumed by the subjects were divided into three categories based on their energy density: 0-7.5 kJ/g, 7.5-15.0 kJ/g, and 15.0-22.5 kJ/g (figure 7.6). The first category contained food items that mainly consist of water (fruit, vegetables, drinks), the second category contained food items that mainly consist of carbohydrate (bread, potatoes, rice, spaghetti), and the third category contained food items that mainly consist of fat (chocolate, cake). Additionally, the nonobese as well as the obese women showed a clear pattern of energy and macronutrient intake during the day. Energy intake as

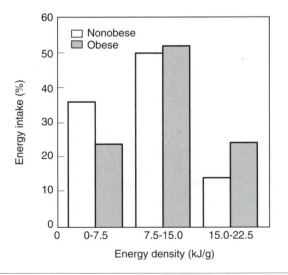

Figure 7.6   Energy intake from three energy density categories in nonobese and obese women. Data from (28).

well as fat intake increased during the day and carbohydrate intake decreased, although the percent energy contribution from fat and carbohydrate (en% fat and en% carbohydrate) at the start and at the end of the day were different between the nonobese and the obese subjects. From these analyses some suggestions for weight reduction diets were made. First, consumption of high energy density food could be replaced by consumption of low energy density food by switching from the highest to the lowest category of foods in order to achieve the energy density distribution that is shown by the nonobese. Second, considering the daily energy and macronutrient intake pattern, these interventions would have a greater effect in the afternoon and evening, when the foods with higher energy density are generally consumed.

Macronutrient Intake and Energy Balance

Several studies have shown that energy intake can be manipulated by changing the energy density of the diet. Unlike animals such as rats, humans do not seem to fully adjust the amount of food consumed when the energy density is varied. However, human studies have been criticized because of methodological limitations. In these studies, food intake was restricted by temporal constraints, such as limited meal times, or by a limited choice of experimental diets. Some typical studies will be described in the following paragraphs.

A typical example of an early study on the effect of dietary fat on the regulation of energy intake is the experiment of Lissner et al. (14). They measured the effect of dietary fat on spontaneous food consumption in 24 women by manipulating a conventional diet with similar foods. Three experimental diets were formulated, consisting of 20 food items each, containing low (15-20 en%), medium (30-35 en%), or high (45-50 en%) levels of fat. Diets were consumed in a random order for 2 weeks each. Relative to their energy consumption on the medium-fat diet, subjects on the low-fat diet spontaneously consumed 11% less energy and those on the high-fat diet consumed 15% more energy. These responses produced a weight change of -0.4 kg and +0.3 kg, respectively, over the 2-week duration of the diet; this is consistent with the variations in energy intake. Changes in body weight were related to adiposity, such that the leanest subjects showed the largest energy compensation. It was concluded that habitual, unrestricted consumption of low-fat diets may be an effective approach to weight control.

The effect of a manipulation of energy density on daily food intake in relation to meal times was studied by Westerterp-Plantenga et al. (29). They compared subjects with contrasting meal patterns, specifically, "gorgers" (a low-frequency pattern) and "nibblers" (a high-frequency pattern). Gorgers showed a clear periodicity of nutrient utilization. Carbohydrate utilization was elevated after meals and, to cover energy needs, was compensated by an increase in fat oxidation during the fasting periods. Nibblers have almost synchronized food quotients and respiratory quotients throughout the day, continuously using the nutrients they have ingested.

Twenty women, categorized as nibblers (those who habitually eat between meals) or gorgers (10 per group), were offered energy-reduced or normal-energy lunches with snacks, evening meals, and their own standard breakfasts. Compensatory energy intake occurred in the nibblers within 5 hours of the "light" lunch. In the gorgers, compensation of energy intake was not reached within 48 hours. Westerterp-Plantenga et al. (29) concluded that differences in short-term compensation of intake can arise from habitual snacking or its absence.

Finally, Westerterp et al. (26) did a long-term intervention study on the effect of dietary fat on body fat. A group of 108 women and 109 men, equally distributed over the age range of 19 to 55 years and with body mass indexes between 21 and 30, were, after a baseline measurement, randomly assigned to either a group consuming reduced-fat products or a group consuming full-fat products. Figure 7.7 shows the change in body mass for a subgroup of subjects where measurements covered a full year from June to June, including the diet intervention (full-fat versus reduced-fat products) that took place from August to February. The change in fat content of the diet was positively related to a change in energy intake, the latter explaining 5% of the variation in the change in body-fat mass. Subjects changing the fat content of the diet showed a consequent change in body-fat mass only when energy intake changed as well. It is often suggested that the influence of fat on energy balance is independent of energy intake. The study described clearly showed that the fat content of the diet had an effect on body fat as a function of the effect of dietary fat on energy intake.

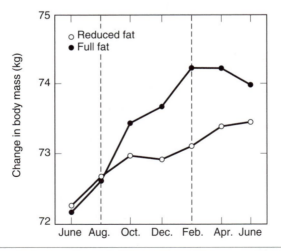

Figure 7.7   Mean body mass change in 36 women and 33 men, age 36 ± 11 y (range 19-55) and body mass index 24 ± 2 kg/m² (range 21-30), equally distributed over a group consuming reduced-fat products and a group consuming full-fat products from August through February. Data from (26).

Conclusions

The assessment of energy and nutrient intake is very difficult, especially in the obese. Reference methods, such as simultaneous measurement of urine nitrogen excretion and total energy expenditure, have shown that subjects systematically underreport their energy intake. Discrepancies between reported intake and estimated requirements range between 10 and 40%. Additionally, there are indications that obese subjects preferentially omit reporting fatty foods and foods rich in carbohydrates, the snack-type foods. Experimental studies have shown that the fat content of the diet is positively related to energy intake, and thus to energy balance.

References

1. Acheson, K.J., I.T. Campbell, O.G. Edholm, D.S. Miller, and M.J. Stock. 1980. The measurement of food and energy intake in man—an evaluation of some techniques. *American Journal of Clinical Nutrition* 33: 1147-1154.
2. Basiotis, P.P., S.O. Welsh, F.J. Cronin, J.L. Kelsay, and W. Mertz. 1987. Number of days of food intake records to estimate individual and group nutrient intakes with defined confidence. *Journal of Nutrition* 117: 1638-1641.
3. Bingham, S.A., A. Cassidy, T.J. Cole, et al. 1995. Validation of weighed dietary records and other methods of dietary assessment using the 24 h urine technique and other biological markers. *British Journal of Nutrition* 73: 531-550.
4. Bingham, S.A., and J.H. Cummings. 1985. Urine nitrogen as an independent validatory measure of dietary intake: a study of nitrogen balance in individuals consuming their normal diet. *American Journal of Clinical Nutrition* 42: 1276-1289.
5. Bouten, C.V.C, W.P.H.G. Verboeket-van de Venne, K.R. Westerterp, M. Verduin, and J.D. Janssen. 1996. Daily physical activity assessment: comparison between movement registration and doubly labelled water. *Journal of Applied Physiology* 81: 1019-1026.
6. De Vries, J.M., P.L. Zock, R.P. Mensink, and M.B. Katan. 1994. Underestimation of energy intake by 3-d records compared with energy intake to maintain body weight in 269 nonobese adults. *American Journal of Clinical Nutrition* 60: 855-860.
7. Dreon, D.M., B. Frey-Hewitt, N. Ellsworth, P.T. Williams, R.B. Terry, and P.D. Wood. 1988. Dietary fat:carbohydrate ratio and obesity in middle-aged men. *American Journal of Clinical Nutrition* 47: 995-1000.
8. Garrow, J.S. 1981. *Treat obesity seriously, a clinical manual.* Churchill Livingstone, Edinburgh.

9. Goldberg, G.R., A.E. Black, S.A. Jebb, T.J. Cole, P.R. Murgatroyd, W.A. Coward, and A.M. Prentice. 1991. Critical evaluation of energy intake data using fundamental principles of energy physiology: 1. Derivation of cut-off limits to identify under-recording. *European Journal of Clinical Nutrition* 45: 569-582.

10. Goris, A.H.C., M.S. Westerterp-Plantenga, and K.R. Westerterp. 2000. Undereating and underrecording of habitual food intake in obese men: selective underreporting of fat intake. *American Journal of Clinical Nutrition* 71: 130-134.

11. Goris, A.H.C., and K.R. Westerterp. 1999. Underreporting of habitual food intake explained by undereating in highly motivated lean women. *Journal of Nutrition* 129:878-882.

12. Heitmann, B.L., and L. Lissner. 1995. Dietary underreporting by obese individuals—is it specific or non-specific? *British Medical Journal* 311: 986-989.

13. Jiang, C.L., and J.N. Hunt. 1983. The relation between freely chosen meals and body habitus. *American Journal of Clinical Nutrition* 38: 32-40.

14. Lissner, L., D.A. Levitsky, B.J. Strupp, H.J. Kalkwarf, and D.A. Roe. 1987. Dietary fat and the regulation of energy intake in human subjects. *American Journal of Clinical Nutrition* 46: 886-892.

15. Meijer, G.A.L., K.R. Westerterp, A.M.P. van Hulsel, and F. ten Hoor. 1992. Physical activity and energy expenditure in lean and obese adult human subjects. *European Journal of Applied Physiology* 65: 525-528.

16. Miller, W.C., A.K.Lindeman, J. Wallace, and M. Niederpruem. 1990. Diet composition, energy intake, and exercise in relation to body fat in men and women. *American Journal of Clinical Nutrition* 52: 426-430.

17. Pannemans, D.L.E., and K.R. Westerterp. 1993. Estimation of energy intake to feed subjects at energy balance as verified with doubly labelled water: a study in the elderly. *European Journal of Clinical Nutrition* 47: 490-496.

18. Prentice, A.M., A.E. Black, W.A. Coward, H.L. Davies, G.R. Goldberg, P.R. Murgatroyd, J. Ashford, M. Sawyer, and R.G. Whitehead. 1986. High levels of energy expenditure in obese women. *British Medical Journal* 292: 983-987.

19. Sjödin, A., A. Andersson, J. Högberg, and K. Westerterp. 1994. Energy balance in cross country skiers. A study using doubly labeled water and dietary record. *Medicine and Science in Sports and Exercise* 26: 720-724.

20. Tremblay, A., G. Plourde, J.P. Despres, and C. Bouchard. 1989. Impact of dietary fat content and fat oxidation on energy intake in humans. *American Journal of Clinical Nutrition* 49: 799-805.

21. Velthuis-te Wierik, E.J.M., K.R. Westerterp, and H. van den Berg. 1995. Impact of a moderately energy-restricted diet on energy metabolism and body composition in non-obese men. *International Journal of Obesity* 19: 318-324.

22. Warnold, I., G. Carlgren, and M. Krotkiewski. 1978. Energy expenditure and body composition during weight reduction in hyperplastic obese women. *American Journal of Clinical Nutrition* 31: 750-763.

23. Westerterp, K.R., J. Donkers, E.W.H.M. Fredrix, and P. Boekhoudt. 1995. Energy intake, physical activity and body weight; a simulation model. *British Journal of Nutrition* 73: 337-347.

24. Westerterp, K.R., G.A.L. Meijer, E.M.E. Janssen, W.H.M. Saris, and F. ten Hoor. 1992. Long term effect of physical activity on energy balance and body composition. *British Journal of Nutrition* 68: 21-30.

25. Westerterp, K.R., W.H.M. Saris, M. van Es, and F. ten Hoor. 1986. Use of the doubly labeled water technique in man during heavy sustained exercise. *Journal of Applied Physiology* 61: 2162-2167.

26. Westerterp K.R., W.P.H.G. Verboeket-van de Venne, M.S. Westerterp-Plantenga, E.J.M. Velthuis-te Wierik, C. de Graaf, and J.A. Weststrate. 1996. Dietary fat and body fat: an intervention study. *International Journal of Obesity* 20: 1022-1026.

27. Westerterp, K.R, L. Wouters, and W.D. van Marken Lichtenbelt. 1995. The Maastricht protocol for the measurement of body composition and energy expenditure with labeled water. *Obesity Research* 3(supplement 1): 49-57.

28. Westerterp-Plantenga, M.S., W.J. Pasman, M.J.W. Ydema, and N.E.G. Wijckmans-Duijsens. 1996. Energy intake adaptation of food intake to extreme energy densities of food by obese and non-obese women. *European Journal of Clinical Nutrition* 50: 401-407.

29. Westerterp-Plantenga, M.S., N.E.G. Wijckmans-Duijsens, and F. ten Hoor. 1994. Food intake in the daily environment after energy-reduced lunch, related to habitual meal frequency. *Appetite* 22: 173-182.

CHAPTER 8

Human Energy and Nutrient Balance

Isabelle Dionne, PhD, and Angelo Tremblay, PhD
Division of Kinesiology, Physical Activity Sciences Laboratory,
Laval University, Ste-Foy, Québec, Canada

The last century has been marked by a significant decrease in human energy needs and an increasing prevalence of obesity in affluent societies. This situation might be the outcome of the major changes that occurred in energy metabolism with industrialization, as energy expenditure and thus needs have seriously decreased. In the United States, the prevalence of obesity in the adult population has surged from 24.9% in 1976 to 33.4% in 1991 (71). The situation is somewhat identical in Finland (47), Australia (88), Sweden (72, 73), Great Britain (107), and Canada (51).

Improvements in general knowledge and the development of technology have led to major changes in the human lifestyle. During the first stages of humanity, man had to hunt and fish for foods. More recently, man became able to cultivate and breed his foods. Nowadays, new technology has enhanced the availability of food and facilitated its preparation. Foods are now much more palatable and plentiful (46). In the meantime, the fast pace and variety of social events in the modern lifestyle have directed the population to select fast meals, often rich in fat and simple sugars, frequently accompanied by alcohol consumption (144). Taken together, these changes are very likely to have contributed to an overconsumption of energy-dense foods.

As energy intake relative to energy needs has increased, the modern lifestyle has led to a decrease in overall energy expenditure. Industrialization has caused a decrease in physically demanding work as motorized transportation has largely replaced biking and walking. In addition, leisure time activities now include more sedentary activities such as video games, television, and the movies (81). The resulting positive energy balance emerging from this fattening combination is very likely a significant contributor to the current rise in the prevalence of obesity.

Another factor that has influenced overall energy balance is a diminished number of cigarette smokers in comparison to a generation ago. In the United States, the number of cigarettes smoked every year *per capita* has dropped from 4266 in 1961 to 2800 in 1988-1989 (89). Because cigarette smoking inhibits taste and smell, those who quit smoking are at risk of increasing their total energy intake. Indeed, the cessation of smoking produces a mean body weight gain of 2.7 kg (82); this fact is meaningful when the body weight of an entire population is taken into account. However, the health benefits associated with the cessation of smoking compensate for the harmful effects of a small gain in weight. This potential small weight gain should not influence the decision of smokers who wish to quit.

It is difficult to estimate the extent of the energy balance shift that has taken place during the last century. Energy expenditure measurement techniques were developed and became widely available only a few decades ago, and the obesity issue was probably not preoccupying scientists as it was not yet a problem at the beginning of the 20th century. With the advancement of study in the genetics of obesity, it is now apparent that the increase in the prevalence of obesity is only partly due to genetically determined factors (23), and that the modern lifestyle is responsible for much of the positive energy balance of the last decades. Epidemiological studies of societies with a more traditional lifestyle have demonstrated that energy intake and especially energy expenditure are far different from that found in affluent societies. For example, Gambian women, who maintain an agricultural lifestyle, expend about 2.35 times their resting metabolic rate (RMR) (129). In contrast, women living in an industrialized context, expend between 1.46 and 1.9 times their RMR (16, 78, 83, 107, 130) (table 8.1). The discrepancy in energy expenditure between these two current types of societies is probably comparable to the difference in energy expenditure between the past and present lifestyles for women in industrialized society.

We can now hypothesize that a substantial fat gain was essential to establish a new steady state of energy balance in the general population. However, even though fat-mass accretion allows increases in energy expenditure and fat oxidation that progressively match the respective intakes, it is also responsible for numerous health problems, such as cardiovascular disease, insulin resistance, diabetes mellitus, cancer, etc. (26). Thus, alternatives to fat gain should be promoted to facilitate body weight stability in the context of a modern lifestyle.

Table 8.1   Mean Daily Activity Level in Women

Population	Lifestyle	X • RMR
Gambian	Traditional	2.35*
North American	Sedentary	1.4-1.7

* Energy expenditure for 48-kg subjects. At this body weight, the difference between the two lifestyles corresponds to 800-1000 kcal/day.
Adapted from Singh et al. (129).

Definition and Types of Obesity

The interest of scientists in obesity has developed since a few health problems were found to be associated with fat mass accretion, particularly in the trunk area. Obesity is classified according to fat distribution (table 8.2). Each type of obesity differs somewhat in its morbidity and mortality rates; however, the classification is solely based on the pattern of fat distribution (20-22).

Table 8.2   The Types of Obesity Phenotypes in an Anatomical Perspective

Type I	Excess body mass or percent fat
Type II	Excess subcutaneous truncal-abdominal fat (android)
Type III	Excess abdominal visceral fat
Type IV	Excess gluteofemoral fat (gynoid)

From Bouchard (21)

Type I obesity is characterized by an overall excess of body mass or percent of body fat distributed all over the body. In contrast, the three other types of obesity are characterized by fat deposition in a particular area. Type II obesity constitutes android obesity, which is characterized by an accumulation of fat at the trunk, particularly in the abdomen. This type of obesity is mainly found in males and has some important association with hypertension and diabetes mellitus (26). Type III obesity is defined by an excess accumulation of fat in the visceral compartment and is commonly called visceral obesity. Visceral fat is the fat contained in the deep part of the abdomen (in the visceres), whereas abdominal fat is located all around the trunk and includes visceral fat. Finally, type IV obesity is the female type of obesity with the fat accumulation concentrated in the lower body, or the gluteofemoral area. Based on these classifications, it is obvious that a given amount or percent of body fat can result in various anatomical characterizations.

In order to assess the level of obesity, body mass index (BMI) is usually measured. A BMI over 30 kg/m^2 is normally considered as obesity, while that below 27 kg/m^2 is thought to be normal. However, BMI is more representative of body weight than fat mass *per se,* and it should be used with caution, especially in individuals displaying a large muscle mass. The percentage of body fat is more precise in indicating the level of obesity, but its measurement implies specific techniques such as underwater weighing, plethysmography, electrical impedance, or isotope dilution—techniques that are usually not available in clinical environments. Anthropometric measurements such as skinfold thickness and abdominal circumference are both commonly used in clinics because of the ease with which they can be measured.

The waist-to-hip ratio is also a very useful and easy-to-obtain estimation of abdominal obesity. Its use provides practical information about the distribution of fat accretion.

The Interaction of Genetics and Environment

The individual variations in body fat are caused by a complex interaction between genetic, psychological, social, nutritional, and physical activity factors. It is thus very difficult to establish the exact cause underlying the onset of obesity or to implement an effective treatment aimed at prevention and cure. Research into the genetics of obesity has progressed during the last decade, but the polygenic aspect of obesity makes it difficult to pinpoint the exact cluster of genes and/or mutations responsible for the disorder.

Despite the ignorance of the precise defects underlying obesity, important knowledge about energy balance is now available that provides some information about how obesity takes place and how it could be counteracted. Each component of energy balance, i.e., energy intake and energy expenditure, and their respective known regulatory mechanisms are addressed in this chapter. We will also focus on macronutrient balance, because we now know that the regulation of each energy substrate balance must be considered in an attempt to influence overall energy balance. Finally, the issue of suitable behavioral changes in prone-to-obesity individuals, for example dieting and exercising, will also be discussed.

Energy and Substrate Balance

When a stable body weight is maintained over a long period of time, it is assumed that energy balance and macronutrient balance are also steady. The short-term balance is somewhat meaningless, because over a period of a few days most consumed nutrients are generally oxidized. Natural adjustment in preferential use of one type of substrate can compensate for day-to-day variations in macronutrient and energy intake. In the case of a sustained imbalance between intake and expenditure, body weight fluctuations will occur.

Fuel Oxidized vs. Macronutrient Content of the Diet

Energy balance means that the amount of energy consumed is equivalent to the amount of energy expended and that there is no storage or depletion of the body's reserves. An energy imbalance has an impact on body reserves and substrate balance. To maintain a stable body weight, the fuel mix oxidized has to closely match the macronutrient content of the diet. This concept is known as Flatt's RQ/FQ concept (RQ = respiratory quotient = $\dot{V}CO_2/\dot{V}O_2$; FQ = food quotient). The RQ to FQ ratio is directly related to energy balance. A ratio higher than 1.0 implies a positive energy balance, whereas a ratio lower than 1.0 reflects a negative energy balance. When RQ/FQ is close to 1.0, we assume that protein, carbohydrate, and fat are in balance and that body weight is maintained (44).

When excess foods are ingested compared to the expenditure, the organism's tendency to adjust glucose oxidation to carbohydrate intake in order to maintain stable glycogen stores will elevate the RQ, which reflects a diminution of fat oxidation (44). The excess energy is then stored as adipose tissue. Conversely, when energy intake is lower than the expenditure, the resulting deficit will be fulfilled by the oxidation of body reserves, mainly fat stores. This mechanism will preserve the glycogen stores and implies a decrease of body weight in the long term.

Substrate Balance

It has long been recognized that body nitrogen reserves tend to be maintained, even when large deviations in protein intake are observed. Substantial changes in protein intake induce small gains or losses of body protein stores because the natural adjustment of oxidation to intake rapidly takes place. Because protein balance tends to acutely maintain itself and because energy ingested from protein accounts for a minor fraction of the total energy intake, it has been postulated that variations in body weight are regulated by carbohydrate and fat metabolism (45).

Carbohydrates, as glucose, and fat, as fatty acids, are the major fuels used for metabolism and energy production. Dietary energy is mainly provided in the form of starch and sugars as carbohydrates and triglycerides as fat. The body's capacity to store carbohydrate is quite limited. These reserves are usually maintained within a very small, relatively stable range (59). This range is maintained because a constant availability of glucose is of paramount importance in assuring an adequate supply for the brain. In addition, carbohydrates are the most efficient energy producer of all the macronutrients because of their rapid and low-energy demanding oxidation. When excess carbohydrates are ingested while glycogen reserves are already saturated, they are stored as adipose tissue by entering *de novo lipogenesis*. *De novo lipogenesis* is marked by RQ higher than 1.0, indicating that fat storage is greater than concomitant fat oxidation. However, this was demonstrated only as a result of massive carbohydrate overfeeding (3, 124). The normal consumption of a mixed diet containing variable amounts of carbohydrates does not induce an important rate of *de novo lipogenesis* and fat gain as demonstrated by a value of RQ that normally fluctuates between 0.75 and 0.95 (2). These findings indicate that carbohydrate balance is maintained relatively stable by short-term adjustments.

Because protein and carbohydrate balances are naturally stable on a short-term basis, it appears that energy balance corresponds to fat balance. In this regard, acute overfeeding that provides 33% in excess energy (61), or a massive fat load ingested in the course of a breakfast that contains 35% carbohydrates and 50% fat (43), did not lead to a post-prandial decreased RQ. Indeed, increased carbohydrate ingestion, rather than the fat content of the meal, induced an increase in RQ. Thus, fat intake does not determine fat oxidation rate. Fat oxidation seems to be regulated by the state of carbohydrate ingestion and reserves that direct carbohydrate availability for production of energy. The adjustment of fat oxidation is consequently imprecise.

An enlargement of the adipose tissue mass of the body leads to an increase in plasma availability of FFA (14). This affects the relative contribution of fat to the oxidized fuel mix. In fact, oxidation of fat is directly affected by a gain of fat mass (125) and favors the enhancement of FFA availability and consequently fat oxidation. This phenomenon probably promotes the restoration of fat and energy balance (figure 8.1).

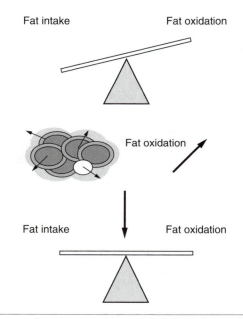

Figure 8.1   Long-term adaptation in energy and fat balance induced by fat-mass accretion.

Energy Intake and Composition of the Diet

The energy intake of individuals is highly variable and is determined by social status, although the energy and macronutrient intake regulatory mechanisms are partly inherited. Familial resemblance was shown to drive the preference for specific macronutrients. In this regard, cultural inheritance is more significant than genetic inheritance in affecting energy and macronutrient intake (93), although genetic factors are also important.

Dietary proteins, carbohydrates, and lipids constitute the three macronutrients of foods. As discussed in the next subsections, each macronutrient has a different potential to affect both energy intake and expenditure. Alcohol, which is not a macronutrient, also plays a role in energy balance and is therefore discussed in this section.

Proteins

Proteins are used mainly for tissue constitution and contribute to only a small fraction of total energy intake. Many studies have reported that protein ingestion favors satiety and may reduce energy intake (9, 55, 56, 147). In addition, protein's thermogenic effect is higher than that of fats and carbohydrates (84). Proteins are thus low-energy dense nutrients. Because of their high satiating and thermogenic capacity, as well as their important role in fat-free mass constitution, their inclusion in the diet, although smaller than for other macronutrients, is of major importance.

Carbohydrates

Carbohydrate energy density is lower than that of fats and alcohol. The satiating effect of carbohydrates is higher than that of fats (18, 117, 131) and lower than that of proteins. The satiating potential of carbohydrates has been related to the associated postprandial rise in insulin concentration (58, 108).

The thermic effect of carbohydrates is about 10% of their energy content (43). This includes a facultative thermogenesis mediated by the sympathetic nervous system (2). It appears concordant with the fact that men who are vegetarians and who consume a high-carbohydrate diet display a higher resting energy expenditure than omnivorous males (89, 99). As shown in table 8.3, our integration of these observations leads us to consider carbohydrates as intermediate between proteins and fat with regard to their potential to favor a negative energy balance.

Table 8.3   Comparison of Energy Substrates in Their Potential to Alter Energy Balance-Related Variables

	Energy substrates			
Potential to:	Proteins	Carbohydrates	Alcohol	Fat
Increase energy expenditure	Highest	Intermediate		Lowest
Favor satiety at low energy intake	Highest	High	Low	Lowest

Fat

Fat is a high-energy dense macronutrient that is characterized by a weak potential to favor negative energy balance. Because of the poor capacity of high-fat foods to favor satiety (18, 75, 118, 131) and to decrease spontaneous overfeeding (118, 131, 135, 138), dietary fat intake has been associated with body-fat gain (37, 70, 119, 135). Interestingly, obese and post-obese individuals display an increased preference for dietary fat relative to lean controls (38, 39).

Alcohol

Alcohol generally does not contribute to a large proportion of total energy intake. For example, in the United States, the caloric content of alcohol accounts for 5% of the total calorie intake (17). However, in a number of studies, calories derived from alcohol consumption represented a surplus that was added to the remaining energy intake. Therefore, there was no compensation by a decrease in the intake of other macronutrients (33, 49, 62, 142). In contrast, in binge drinkers, calories ingested from alcohol replaced caloric intake from other macronutrients (49). We can thus hypothesize that social and binge drinking do not induce the same effect on energy metabolism and should be treated differently. In this chapter, our discussion of alcohol consumption will be limited to social drinking, which represents a few drinks per week, usually ingested before and during meals.

Our research group investigated the effect of the combination of alcohol and a high-fat diet on spontaneous energy intake. Specifically, the effect of the two following types of appetizers on subsequent energy intake were compared: alcohol with high-fat food and no alcohol with high-carbohydrate food. Even if the energy density of appetizers was comparable, *ad libitum* (all you can eat) energy intake for the rest of the meal was significantly higher after the alcohol with a high-fat appetizer (143). This confirms the low potential of the combination of high-fat diet and alcohol to favor satiety with low energy intake. Similar findings were also reported by others. It was shown that alcohol ingested with carbohydrates did not induce any change in hunger or spontaneous intake compared to carbohydrates alone (105). It was also reported that alcohol intake at different times of the day did not favor a substantial compensation in the intake of macronutrients (48). Taken together, these results obtained in an experimental context highlight the fact that alcohol may promote overfeeding and a long-term weight gain.

Hellerstedt et al. (52) reviewed the literature pertaining to alcohol consumption and body-fat mass. They found equal numbers of studies that reported negative, positive, and neutral relationships between alcohol consumption and body-fat gain. Since the publication of this paper, Tremblay et al. (141) and Cigolini et al. (31) demonstrated a positive association between usual alcohol ingestion and body-fat accumulation at the trunk. However, Liu et al. (77) found no such relationship.

We can propose a few hypotheses regarding the discrepancy between findings that concern the issue of alcohol. Other factors such as cigarette smoking, physical activity participation, genetic predisposition to burn ethanol, or diet composition could be varying with alcohol ingestion and thus influence body-fat gain. Furthermore, as discussed previously, the type of drinking is probably of major importance and should be considered, inasmuch as binge and social drinkers do not display the same food habits.

Overall energy and particularly macronutrient intake thus have an influence on energy balance. The composition of the diet, including the ingestion of alcohol, affects overall balance by its energy content as well as by its specific aftereffect on dietary intake and energy expenditure.

Exercise, Energy Expenditure, and Substrate Balance

Energy Expenditure Components

Physiologists subdivide total energy expenditure into three components: basal metabolic rate (BMR), diet-induced thermogenesis (DIT), and physical activity energy expenditure (PAEE). These components are influenced by both genetic and environmental factors.

Resting metabolic rate (RMR) is the minimal amount of energy expended necessary to maintain life in different tissues and organs and constitutes the greatest part of total daily energy expenditure. Resting metabolic rate is primarily influenced by fat-free mass (85, 90, 91, 112, 114), which likely explains why obese people display higher RMR than nonobese individuals (111). There is also evidence that low RMR is associated with a greater risk of weight gain (113). However, other recent data do not support this concept. RMR normally decreases with age and has also been shown to be lower in females (5).

Thermogenesis includes the energy necessary to digest, transform, and store macronutrients and accounts for the smaller part of daily energy expenditure (about 10%). It is suggested that obese people are characterized by a low thermogenesis. However, because it is the least variable component of total energy expenditure, its role in obesity and weight management is not of major importance.

Physical activity energy expenditure includes the surplus of energy expended for the slightest movement to the most strenuous exercise. Therefore, PAEE is the most variable type of energy expenditure. These variations account for a substantial part of the fluctuations in total daily energy expenditure. PAEE is thus mainly influenced by voluntary participation in different physical activities. However, the energetic cost of activities (i.e., the energy expended for a given workload) is partly affected by heredity (19, 98).

Exercise thus exerts a beneficial effect on energy balance by providing an additional source of energy expenditure. One single bout of exercise generally expends about 200-500 kcal (or more), depending on the duration and the intensity of the exercise session. Theoretically, exercise alone can induce a negative energy balance that is sufficiently important to induce a significant weight loss. In addition, the positive influence of exercise on energy expenditure is prolonged for many hours after exercise in that it causes a further increase in energy expenditure and fat oxidation. Exercise can be useful as part of a weight-reducing treatment in association with other modes of intervention.

Exercise, Oxygen Consumption, and Substrate Oxidation

Bielinski et al. (12) were the first to monitor post-exercise energy expenditure and substrate oxidation for a period of many hours. They demonstrated that after 3 hours of exercise at 50% of $\dot{V}O_2$max, there was an increase in resting energy expenditure as well as a significant decrease in RQ 17 hr after the end of the exercise session. Other studies also showed a persistent elevation of oxygen consumption as well as a shift to fat oxidation after exercise of different durations and intensities (127, 145). On the other hand, Weststrate et al. (150) did not find any effect of exercise on resting metabolic rate 12 hours after an exercise bout of 90 minutes at about 25-35% of $\dot{V}O_2$max. However, it was previously demonstrated that such a low-intensity exercise may fail to produce any excess post-exercise oxygen consumption (EPOC) (127).

During exercise, predominantly carbohydrates are oxidized. This is followed by a shift to fat oxidation while glycogen stores are progressively being depleted, depending on the duration and intensity of exercise (65, 66). This oxidation seems to be prolonged after the end of exercise, as reflected by a decrease in RQ (12, 80, 145, 150). This increase in fat oxidation following exercise is possibly secondary to the glycogen depletion and to the acute negative energy balance induced by exercise. To verify this hypothesis, Calles-Escandon et al. (29) submitted 21 subjects to one of four 10-day treatments: (1) control, (2) overfed, (3) exercised, and (4) overfed and exercised. They found a similar influence of exercise on fat oxidation with or without caloric compensation, and they concluded that after an exercise session of 50% of $\dot{V}O_2$max representing 50% of resting daily energy expenditure, fat oxidation increased at rest, independently of energy balance.

It was recently demonstrated that fatty acid oxidation is regulated by carbohydrate metabolism via insulinemia during exercise (32). Thus, during the recovery, the exercise-induced glycogen depletion and the subsequent decrease in circulating insulin could favor an increase in fat oxidation by enhancing fatty acid metabolism probably until carbohydrate stores are replenished. A recent study conducted in our laboratory reinforced these assumptions. We submitted eight young men to an exercise bout of 60 minutes at 55% of $\dot{V}O_2$max before they stayed in a respiratory chamber for 62 consecutive hours. The specificity of this study was that the exercise bout was immediately followed by a dietary compensation that matched the energy expended and the substrate composition of the fuel mix oxidized during exercise. We found no difference in post-exercise energy expenditure or in substrate oxidation between the post-exercise condition and a control resting state session (36). During exercise recovery, energy expenditure and substrate oxidation were similar as during a control condition. These results suggest that the post-exercise shift from carbohydrate use to fat oxidation depends on the post-exercise energy and macronutrient balance. They also extend the classical concept of Randle et al. (110) by showing that an increase in fat oxidation is not detrimental for glucose homeostasis when glucose storage space is increased by exercise.

Exercise Intensity and Duration

The intensity and duration of a single bout of exercise has an important influence on post-exercise energy expenditure and substrate oxidation. A few studies compared exercise of different intensities but of the same duration. They showed that high-intensity exercise promotes a more important Excess Post-exercise Oxygen Consumption (EPOC) and decreased RQ for the 24 hr following exercise compared to an exercise of moderate intensity (30, 127). However, one study revealed that low-intensity exercise produced a greater 3-hr total EPOC (94). The authors attributed these findings to an increased fat oxidation during and following exercise, but this is not likely to be the cause inasmuch as other studies have not resulted in higher RQs after high-intensity exercise. One possible explanation of the finding concerns the duration of the measurement of post-exercise oxygen consumption that lasted for a period of only 3 hr. In this regard, Sedlock et al. (127) reported that the intensity of exercise has an impact on the magnitude as well as on the duration of EPOC.

The effect of exercise duration has not been widely studied. However, one group of researchers compared two exercise bouts of the same intensity but of different duration; they reported that the longer the duration, the longer the post-exercise effect on oxygen consumption (127).

Exercise and Subsequent Nutritional Behavior

It is important to prevent a substantial compensation in energy and fat intake in the post-exercise state to profit from the negative balance induced by exercise. In this regard, we compared two conditions defined as exercise/low-fat diet and rest/mixed diet to quantify the acute impact of exercise and a low-fat diet on energy balance (35). We found a difference in 24-hr energy balance of as much as 1740 kcal/day, only by submitting heavy men to a 60-min exercise bout at 50% of $\dot{V}O_2$max and by enhancing the diet FQ from 0.85 to 0.89 (table 8.4). This important contribution of diet composition was observed in a context where changes in appearance and palatability of food were minimal. The results of this study provide a good example of the potential of exercise to alter energy balance when the nutrient balance conditions are optimized. This observation is concordant with two recent studies

Table 8.4   Difference in Energy Balance Between an Exercise/Low-Fat Diet Condition and a Control Condition

a) Surplus of energy above resting level due to energy of exercise	910 kcal
b) Difference in post-exercise daily energy expenditure	120 kcal
c) Difference in daily energy intake between the two conditions	717 kcal
Difference in energy balance (a + b + c)	**1747 kcal**

From Dionne et al., 1997

that show that modification of the post-exercise dietary fat intake plays a major role in variations in post-exercise energy intake and balance (68, 140).

Acute exercise is suspected to exert some effects on subsequent appetite and feeling of hunger. A few studies found a decreased post-exercise feeling of hunger (69, 132) after strenuous exercise (69). In a study recently performed in our laboratory (60), the effect of two levels of exercise intensity (35% vs. 72% of $\dot{V}O_2$max) on subsequent subjective feeling of hunger and energy intake were compared. Exercise intensity had no influence on post-exercise voluntary energy and macronutrient intake or on appetite and feelings of hunger. However, the excess of energy ingested after exercise relative to the amount of energy expended during exercise was greater after low- than high-intensity exercise.

Variations in exercise, substrate oxidation, and post-exercise spontaneous macronutrient intake have also been investigated. Several studies have shown that exercise is followed by an increase in carbohydrate intake (132, 148), whereas other studies have not found such an effect (67). The post-exercise spontaneous macronutrient intake could be influenced by variations in substrate oxidation during exercise. Likewise, it was previously demonstrated that following an exercise session, exercisers with high RQ had a higher post-exercise energy intake relative to expenditure (4), maybe to facilitate glycogen resynthesis. Accordingly, when the exercise RQ was low, which indicates increased fat oxidation and lower carbohydrate use, the subsequent intake relative to expenditure was lower.

Post-exercise eating behavior is thus of major importance when considering the effect of physical activity on energy balance. Scientific evidence strongly suggests that exercise should be followed by a low-fat, high-carbohydrate diet in order to allow exercise to induce a negative energy and fat balance.

The Effect of Physical Training on Substrate Balance

Exercise produces both an acute and chronic effect on substrate balance. An improved capacity to mobilize and oxidize lipids was found to occur after a certain period of sustained physical training. The lipolytic response of adipose tissue to catecholamines is increased (34, 116), and the muscle lipoprotein lipase activity was reported to be positively related to the state of training (66). Taken together, these changes appear to improve the rate of fat oxidation both during exercise and in post-exercise resting state, as demonstrated by a decreased mean daily RQ following a training program (27, 102). Contradictory results were obtained in other studies (74, 123), which could be related to a higher carbohydrate intake that naturally occurs in trained individuals (148).

Highly trained individuals thus exhibit a good oxidative and lipolytic potential, despite a low adiposity. This was ascertained by a demonstration showing that trained individuals present a higher fat-oxidation rate for an equivalent free fatty acids (FFA) concentration (28, 146). It seems that physical training has the capacity to exert the same effect as body-fat-mass accretion. In fact, as reported earlier, fat-mass gain exerts an increasing effect on fat oxidation by elevating FFA concentration in circulation, whereas exercise increases the efficiency of FFA oxidation. It thus seems that both exercise and fat gain have the potential to favor the restoration of fat balance. However, the negative health effects associated with an excess of body-fat mass and the subsequent elevation of circulating FFA can be prevented by using exercise to reinduce fat balance.

Regulation of Energy Balance

A sustained energy and fat imbalance leads to a reaction of the sympathetic nervous system (SNS), the system that allows the control of substrate and energy fluxes. Acute energy restriction that leads to a marked negative energy balance normally induces a reduction in sympathoadrenal drive, which reduces energy expenditure. Conversely, the stimulating effect of overfeeding and a positive energy balance on SNS results in a decrease in energy intake (25, 57) and an increase in energy expenditure (79, 87, 128). The elevation of energy expenditure in response to an increase in SNS activity can be attributed to an increased level of resting metabolic rate (121, 133) and thermogenesis (25, 57). An important role of SNS is thus to reestablish energy balance by exerting an effect on energy intake and expenditure.

Because fat balance is sensitive to fluctuations in SNS (3), we suggest that the reequilibrium of energy balance also depends on the response of fat oxidation to changes in SNS activity. Fat oxidation is enhanced by a stimulation of SNS (139), and low sympathetic activity is suspected in obesity-prone individuals and in a reduced potential for diet-induced weight loss (7). Accordingly, SNS activity is positively correlated to daily energy expenditure adjusted for body size and composition (121, 133). However, reduced SNS activity seems to be normalized with body weight and fat-mass gain as shown by the positive relationship between fat mass and SNS activity measured by catecholamines (103) or muscle sympathetic nerve activity (93, 103, 122). All together, these studies indicate that overall SNS activity may be an important factor in the metabolic defects underlying obesity.

Exercise has the potential to cause an increase of SNS activity and, consequently, to induce anorectic effects (101, 139). As with fat gain, an elevation of SNS activity leads to an increase in fat oxidation and energy expenditure. However, the effect of exercise on SNS activity was suggested to be concentrated at the muscle β-adrenergic receptors (95, 96) and is therefore less expected to enhance the risk of hypertension and cardiovascular diseases than does fat-mass gain.

The effect of exercise on energy balance seems to be partly related to the corticotrophin-releasing hormone (CRH), which is known to reduce energy intake (10, 76) and promote energy expenditure (120). In this regard, CRH has been demonstrated to play a major role in the anorectic effect of exercise (115). Furthermore, it also has a negative effect on the preference for dietary fat (152).

In addition to exercise, dieting is a common way to lose body weight. However, dietary restriction has been demonstrated to decrease SNS activity (126). During weight loss treatment, obese subjects would thus benefit from an exercise treatment that has the potential to compensate for the decrease in SNS activity, as well as providing all the other benefits of regular physical activity participation.

Sympathomimetic agents such as caffeine and capsaicin may also have a significant impact on energy expenditure and intake. It was shown that caffeine stimulates SNS (11), thereby affecting energy expenditure (1, 6, 26, 40, 97, 153) and fat oxidation (6, 24) for up to 12 hours after its ingestion. Capsaicin, a pungent product of red pepper, also induces acute effects on SNS activity (149). The ingestion of capsaicin produces an increase in energy expenditure (53, 154) and fat oxidation (63, 154). However, the nonselective stimulating effect of these compounds on the SNS could be harmful, especially regarding hypertension. Their use as thermogenic agents should thus be considered with caution.

Weight Loss Treatments

Weight loss will occur only if energy expenditure exceeds energy intake. Moreover, according to Flatt's theory (45), fat oxidation also has to be higher than fat intake to favor negative energy balance. Considering these concepts, treatments aimed at producing body weight loss should induce a negative energy and fat balance, whether by decreasing energy and fat intake and/or by increasing energy expenditure and fat oxidation.

Exercise

A negative relationship exists between physical activity participation and body weight, with active people being generally leaner than sedentary individuals (136). Endurance training and its effect on daily energy expenditure have been widely investigated. Many intervention studies have reported that endurance training from

4 to 20 weeks has significantly increased daily energy expenditure in young males (13, 15, 83) from 301 to 956 kcal/day. However, no studies reported such augmentation in females (13, 83, 100, 109). The difference between results of these various studies could depend on differences in intensity and duration of the training, possibly because females seem to naturally train at a lower intensity, expending less energy for the same period of time.

The effects of exercise intensity on body weight and fat-mass changes have been investigated in our laboratory. Experimental evidence indicates that higher intensity exercise decreases subcutaneous adiposity for a given energy expenditure in leisure-time physical activities (136), as well as favoring other benefits (table 8.5). A high-intensity interval training program combined with endurance training was more effective in body weight reduction than endurance training alone. Moreover, interval training combined with endurance training produces changes in the oxidative potential of the skeletal muscle, meaning that this type of training may also improve the capacity to oxidize fat (140).

The duration and frequency of exercise are also of major importance in body weight and fat loss, because both parameters increase PAEE. Cross-sectional data have demonstrated positive and significant correlations between frequency and duration of exercise in relation to body weight and fat-mass changes in men (8). Concordant data obtained in an experimental protocol demonstrated that increasing the frequency of training further increased weight loss and subcutaneous adipose tissue reduction (104).

Exercise training is thus useful in producing weight loss. A single bout of exercise produces a surplus of energy expended as well as an increase in the post-exercise energy expenditure and fat oxidation. It is noteworthy that higher intensity training is more potent in elevating energy expenditure and has been shown to be more efficient in inducing weight loss and fat-mass loss. However, frequency and duration are also of major importance, and their respective positive relationship with body weight and fat-mass changes indicate that regularity of exercise is an important factor when exercise is considered as a weight loss treatment.

Table 8.5   Effects of High-Intensity Exercise on Weight-Loss-Related Variables

Effects of intensity	Reference
1. Decreases subcutaneous adiposity for a given energy expenditure in leisure-time physical activities	136
2. Increases subcutaneous fat-mass loss	140
3. Increases muscular lipolytic capacity	140
4. Decreases post-exercise energy intake compensation	60
5. Enhances the potential of exercise to increase post-exercise resting metabolic rate and fat oxidation	155

The Combination of Diet and Exercise as a Treatment for Weight Loss

Previous studies compared the effect of exercise or dieting alone on weight loss to the use of both treatments combined. As expected, subjects receiving a combination of the two interventions were more successful in losing weight and maintaining the weight loss for one year (151). Our laboratory submitted obese women to a weight-reducing program consisting of aerobic exercise followed by the addition of a low-fat diet (137). Exercise induced a substantial weight loss and adding a low-fat diet improved the body weight reduction process. After 21 months of treatment, weight loss leveled off, indicating a new steady state in energy balance at a reduced body weight. Moreover, the metabolic profile of these women was significantly improved even though some of the subjects were still considered obese.

Available evidence suggests that fat oxidation tends to decline with the decrease in body-fat mass because of a decrease in fat stores and FFA availability (134) that diminishes the fat content of the fuel mix oxidized relative to carbohydrates (125). The addition of a low-fat diet to the weight loss program thus allowed the decrease of fat intake to match fat oxidation at a lower body weight. As illustrated in figure 8.2, this permits achievement of energy and fat balance even if the contribution of fat mass to fat balance regulation is reduced.

Weight Maintenance in the Reduced-Obese State

When the patient has lost the desired amount of weight, the combination of a healthy diet and exercise seems to be the best approach to maintain body weight reduction. In fact, it was reported that 90% of reduced-obese women who were successfully maintaining weight loss were exercising regularly, whereas only 34% of those who relapsed were active (64). Moreover, the highly active women regained significantly less weight than moderately active and sedentary subjects after one and two years of a weight maintenance follow-up (41, 54). In addition, post-obese men who exercised at a higher intensity level regained substantially less weight than their counterparts who exercised at low levels, two to three years after a mean body weight loss of 27 kg (50).

Exercise in combination with nutritional changes is thus an important mode of treatment to help maintain weight loss. Because dietary restriction causes decreased SNS activity and energy expenditure, achieving and maintaining negative energy balance with dieting alone is difficult. However, exercise is an efficient mode of counteracting this difficulty because it has the potential to favor an elevation of SNS activity, to increase energy expenditure and fat oxidation, and to induce behavioral changes regarding macronutrient preferences and energy intake habits. All together, these adaptations have the potential to lead to the reoccurrence of a stable energy and fat balance in addition to an improved metabolic profile at a lower body weight and fat mass.

Exercise and fat oxidation

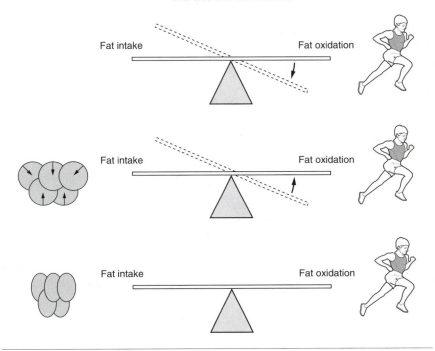

Figure 8.2   Long-term adaptation in energy and fat balance following the adherence to an exercise program.

Conclusion

Energy and macronutrient balance need to be closely regulated in order to maintain a stable body weight. In the case of a sustained negative or positive imbalance, body reserves will fluctuate to compensate for the deficit or excess in energy; therefore, body weight and fat mass will be subjected to changes. When some defects impair the regulation of energy and macronutrient balance, it appears that the subsequent body-fat tissue accretion generates an increase in fat oxidation and energy expenditure that can compensate for the impairment. However, various metabolic and health problems inherent to an excess of body-fat mass can be prevented by adhering to a healthy lifestyle.

Exercise is considered to be one of the most effective treatments against excess body weight. Exercise seems to promote the restoration of fat and energy balance by increasing energy expenditure and fat oxidation at rest and during exercise; it also appears to improve nutritional preferences, enhance the sympathoadrenal drive, and improve the metabolic profile. Exercise, as part of a healthy lifestyle, is thus a useful and beneficial mode of intervention to reestablish energy and macronutrient balance at a lower body weight.

References

1. Acheson, K.J., B. Zahorska-Markiewicz, P. Pittet, K. Anantharaman, and E. Jéquier. 1980. Caffeine and coffee: their influence on metabolic rate and substrate utilization in normal-weight and obese individuals. *American Journal of Clinical Nutrition* 33: 989-997.
2. Acheson, K.J., Y. Schutz, T. Bessard, E. Ravussin, E. Jéquier, and J.P. Flatt. 1984. Nutritional influences on lipogenesis and thermogenesis after a carbohydrate meal. *American Journal of Physiology* 246: E62-E70.
3. Acheson, K.J., Y. Schutz, T. Bessard, K. Anantharaman, J.P. Flatt, and E. Jéquier. 1988. Glycogen storage capacity and de novo lipogenesis during massive carbohydrate overfeeding in man. *American Journal of Clinical Nutrition* 48: 240-248.
4. Alméras, N., N. Lavallée, J.P. Després, C. Bouchard, and A. Tremblay. 1995. Exercise and energy intake: effect of substrate oxidation. *Physiology and Behavior* 57: 995-1000.
5. Arciero, P.J., M.I. Goran, and E.T. Poehlman. 1993. Resting metabolic rate is lower in women than in men. *Journal of Applied Physiology* 75: 2514-2520.
6. Astrup, A., S. Toubro, S. Cannon, P. Hein, L. Breum, and J. Madsen. 1990. Caffeine: a double-blind placebo-controlled study of its thermogenic, metabolic, and cardiovascular effects in healthy volunteers. *American Journal of Clinical Nutrition* 51: 759-767.
7. Astrup, A., and I.A. Macdonald. 1997. Sympathoadrenal system and metabolism. In *Handbook of Obesity*, ed. Bray, G.A., C. Bouchard, and W.P.T. James. New York: Marcel Dekker, Inc.
8. Ballor, D.L., and R.E. Keesey. 1991. A meta-analysis of the factors affecting exercise-induced changes in body mass, fat mass and fat-free mass in males and females. *International Journal of Obesity* 15: 717-726.
9. Barkeling, B., S. Rössner, and H. Bjorvell. 1990. Efficiency of a high-protein meal (meat) and a high-carbohydrate meal (vegetarian) on satiety measured and by automated computerized monitoring of subsequent food intake, motivation to eat and food preferences. *International Journal of Obesity* 14: 743-751.
10. Beck, B. 1992. Cholecystokinin, neurotensin and corticotropin release factor—3 important anorexic peptides. *Annals of Endocrinology* 53: 44-56.

11. Bellet, S., A. Kershbaum, and E.M. Finck. 1968. Response of free fatty acids to coffee and caffeine. *Metabolism* 17: 702-707.

12. Bielinski, R., Y. Schutz, and E. Jéquier. 1985. Energy metabolism during the postexercise recovery in man. *American Journal of Clinical Nutrition* 42: 69-82.

13. Bingham, S.A., G.R. Goldberg, W.A. Coward, A.M. Prentice, and J.H. Cummings. 1989. The effect of exercise and improved physical fitness on basal metabolic rate. *British Journal of Nutrition* 61: 155-173.

14. Bjorntorp, P., H. Bergman, E. Varnauskas, and B. Lingholm. 1969. Lipid mobilization in relation to body composition. *Metabolism* 18: 841-851.

15. Blaak, E.E., K.R. Westerterp, O. Bar-Or, L.J. M. Wouters, and W.H.M. Saris. 1992. Total energy expenditure and spontaneous activity in relation to training in obese boys. *American Journal of Clinical Nutrition* 55: 777-782.

16. Black, A.E., W.A. Coward, T.J. Cole, and A.M. Prentice. 1996. Human energy expenditure in affluent societies: an analysis of 574 doubly-labelled water measurements. *European Journal of Clinical Nutrition* 50: 72-92.

17. Block, G., C.M. Dresser, A.M. Hartman, and M.D. Carroll. 1985. Nutrient sources in the American diet: quantitative data from the NHANES II survey. Macronutrients and fats. *American Journal of Epidemiology* 122: 27-40.

18. Blundell, J.E., V.J. Burley, J.R. Cotton, and C.L. Lawton. 1993. Dietary fat and control of energy intake: evaluating the effects of fat on meal size and postmeal satiety. *American Journal of Clinical Nutrition* 57 (Suppl): 772S-778S.

19. Bouchard, C., A. Tremblay, J. Nadeau, J.P. Després, G. Thériault, M. Boulay, G. Lortie, C. Leblanc, and G. Fournier. 1989. Genetic effect in resting and exercise metabolic rates. *Metabolism* 38: 364-370.

20. Bouchard, C. 1990. Variation in human body fat: The contribution of the genotype. In *Obesity: Towards a molecular approach,* eds. Bray, G., D. Ricquier, and B. Spiegelman. New York: Alan R. Liss.

21. Bouchard, C. 1992. Human obesities: chaos or determinism? Obesity in Europe 91. *Proceedings of the 3rd European Congress on Obesity.* eds. Ailhaud G., B. Guy-grand, M. Lafontan, and D. Ricquier. Paris: John Libbey.

22. Bouchard, C. 1992. La génétique des obésités humaines. *Annales de Médecine Interne* 143: 463-471.

23. Bouchard, C., J.P. Després, and P. Mauriège. 1993. Genetic and nongenetic determinants of regional fat distribution. *Endocrinology Review* 14: 72-93.

24. Bracco, D., J.M. Ferrarra, M.J. Arnaud, E. Jéquier, and Y. Schutz. 1995. Effects of caffeine on energy metabolism, heart rate, and methylxanthine metabolism in lean and obese women. *American Journal of Physiology* 32: E671-E678.

25. Bray, G.A. 1990. A state of reduced sympathetic activity and normal or high adrenal activity: the autonomic and adrenal hypothesis revisited. *International Journal of Obesity* 14: 77-92.

26. Bray, G.A. 1991. Treatment for obesity: A nutrient balance/nutrient partition approach. *Nutrition Review* 49: 33-45.

27. Buemann, B., A. Astrup, and N.J. Christensen. 1992. Three months aerobic training fails to affect 24-hour energy expenditure in weight-stable, post-obese women. *International Journal of Obesity* 16: 809-816.

28. Calles-Escandon, J., and P. Driscoll. 1994. Free fatty acid metabolism in aerobically fit individuals. *Journal of Applied Physiology* 77: 2374-2379.

29. Calles-Escandon, J., M.I. Goran, M. O'Connell, K.S. Nair, and E. Danforth. 1996. Exercise increases fat oxidation at rest unrelated to changes in energy balance or lipolysis. *American Journal of Physiology* 270 (Endocrinol. Metab. 33): E1009-E1014.

30. Chad, K.E., and B.M. Quigley. 1991. Exercise intensity: effect of postexercise O_2 uptake in trained and untrained women. *Journal of Applied Physiology* 70: 1713-1719.

31. Cigolini, M., G. Targher, A.I.A. Bergamo, M. Tonoli, F. Filippi, M. Muggeo, and G. De Sandre. 1996. Moderate alcohol consumption and its relation to visceral fat and plasma androgens in healthy women. *International Journal of Obesity* 20: 206-212.

32. Coyle, E.F., A.E. Jeukendrup, A.J.M. Wagenmakers, and W.H.M. Saris. 1997. Fatty acid oxidation is directly regulated by carbohydrate metabolism during exercise. *American Journal of Physiology* 273 (Endocrinol. Metab. 36): E268-E275.

33. De Castro, J.M., and S. Orozco. 1990. Moderate alcohol intake on the spontaneous eating patterns of humans: evidence of unregulated supplementation. *American Journal of Clinical Nutrition* 52: 246-253.

34. Després, J.P., C. Bouchard, S. Savard, A. Tremblay, M. Marcotte, and G. Thériault. 1984. Level of physical fitness and adipocytes lipolysis in human. *Journal of Applied Physiology* 56: 1157-1161.

35. Dionne, I., M. Johnson, M.D. White, S. St.-Pierre, and A. Tremblay. 1997. Acute effect of exercise and low-fat diet on energy balance. *International Journal of Obesity* 21: 413-416.

36. Dionne, I., S. Van Vugt, and A. Tremblay. 1999. Postexercise substrate oxidation: a factor dependent on postexercise macronutrient intake. *American Journal of Clinical Nutrition* 69: 927-930.

37. Dreon, D.M., B. Frey-Hewitt, N. Ellsworth, P.T. Williams, R.B. Terry, and P.D. Wood. 1988. Dietary fat:carbohydrate ratio and obesity in middle-aged men. *American Journal of Clinical Nutrition* 47: 995-1000.

38. Drewnowski, A., and M.R.C. Greenwood. 1983. Cream and sugar: human preferences for high-fat foods. *Physiology and Behavior* 30: 629-633.

39. Drewnowski, A., C. Kurth, J. Holden-Wiltse, and J. Saari. 1992. Food preferences in human obesity: carbohydrates versus fats. *Appetite* 18: 207-221.

40. Dulloo, A.G., and C.A. Geissler. 1989. Normal caffeine consumption: influence on thermogenesis and daily energy expenditure in lean and

postobese human volunteers. *American Journal of Clinical Nutrition* 49: 44-50.

41. Ewbank, P.P., L.L. Darga, and C.P. Lucas. 1995. Physical activity as a predictor of weight maintenance in previously obese subjects. *Obesity Research* 3: 257-263.

42. Ferraro, R., S. Lillioja, A.M. Fontvieille, R. Rising, C. Bogardus, and E. Ravussin. 1992. Lower sedentary metabolic rate in women compared with men. *Journal of Clinical Investigation* 90: 780-784.

43. Flatt, J.P., E. Ravussin, K.J. Acheson, and E. Jéquier. 1985. Effects of dietary fat on post-prandial substrate oxidation and on carbohydrate and fat balances. *Journal of Clinical Investigation* 76: 1119-1124.

44. Flatt, J.P. 1987. Dietary fat, carbohydrate balance, and weight maintenance: effects of exercise. *American Journal of Clinical Nutrition* 45: 296-306.

45. Flatt J.P., 1988. Importance of nutrient balance in body weight regulation. *Diabetes* 4: 571-581.

46. Flatt, J.P. 1995. Diet, lifestyle, and weight maintenance. Mc Collum Award Lecture. *American Journal of Clinical Nutrition* 62: 820-836.

47. Fogelholm, M., S. Männistö, E. Vartiainen, and P. Pietinen. 1996. Determinants of energy balance and overweight in Finland 1982 and 1992. *International Journal of Obesity* 20: 1097-1104.

48. Foltin, R.W., T.H. Kelly, and M.W. Fischman. 1993. Ethanol as an energy source in humans: comparison with dextrose containing beverages. *Appetite* 20: 95-110.

49. Grucshow, H.W., K.A. Sobocinski, J.J. Barboriak, and J.G. Scheller. 1985. Alcohol consumption, nutrient intake and relative weight among US adults. *American Journal of Clinical Nutrition* 42: 289-295.

50. Hartman, W.M., M. Stroud, D.M. Sweet, and J. Saxton. 1993. Long-term maintenance of weight loss following supplemented fasting. *International Journal of Eating Disorders* 14: 87-93.

51. Health Canada. 1995. *Canadians and heart health: reducing the risk.*

52. Hellerstedt, W.L., R.W. Jeffery, and D.M. Murray. 1990. The association between alcohol intake and adiposity in the general population. *American Journal of Epidemiology* 132: 594-611.

53. Henry, C.J.K., and B. Emery. 1985. Effects of spiced food on metabolic rate. *Human Nutrition* 40C: 165-168.

54. Hensrud, D.D., R.L. Weinsier, B.E. Darnell, and G.R. Hunter. 1994. A prospective study of weight maintenance in obese subjects reduced to normal body weight without weight-loss training. *American Journal of Clinical Nutrition* 60: 688-694.

55. Hill, A.J., and J.E. Blundell. 1986. Macronutrients and satiety: the effects of high-protein or high-carbohydrate meal on subjective motivation to eat and food preferences. *Nutrition and Behavior* 3: 133-144.

56. Hill, A.J., and J.E. Blundell. 1990. Comparison of the action of macronutrients on the expression of appetite in lean and obese humans. *Annals of the New York Academy of Sciences* 597: 529-531.

57. Himms-Hagen, J. 1984. Thermogenesis in brown adipose tissue as an energy buffer: implication for obesity. *New England Journal of Medicine* 311: 1549-1558.

58. Holt, S.H.A., J.C. Brand Miller, and P. Petocz. 1996. Interrelationships among postprandial satiety, glucose and insulin responses and changes in subsequent food intake. *European Journal of Clinical Nutrition* 50: 788-797.

59. Hultman, E., and L.H. Nilsson. 1975. Factors influencing carbohydrate metabolism in man. *Nutrition and Metabolism* 18 (Suppl 1): 45-64.

60. Imbeault, P., S. Saint-Pierre, N. Alméras, and A. Tremblay. 1997. Acute effects of exercise on energy intake and feeding behaviour. *British Journal of Nutrition* 77: 511-521.

61. Jebb, S.A., A.M. Prentice, G.R. Goldberg, P.R. Murgatroyd, A.E. Black, and W.A. Coward. 1996. Changes in macronutrient balance during over- and underfeeding assessed by a 12-d continuous whole-body calorimetry. *American Journal of Clinical Nutrition* 64: 259-266.

62. Jones, B.R., E. Barrett-Connor, M.H. Criqui, and M.J. Holbrook. 1982. A community study of calorie and nutrient intake in drinkers and non-drinkers of alcohol. *American Journal of Clinical Nutrition* 35: 135-139.

63. Kawada, T., K.I. Hagihara, and K. Iwai. 1986. Effects of capsaicin on lipid metabolism in rats fed high fat diet. *Journal of Nutrition* 116: 1272-1278.

64. Kayman, S., W. Bruvold, and J.S. Stern. 1990. Maintenance and relapse after weight loss in women. *American Journal of Clinical Nutrition* 52: 800-807.

65. Kiens, B., B. Essen-Gastavsson, and H. Lithell. 1987. Lipoprotein lipase activity and intramuscular triglyceride stores after long-term high-fat and high-carbohydrate diets in physically trained men. *American Journal of Physiology* 7: 1-9.

66. Kiens, B., and H. Lithell. 1989. Lipoprotein metabolism influenced by training-induced changes in human skeletal muscle. *Journal of Clinical Investigation* 83: 558-564.

67. King, N.A., and J.E. Blundell. 1994. Exercise-induced suppression of appetite: effects on food intake and implications for energy balance. *European Journal of Clinical Nutrition* 48: 715-724.

68. King, N.A., and J.E. Blundell. 1995. High fat foods overcome the energy expenditure due to exercise after cycling and running. *European Journal of Clinical Nutrition* 49: 114-123.

69. Kissileff, H.R., F.X. Pi-Sunyer, K. Segal, S. Meltzrer, and P.A. Foelsch. 1990. Acute effects of exercise on food intake in obese and nonobese women. *American Journal of Clinical Nutrition* 52: 715-724.

70. Klesges, R.C., L.M. Klesges, C.K. Haddock, and L.H. Eck. 1992. A longitudinal analysis of the impact of dietary intake and physical activity on weight change in adults. *American Journal of Clinical Nutrition* 55: 818-822.

71. Kuczmarski, R.J., K.M. Flegal, S.M. Campbell, and C.L. Johnson. 1994. Increasing prevalence of overweight among US adults. *Journal of the American Medical Association* 272: 205-211.

72. Kuskowska, W.A., and R. Bergstrom. 1993. Trends in body mass index and prevalence of obesity in Swedish men 1980-89. *Journal of Epidemiology and Community Health* 47: 103-108.

73. Kuskowska, W.A., and R. Bergstrom. 1993. Trends in body mass index and prevalence of obesity in Swedish Women 1980-89. *Journal of Epidemiology and Community Health* 47: 195-199.

74. Lawson, S., J.D. Webster, and J.D. Pacy. 1987. Effect of a 10-week aerobic exercise program on metabolic rate, body composition and fitness in lean sedentary females. *British Journal of Clinical Practice* 41: 684-688.

75. Lawton, C.L., V.J. Burley, J.K. Wales, and J.E. Blundell. 1993. Dietary fat and appetite control in obese subjects. *International Journal of Obesity* 17: 409-416.

76. Levine, A.S., and C.J. Billington. 1991. Stress, peptides, and regulation of ingestive behavior. In *Stress, Neuropeptides, and Systemic Diseases,* eds. McCubbin, J.A., P.G. Kaufmann, and C.B. Nemeroff. San Diego: Academic, 327-339.

77. Liu, S., M.K. Serdula, D.F. Williamson, A.H. Mokdad, and T. Byers. 1994. A prospective study of alcohol intake and change in body weight among US adults. *American Journal of Epidemiology* 140: 912-920.

78. Livingstone, M.B.E., J.J. Strain, A.M. Prentice, W.A. Coward, G.B. Nevin, M.E. Barker, R. Hickey, P.G. McKenna, and R.G. Whitehead. 1991. Potential contribution of leisure time activity to the energy expenditure patterns of sedentary populations. *British Journal of Nutrition* 65: 145-155.

79. Macdonald, I.A., and J. Webber. 1995. Feeding, fasting and starvation: factors affecting fuel utilization. *Proceedings of the Nutrition Society* 54: 267-274.

80. Maehlum, S., M. Grandmontagne, E.A. Newshome, and O.M. Sejersted. 1986. Magnitude and duration of excess postexercise oxygen consumption in healthy young subjects. *Metabolism* 35: 425-429.

81. McGinnis, J.M. 1992. The public health burden of a sedentary lifestyle. *Medicine and Science in Sports and Exercise* 24 (Suppl 6): S196-S200.

82. McGinnis, J.M., and W.H. Foege. 1993. Actual causes of death in the United States. *Journal of the American Medical Association* 270: 2207-2212.

83. Meijer, G.A.L., K.R. Westerterp, A.M.P. Van Hulsel, and F. Ten Hoor. 1992. Physical activity and energy expenditure in lean and obese adult human subjects. *European Journal of Applied Physiology* 65: 525-528.

84. Nair, K.S., D. Halliday, and J.S. Garrow. 1983. Thermic response to isoenergetic protein, carbohydrate or fat meals in lean and obese subjects. *Clinical Sciences* 65: 307-312.

85. Nelson, K.M., R.L. Weinsier, C.L. Long, and Y. Schutz. 1992. Prediction of resting energy expenditure from fat-free mass and fat mass. *American Journal of Clinical Nutrition* 56: 848-856.

86. Oberlin, P., C.L. Melby, and E.T. Poehlman. 1990. Resting energy expenditure in young vegetarians and non-vegetarians. *Nutrition Research* 10: 39-49.

87. O'Dea, K., M.D. Ester, P. Leonards, J.R. Stockigt, and P. Nestel. 1982. Noradrenaline turnover during under- and overeating in normal-weight subjects. *Metabolism* 31: 95-100.

88. O'Dea, K. 1984. Marked improvement in carbohydrate and lipid metabolism in diabetic Australian aborigines after temporary reversion to traditional lifestyle. *Diabetes* 33: 596-603.

89. Office of Smoking and Health. 1989. *Reducing the health consequence of smoking: 25 years of progress.*

90. Owen, O.E., E. Kavle, R.S. Owen, M. Polansky, S. Caprio, M.A. Mozzoli, Z.V. Kendrick, M.C. Bushman, and G. Boden. 1986. A reappraisal of caloric requirements in healthy women. *American Journal of Clinical Nutrition* 44: 1-19.

91. Owen, O.E., J.L. Holup, D.A. D'Alessio, E.S. Craig, M. Polansky, K.J. Smalley, E.C. Kavle, M.C. Bushman, L.R. Owen, and M.A. Mozzoli. 1987. A reappraisal of the caloric requirements of men. *American Journal of Clinical Nutrition* 46: 875-885.

92. Parker, J.P., S. Snitker, J.S. Skinner, and E. Ravussin. 1996. Gender differences in muscle sympathetic nerve activity: effect of body fat distribution. *American Journal of Physiology* 270: E363-E366.

93. Pérusse, L., A. Tremblay, C. Leblanc, C.R. Cloninger, T. Reich, J. Rice, and C. Bouchard. 1988. Familial resemblance in energy intake: contribution of genetic and environmental factors. *American Journal of Clinical Nutrition* 47: 629-635.

94. Phelain, J.F., E. Reinke, M.A. Harris, and C.L. Melby. 1997. Postexercise energy expenditure and substrate oxidation in young women resulting from exercise bouts of different intensities. *Journal of American College of Nutrition* 16: 140-146.

95. Plourde, G., S. Rousseau-Migneron, and A. Nadeau. 1991. β-adrenoreceptor adenylate cyclase system adaptation to physical training in rat ventricular tissue. *Journal of Applied Physiology* 70: 1633-1638.

96. Plourde, G., S. Rousseau-Migneron, and A. Nadeau. 1993. Effect of endurance training on beta-adrenergic system in three different skeletal muscles. *Journal of Applied Physiology* 74: 1641-1646.

97. Poehlman, E.T., J.P. Després, H. Bessette, E. Fontaine, A. Tremblay, and C. Bouchard. 1985. Influence of caffeine on the resting metabolic rate of exercise-trained and inactive subjects. *Medicine and Science in Sports and Exercise* 17: 689-694.

98. Poehlman, E.T., A. Tremblay, A. Nadeau, J. Dussault, G. Thériault, and C. Bouchard. 1986. Heredity and changes in hormones and metabolic rates with short-term training. *American Journal of Physiology* 250: E711-E717.

99. Poehlman, E.T., P.J. Arciero, C.L. Melby, and D.M.V. Badylak. 1988. Resting metabolic rate and postprandial thermogenesis in vegetarians and non-vegetarians. *American Journal of Clinical Nutrition* 48: 209-213.

100. Poehlman, E.T., C.L. Melby, S.F. Badylak, and J. Calles. 1989. Aerobic fitness and resting energy expenditure in young adult males. *Metabolism* 38: 85-90.

101. Poehlman, E.T., and E. Danfort. 1991. Endurance training increases metabolic rate and norepinephrine appearance rate in older individuals. *American Journal of Physiology* 261: E233-E239.

102. Poehlman, E.T., A.W. Gardner, P.J. Arciero, M.I. Goran, and J. Calles-Escandon. 1994. Effects of endurance training on total fat oxidation in elderly persons. *Journal of Applied Physiology* 76: 2281-2287.

103. Poehlman, E.T., A.W. Gardner, M.I. Goran, P.J. Arciero, M.J. Toth, P.A. Ades, and J. Calles-Escandon. 1995. Sympathetic nervous system activity, body fatness, and body fat distribution in younger and older males. *Journal of Applied Physiology* 78: 802-806.

104. Pollock, M.L., H.S. Miller, A.C. Linnerud, and K.H. Cooper. 1975. Frequency of training as a determinant for improvement in cardiovascular function and body composition of middle-aged men. *Archives of Physical Medicine and Rehabilitation* 56: 141-144.

105. Poppitt, S.D., J.W. Eckhardt, J. McGonagle, P.R. Murgatroyd, and A.M. Prentice. 1996. Short-term effects of alcohol consumption on appetite and energy intake. *Physiology and Behavior* 60: 1063-1070.

106. Prentice, A.M., and S.A. Jebb. 1995. Obesity in Britain: gluttony or sloth? *British Journal of Medicine* 311: 437-439.

107. Prentice, A.M., A.E. Black, W.A. Coward, and T.J. Cole. 1996. Energy expenditure in overweight and obese adults in affluent societies: an analysis of 319 doubly-labelled water measurements. *European Journal of Clinical Nutrition* 50: 93-97.

108. Raben, A., and A. Astrup. 1996. Manipulating carbohydrate content and sources in obesity prone subjects: effect on energy expenditure and macronutrient balance. *International Journal of Obesity* 20 (Suppl. 2): S24-S30.

109. Racette, S.B., D.A. Schoeller, R.F. Kusher, K.M. Neil, and K. Herling-Iaffaldano. 1995. Effects of aerobic exercise and dietary carbohydrate on energy expenditure and body composition during weight reduction in obese women. *American Journal of Clinical Nutrition* 61: 486-494.

110. Randle, P.J., P.B. Garland, C.N. Hales, and E.A. Newsholme. 1963. The glucose-fatty-acid cycle. Its role in insulin sensitivity and the metabolic disturbances of diabetes mellitus. *The Lancet* 1: 785-789.

111. Ravussin, E., B. Burnand, Y. Schutz, and E. Jéquier. 1982. Twenty-four-hour energy expenditure and resting metabolic rate in obese, moderately obese, and control subjects. *American Journal of Clinical Nutrition* 35: 566-573.

112. Ravussin, E., S. Lillioja, T.E. Anderson, L. Christin, and C. Bogardus. 1986. Determinants of 24-hour energy expenditure in man. *Journal of Clinical Investigation* 78: 1568-1578.

113. Ravussin, E., S. Lillioja, W.C. Knowler, L. Christin, D. Freymond, W.G.H. Abbot, V. Boyce, B.V. Howard, and C. Bogardus. 1988. Reduced rate of energy expenditure as a risk factor for body weight gain. *New England Journal of Medicine* 318: 467-472.

114. Ravussin, E., and C. Bogardus. 1989. Relationships of genetics, age and physical fitness to daily energy expenditure and fuel utilization. *American Journal of Clinical Nutrition* 49: 968-975.

115. Rivest, S., and D. Richard. 1990. Involvement of corticotropin-releasing factor in the anorexia induced by exercise. *Brain Research Bulletin* 25: 169-172.

116. Rivière, D., F. Crampes, M. Beauville, and M. Garrigues. 1989. Lipolytic response of fat cells to catecholamines in sedentary and exercise-trained women. *Journal of Applied Physiology* 66: 330-335.

117. Rolls, B.J., M. Hetherington, and V. Burling. 1988. The specificity of satiety: the influence of foods of different macronutrient content on the development of satiety. *Physiology and Behavior* 43: 145-153.

118. Rolls, B.J., S. Kim-Harris, M.W. Fischman, R.W. Foltin, T.H. Moran, and S.A. Stoner. 1994. Satiety after preloads with different amounts of fat and carbohydrate: implications for obesity. *American Journal of Clinical Nutrition* 60: 476-487.

119. Romieu, I., W.C. Willett, M.J. Stampfer, G.A. Colditz, L. Sampson, B. Rosner, C.H. Hennekens, and F.E. Speizer. 1988. Energy intake and other determinants of relative weight. *American Journal of Clinical Nutrition* 47: 406-412.

120. Rothwell, N.J. 1990. Central effect of CRF on metabolism and energy balance. *Neuroscience and Biobehavioral Reviews* 14: 263-271.

121. Saad, M.F., S.A. Alger, F. Zurlo, J.B. Young, C. Bogardus, and E. Ravussin. 1991. Ethnic differences in sympathetic nervous system-mediated energy expenditure. *American Journal of Physiology* 261: E789-E794.

122. Scherrer, U., D. Randin, L. Tappy, P. Vollenweider, E. Jéquier, and P. Nicod. 1994. Body fat and sympathetic nerve activity in healthy subjects. *Circulation* 89: 2634-2640.

123. Schulz, L.O., B.L. Nyomba, S. Alger, T.E. Anderson, and E. Ravussin. 1991. Effect of endurance training on sedentary energy expenditure measured in a respiratory chamber. *American Journal of Physiology* 260: E257-E261.

124. Schutz, Y., K.J. Acheson, and E. Jéquier. 1985. Twenty-four hour energy expenditure and thermogenesis response to progressive carbohydrate overfeeding in man. *International Journal of Obesity* 9 (Suppl 2): 111-114.

125. Schutz, Y., A. Tremblay, R.L. Weinsier, and K.M. Nelson. 1992. Role of fat oxidation in the long-term stabilization of body weight in obese women. *American Journal of Clinical Nutrition* 55: 670-674.

126. Schwartz, R.S., L.F. Jaeger, R.C. Weight, and S. Lakshminaratan. 1990. The effect of the diet or exercise on plasma norepinephrine kinetics in moderate obese young men. *International Journal of Obesity* 14: 1-11.

127. Sedlock, D.A., J.A. Fissinger, and C.L. Melby. 1989. Effect of exercise intensity and duration on postexercise energy expenditure. *Medicine and Science in Sports and Exercise* 21: 662-666.

128. Shetty, P.S., R.T. Jung, and W.P.T. James. 1979. Effect of catecholamine replacement with levodopa on the metabolic response to semi-starvation. *The Lancet* 1: 77-79.

129. Singh, J., A.M. Prentice, E. Diaz, W.A. Coward, J. Ashford, M. Sawyer, and R.G. Whitehead. 1989. Energy expenditure of Gambian women during peak agricultural activity measured by the doubly-labelled water technique. *British Journal of Nutrition* 62: 315-329.

130. Spurr, G.B., D.L. Dufour, and J.C. Reina. 1996. Energy expenditure of urban Colombian women: a comparision of patterns and total daily expenditure by the heart rate and factorial methods. *American Journal of Clinical Nutrition* 63: 870-878.

131. Stubbs, R.J., C.G. Harbron, P.R. Murgatroyd, and A.M. Prentice. 1995. Covert manipulation of dietary fat and energy density: effect on substrate flux and food intake in men eating ad libitum. *American Journal of Clinical Nutrition* 62: 316-329.

132. Thompson, D.A., L.A. Wolfe, and R. Eikelboom. 1988. Acute effects of exercise intensity on appetite in young men. *Medicine and Science in Sports and Exercise* 20: 222-227.

133. Toubro, S., T.I.A. Sorensen, B. Ronn, N.J. Christensen, and A. Astrup. 1996. Twenty-four hour energy expenditure: the role of body composition, thyroid status, sympathetic activity, and family membership. *Journal of Clinical Endocrinology and Metabolism* 81: 2670-2674.

134. Tremblay, A., J.P. Després, and C. Bouchard. 1985. The effects of exercise-training on energy balance and adipose tissue morphology and metabolism. *Sports Medicine* 2: 223-233.

135. Tremblay, A., G. Plourde, J.P. Després, and C. Bouchard. 1989. Impact of dietary fat content and fat oxidation on energy intake in humans. *American Journal of Clinical Nutrition* 49: 799-805.

136. Tremblay, A., J.P. Després, C. Leblanc, C.L. Craig, B. Ferris, T. Stephens, and C. Bouchard. 1990. Effect of intensity of physical activity on body fatness and fat distribution. *American Journal of Clinical Nutrition* 51: 153-157.

137. Tremblay, A., J.P. Després, J. Maheux, C. Pouliot, A. Nadeau, S. Moorjani, P.J. Lupien, and C. Bouchard. 1991. Normalization of the metabolic profile in obese women by exercise and a low fat diet. *Medicine and Science in Sports and Exercise* 23: 1326-1331.

138. Tremblay, A., N. Lavallée, N. Alméras, L. Allard, J.P. Després, and C. Bouchard. 1991. Nutritional determinants of the increase in energy intake associated with a high-fat diet. *American Journal of Clinical Nutrition* 53: 1134-1137.

139. Tremblay, A., S. Coveney, J.P. Després, A. Nadeau, and D. Prud'homme. 1992. Increased resting metabolic rate and lipid oxidation in exercise-trained individuals: evidence for a role of ß-adrenergic stimulation. *Canadian Journal of Physiology and Pharmacology* 70: 1342-1347.

140. Tremblay, A., J.A. Simoneau, and C. Bouchard. 1994. Impact of exercise intensity on body fatness and skeletal muscle metabolism. *Metabolism* 43: 814-818.

141. Tremblay, A., B. Buemann, G. Thériault, and C. Bouchard. 1995. Body fatness in active individuals reporting low lipid and alcohol intake. *European Journal of Clinical Nutrition* 49: 824-831.

142. Tremblay, A., E. Wouters, M. Wenker, S. St-Pierre, C. Bouchard, and J.P. Després. 1995. Alcohol and high-fat diet: a combination favoring overfeeding. *American Journal of Clinical Nutrition* 62: 639-644.

143. Tremblay, A., and S. St-Pierre. 1996. The hyperphagic effect of a high-fat diet and alcohol intake persists after control for energy density. *American Journal of Clinical Nutrition* 63: 479-482.

144. Treno, A.J., R.N. Parker, and H.D. Holder. 1993. Understanding U.S. alcohol consumption with social and economic factors: a multivariate time series analysis, 1950-1986. *Journal of Studies on Alcohol* 54: 146-156.

145. Treuth, M.S., G.R. Hunter, and M. Williams. 1996. Effects of exercise intensity on 24-h energy expenditure and substrate oxidation. *Medicine and Science in Sports and Exercise* 28: 1138-1143.

146. Turcotte, L.P., E.A. Richter, and B. Kiens. 1992. Increased plasma FFA uptake and oxidation during prolonged exercise in trained vs. untrained humans. *American Journal of Physiology* 292: E791-E799.

147. Van Wyk, M.C.W., R.J. Stubbs, A.M. Johnstone, and C.G. Harbon. 1995. Breakfast high in protein, fat and carbohydrate: effect on within-day appetite and energy balance. *International Journal of Obesity* 19: P134.

148. Verger, P., M.T. Lanteaume, and J. Louis-Sylvestre. 1992. Human intake and choice of foods at intervals after exercise. *Appetite* 18: 93-99.

149. Watanabe, T., T. Kawada, M. Yamamoto, and K. Iwai. 1987. Capsaicin, a pungent principle of hot red pepper, evokes catecholamine secretion from the adrenal medulla of anesthetized rats. *Biochemistry and Biophysics of the Research Community* 142: 259-264.

150. Weststrate, J.A., P. Weys, E. Poortvliet, P. Deurenberg, and G.A.J. Hautvast. 1990. Lack of a systematic sustained effect of prolonged exercise bouts on resting metabolic rate in fasting subjects. *European Journal of Clinical Nutrition* 44: 91-97.

151. Wing, R., L.H. Epstein, M. Paterno-Bayles, A. Kriska, M.P. Nowalk, and W. Gooding. 1988. Exercise in a behavioural weight control program for obese patients with type 2 (non-insulin dependent) diabetes. *Diabetologia* 31: 902-909.

152. York, D.A. 1992. Central regulation of appetite and autonomic activity by CRH, glucocorticoids and stress. *Progress in Neurology, Endocrinology and Immunology* 5: 153-165.

153. Yoshida, T., N. Sakane, T. Umekawa, and M. Kondo. 1994. Relationship between basal metabolic rate, thermogenesis response to caffeine, and body weight loss following combined low calorie and exercise treatment in obese women. *International Journal of Obesity* 18: 345-350.

154. Yoshida, T., T. Umekawa, N. Sakane, K. Yoshimoto, and M. Kondo. 1996. Effect of CL316,243, a highly specific ß3-adrenoreceptor agonist, on sympathetic nervous system activity in mice. *Metabolism* 45: 787-791.

155. Yoshioka, M., S. St-Pierre, D. Richard, F. Labrie, and A. Tremblay. 1997. Effect of exercise intensity on postexercise energy metabolism. *FASEB Journal* 10: A375.

CHAPTER 9

Adipose Tissue Metabolism and Obesity

Paul Poirier, MD[1], and Robert H. Eckel, MD[2]

[1]Québec Heart Institute, Ste-Foy, Québec, Canada and [2]University of Colorado Health Sciences Center, Department of Medicine, Division of Endocrinology, Metabolism and Diabetes, Denver, Colorado, U.S.A.

Adipose tissue represents the largest energy reservoir of the body. This energy is stored in fat cells as triacylglycerols (TG). In a normal-weight individual, about 100 to 300 g of TG are hydrolyzed and synthesized each 24 hours in the total fat depot (12). Only 1 kg of adipose tissue TG is sufficient to supply energy for several marathons. It is important to note that adipose tissue, in addition to its storage capacity, also acts as a secretory organ. Among the secreted substances are lipoprotein lipase (LPL), leptin, tumor necrosis factor α (TNF-α), plasminogen activator inhibitor-1 (PAI-1), angiotensinogen, and acylation stimulating protein (ASP), to name a few (30). This chapter reviews quantitative and regulatory aspects of lipid storage and mobilization in human adipose tissue with implications for exercise and physical training in obesity. Some of these aspects of adipose tissue regulation are summarized in table 9.1.

Fat Storage

Fat storage involves multiple steps. Following food ingestion, nutrients are digested, absorbed, and processed in a way to be eventually utilized or stored in adipose tissue. Numerous molecules are involved, including lipoprotein lipase, insulin, and acylation stimulating protein, all to be discussed in further detail in the following sections.

Lipoprotein Lipase

Circulating chylomicrons and very low-density lipoproteins (VLDL) provide the major source of fatty acids for adipocyte TG storage (31, 36, 39, 89). This storage

Table 9.1   Regulation of Adipose Tissue (AT)

Hormone/ parahormone	Role	Mechanism	Comments
Insulin	Increase TG	↑ LPL ↑ glucose uptake	Men: femoral/gluteal = s/c abd. > visceral AT Women: femoral/gluteal > s/c abd. > visceral AT
	Inhibit lipolysis	↓ PKA, ↓ cAMP, ↓ HSL	Men: visceral AT > s/c abd. = femoral/gluteal AT Women: visceral AT > s/c abd. = femoral/gluteal AT
Catecholamines	Increase lipolysis	↑ PKA (↑ cAMP): β-adrenoceptor	Visceral > s/c abd. > femoral/gluteal AT
	Inhibit lipolysis	Inhibit adenylyl cyclase ↓ cAMP: β-2 adrenoceptor	Less-important effect
Adenosine, neuropeptide Y, prostaglandins E-type	Inhibit lipolysis	Inhibit adenylyl cyclase: ↓ cAMP, ↓ HSL	*In vivo* role unknown
Vasopressin, secretin, glucagon, PTH, TSH	Increase lipolysis	↑ PKA, ↑ cAMP, ↑ HSL	Weak effect on AT

process depends on the enzyme LPL. LPL is a secretory glycoprotein synthesized in a number of tissues including adipose tissue, where it plays an important role in the provision of fatty acids for their uptake and storage as TG (32). Expression of LPL in adipose tissue occurs as an early marker of differentiation after preadipocytes are committed to differentiate into adipocytes (90). Following transcription and translation, LPL is progressively glycosylated and trimmed in the endoplasmic reticulum and golgi apparatus, where activation also occurs (84). LPL is then packaged into secretory vesicles from which secretion occurs.

After secretion, the active lipase is transported to the capillary endothelium where it is bound to glycosaminoglycans and made available for the hydrolysis of circulating triglyceride-rich lipoproteins—chylomicrons and VLDL (figure 9.1). Following activation by apolipoprotein C-II and hydrolysis, the resultant lipolysis products, in particular free fatty acids (FFA), are then taken up by fat cells by several putative mechanisms. Although passive diffusion may occur, a specific binding and transport protein has also been described (1, 9). Once taken up by fat cells, FFA are esterified to TG.

Insulin increases adipose tissue LPL (35, 79). Near-maximal stimulation is seen at serum insulin concentrations that are within the physiological range for humans

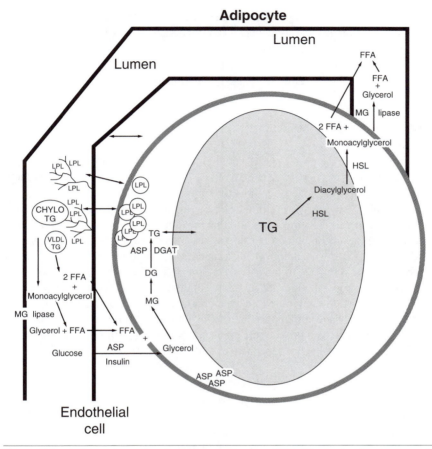

Figure 9.1   Adipose tissue function as a storage organ (TG). The predominant source of FFA for adipose tissue TG is LPL. Glucose incorporation into TG is mostly into the glycerol backbone. ASP facilitates both glucose uptake and TG synthesis (DGAT), whereas insulin increases glucose uptake and adipose tissue LPL. Adipose tissue TG provide FFA to other tissues by HSL-mediated lipolysis. This process is stimulated by cAMP and inhibited by insulin (adapted from reference 30).

($\sim$35 μU/ml) (31). This regulation by insulin is similar between femoral/gluteal and subcutaneous abdominal adipose tissue (97). However, using human omental adipose tissue (visceral) organ cultures (35) or cultured isolated adipocytes (42), the effect of insulin is less apparent. Hence, visceral adipose tissue is less responsive to insulin than subcutaneous adipose tissue. In tissue culture or in animals, catecholamines typically decrease adipose tissue LPL (26, 31, 67, 73). However, it is unclear if this effect is at the level of LPL gene transcription (73, 99). Unlike fatty acids, most of the glucose incorporated into TG is in the glycerol backbone, not the acyl side chains of TG. However, the presence of adipose tissue TG in patients with LPL deficiency (16) strongly suggests that other mechanisms must occur. The synthesis of fatty acids from glucose (lipogenesis) and/or uptake of FFA from the circulation become relevant only in the absence of the lipoprotein lipase enzyme. LPL may have other roles in adipose tissue. Indeed, the provision of fat-soluble vitamin esters and their uptake by adipose tissue may also be facilitated by LPL (13).

Acylation Stimulating Protein

Acylation stimulating protein (ASP) could represent an alternative mechanism to enhance or maintain TG storage (19). ASP is a 76 amino acid basic protein, isolated from human plasma, and is a bioactive fragment of the third component of complement (C3adesArg). Preliminary experiments indicate that ASP binds to specific high affinity receptor sites and that the number of sites increase markedly with differentiation of human preadipocytes to adipocytes (19). The temporal production of ASP during adipocyte differentiation is important because it has been demonstrated that in spite of the early increases in LPL and glycerol-3-phosphate dehydrogenase (key enzymes for substrate supply for TG synthesis), TG synthesis capacity greatly increased following the increase in ASP production by adipocytes. Thus, ASP may have an important role in human adipose tissue physiology. Moreover, chylomicrons produce the greatest increase in ASP production (tenfold), and the increase in ASP is not dependent on LPL-mediated chylomicron triglyceride hydrolysis (18).

Interestingly, the stimulatory effects of ASP are independent of and additive to those of insulin. ASP achieves its effect by two mechanisms: an increase in the activity of diacylglycerol acyltransferase (the last enzyme involved in the synthesis of TG molecules), which generates diglyceride from phosphatidate (94); and an increase in glucose transport (18). Also of interest, defects in ASP action have been implicated in the metabolic syndrome (85).

Fat Mobilization (Lipolysis)

During lipolysis, intracellular TG within fat cells is hydrolyzed by HSL-forming monoacylglycerol and FFA. These products leave the fat cells and are transported by the bloodstream to other tissues to be metabolized (primarily the liver for glycerol, and both liver and muscle for FFA).

Independent of hormones or other regulatory substances, a basal rate of lipolysis from adipocytes exists. When studied *in vitro,* the rate of basal lipolysis varies considerably among species, but is relatively high in human adipose tissue (0.5-1.5 μmol/g tissue/h). An interesting feature of basal lipolysis is its strong relationship to fat cell size, i.e., large fat cells have a much higher rate of basal lipolysis than small ones (12). Indeed, a reduction in fat cell size, e.g., after weight reduction, is followed by a decrease in the basal rate of lipolysis.

Hormone-Sensitive Lipase

HSL is rate limiting for intracellular lipolysis of TG to FFA and monoacylglycerol (MG), and is subject to intense regulation. At the end of the cascade, MG is further hydrolyzed by monoacylglycerol lipase. Monoacylglycerol lipase is abundant in tissues and is not regulated by hormones. The whole process forms 3 moles of FFA and 1 mole of glycerol per mole of completely hydrolyzed TG. If no reesterification occurs, the molar ratio of FFA to glycerol remains 3:1. However, some FFA formed during lipolysis do not leave the fat cell. Instead, they are reesterified and restored as intracellular TG. Because the glycerol formed during lipolysis is not reutilized by adipocytes, glycerol levels constitute a more valid index of lipolysis.

HSL is activated by phosphorylation, and the classical cyclic AMP (cAMP) pathway regulates this enzyme. Lipolysis is stimulated by hormones acting on cell-surface receptors coupled to adenylyl cyclase via Gs (stimulatory) proteins (23). This activates the cyclase, which stimulates the formation of cAMP from ATP. Cyclic AMP activates protein kinase A, which in turn stimulates the phosphorylation of HSL.

Lipolysis can be inhibited by hormones and parahormones via cell-surface receptors, which act by inhibiting adenylyl cyclase in two different ways. Inhibition of adenylyl cyclase via Gi (inhibitory) proteins decreases the production of cAMP. This leads to decreased activation of protein kinase A and decreased phosphorylation of HSL. Another pathway of inhibition is via receptors linked to phosphaditylinositol kinase 3 (PIK-3), e.g., the insulin receptor (52, 66, 71). As for Gi, this leads to the inactivation of protein kinase A, therefore resulting in less phosphorylated HSL.

A number of hormones and parahormones inhibit lipolysis in fat cells. These include insulin and insulin-like growth factors I (IGF-1) and II (IGF-2). Prostaglandins of the E-type, adenosine, and neuropeptide Y, also have inhibitory effects through Gi coupled receptors. In contrast, a number of hormones have lipolytic effects in adipocytes, acting through Gs receptors; these include vasopressin, secretin, glucagon, parathyroid hormone, TSH, and ACTH. In humans, however, glucagon, TSH, and parathyroid hormone have only weak effects. The most important lipolytic hormones in man are the catecholamines. Unlike all other hormones, they have a dual effect on lipolysis; stimulation through the Gs coupled β-adrenoceptors and inhibition via Gi coupled α-2 adrenoceptors (51). In humans, the lipolytic effect of catecholamines is dependent on the balance between α and β adrenergic stimulation.

In vivo administration of high doses of the antilipolytic hormone insulin fails to reduce the rate of lipolysis to zero (20). Moreover, total pharmacologic adrenergic blockade (β and α) does not abolish the basal lipolysis rate (3). Taken together, these data suggest that there is a non-hormonal regulation of adipose tissue lipolysis *in vivo* in humans. A potential regulator of basal lipolysis could be local adenosine production (5). Adenosine has potent antilipolytic effects in human fat cells, and interstitial adenosine concentrations sufficiently high to cause an antilipolytic effect have been demonstrated *in vivo* in human adipose tissue (61). However, the physiological role of adenosine for lipolysis in humans *in vivo* has yet to be completely examined.

Body Fat Distribution and Gender

Site differences in the regulation of lipolysis are frequently demonstrated *in vitro* and *in vivo* in normal-weight subjects (56). Greater resistance to the lipolytic effect of catecholamines and the antilipolytic effect of insulin is found in subjects with abdominal, as compared to peripheral, obesity (20). It is possible that this link between abnormal *in vivo* regulation of lipolysis and upper-body obesity is caused by regional variations in adipocyte regulation. Indeed, catecholamines are less lipolytic in the femoral/gluteal region than the abdominal subcutaneous region; this is due to increased α-2 adrenoceptor function and decreased β-adrenoceptor expression in the gluteal/femoral fat cells (56). The site differences are more marked in women than in men and might explain why women have more fat in peripheral sites than men.

In abdominal obesity, regional variations in lipolysis between visceral and subcutaneous fat cells are further increased. Catecholamine resistance is observed in subcutaneous abdominal fat cells. This is multifactorial in etiology including (1) decreased β-adrenoceptor expression; (2) increased function of α-2 adrenoceptors, and (3) decreased function of HSL (63, 75, 76). In visceral fat, on the other hand, catecholamine action is increased due to an increased β-3 receptor function and decreased α-2 adrenoceptor function (60). These findings taken together suggest that catecholamine-induced lipolysis is decreased in subcutaneous fat but increased in visceral fat. This would cause a relative increase in portal FFA versus peripheral venous FFA and, thus, could be a mechanism for the hepatic insulin resistance of visceral obesity. Furthermore, insulin provides a much greater inhibition of lipolysis in subcutaneous, as compared to visceral, adipocytes (64).

In normal-weight subjects, there are gender-based differences in adipose tissue LPL. In men, adipose tissue LPL activity is higher in the abdominal wall than in the gluteal/femoral region (8); in women, the opposite occurs (8, 10, 74). In subcutaneous sites of normal-weight women, however, the regulation of adipose tissue LPL by insulin is similar between depots (96). In obese women (53, 96) and men (62), fasting adipose tissue LPL activities were not different between the abdominal wall and gluteal regions, and the regulation of the enzyme by insulin (96) and exercise (53) were also similar.

Adipose Tissue Metabolism in Obesity

In obesity, the expanded adipose tissue mass is in a position to supply abundant lipid substrate to meet the needs of increased energy expenditure during exercise.

However, alteration in substrate trafficking to and from adipocytes modifies energy flux in response to such demands.

Abundant literature has documented that adipose tissue LPL is increased in obese subjects (31). In general, the quantity of heparin-releasable adipose tissue LPL relates to body mass index (BMI) and adipocyte size. Here, the manner in which the data are presented is important. If LPL activity is expressed per gram of adipose tissue, increases may not be seen. However, when expressed per adipocyte, increases are typical. The latter has the advantage of relating adipocytes to the source of the lipase.

Most studies on TG turnover in fat cells in obesity have focused on lipolysis. It is well established that circulating FFA levels are increased in obese subjects. On the other hand, when the lipolytic rate is related to the total fat mass, obese subjects have normal or even decreased rates of lipolysis after an overnight fast (20). Using the microdialysis technique, lipolysis per tissue weight was normal in the obese state (40). Thus, increased overall rates of "basal" lipolysis *in vivo,* which causes elevated circulating FFA in obese subjects, is a likely consequence of the enlargement of the fat mass. The basal rate of lipolysis *in vitro* is also increased in adipose tissue of obese subjects, which probably reflects the larger fat-cell size in obesity. When fat-cell size is accounted for, the difference in basal rates of lipolysis *in vitro* between lean and obese subjects disappears (4).

The hormonal regulation of lipolysis—in particular the action of catecholamines—is impaired in obesity. *In vivo* studies have shown blunted catecholamine-induced lipolysis in obese subjects (20). When subcutaneous adipose tissue was examined, the same was found to be true for *in vitro* studies. In part, this resistance appears to be caused by decreased expression and function of β-adrenoceptors and increased function of α-2 adrenoceptors (63, 76).

There are discrepancies in the antilipolytic actions of insulin in fat cells obtained from obese subjects (4). Moreover, *in vivo* data conflict and show normal or decreased antilipolytic action of insulin in the obese state (4, 20). This is in marked contrast to the impact of obesity on insulin-mediated glucose transport in fat cells or in adipose tissue, which invariably shows insulin resistance in obese subjects (17). Preservation or only partial reduction of the antilipolytic effect of insulin could be of importance for the maintenance or acceleration of obesity in overweight subjects who are resistant to other actions of insulin, such as glucose metabolism and LPL (80).

Reduced Obesity

Because of the high recidivism rate, a better understanding of adipose tissue metabolism in weight-reduced obese subjects is important. A reduction of body weight to a level still considered as overweight is accompanied by a decreased basal rate of lipolysis and improved catecholamine-induced lipolysis *in vitro* (4). These data suggest that the abnormalities in basal and catecholamine-induced lipolysis, at least in part, are secondary.

Insulin sensitivity also increases after weight reduction and isocaloric mainte-nance of the reduced-obese state (95, 98). Although most *in vivo* action of insulin is accounted for by insulin action in muscle (24), it is interesting to consider that the ability of increased insulin sensitivity to predict weight regain in reduced-obese subjects (98) relates in part to the many effects of insulin in adipose tissue.

Much more data are available to support the role of adipose tissue LPL overexpression in the recidivism of the obese state. Schwartz, Brunzell, and coworkers (see 32) were the first to report increases in adipose tissue LPL activity in fasted weight-maintaining reduced-obese subjects. Increases in LPL mass and mRNA (43) paralleled the increase in adipose tissue LPL activity. Furthermore, in several studies, adipose tissue LPL was not increased following isocaloric maintenance of the reduced obese state, but the responsiveness of the enzyme to insulin (33), or high fat meals (95), was increased. The concomitant decrease in skeletal muscle LPL would further favor adipose tissue instead of skeletal muscle deposition of lipids if caloric intake increases (34). A plethora of rodent data support this pathophysiology (11).

Adipose Tissue Metabolism and Exercise

Exercise requires energy. The two main sources of energy for muscle contraction are carbohydrate and lipid. The mobilization and oxidation of fat during exercise spares the use of the more-limited carbohydrate stores. The major source of lipid energy is TG stored in adipose tissue but transported to working muscle as FFA. In fact, exercise is one of the more potent physiological stimuli for lipolysis. Peak glycerol rates of appearance (Ra) during exercise in trained subjects are higher than those reported during critical illness (48) or even after 84 hours of starvation (50). Alternative sources of lipids include circulating lipoproteins and TG stores within muscle tissue.

Numerous interventions, such as L-carnitine, caffeine, medium-chain TG feed-ing, and high-fat diets, have been studied with the goal of enhancing fatty acid availability and oxidation during exercise to improve endurance capacity. It appears, however, that only aerobic endurance training has been successful in adapting adipose tissue metabolism to enhance exercise performance (15).

Humoral feedback mechanisms and autonomic nervous system reflexes (spinal reflexes) are not sufficient to establish a balance between substrate mobilization and substrate need during exercise (44, 45). Both neural input from motor centers and feedback from working muscles are important for energy supply in humans. Thus, the relative roles of neural mechanisms (central command and neural reflex) and blood-borne mechanisms for regulation of hormone release and energy substrate turnover during exercise are complementary (44, 45). These processes are balanced by inhibitory factors. As stated previously, insulin is the main antilipolytic hormone. Insulin concentration is lower during exercise because of the inhibitory effect of α-adrenergic tone on insulin secretion. This decrement in insulin levels facilitates FFA availability.

The methodology used to evaluate whole-body lipolytic sensitivity during exercise is an important consideration. It is possible to directly investigate subcutaneous adipose tissue metabolism in humans using the Fick or microdialysis technique. Conclusions regarding whole-body lipolysis based solely on the measurement of plasma FFA or glycerol concentrations can be misleading, because changes in plasma concentrations may not necessarily reflect changes in the whole-body lipolytic rate. The most reliable method for evaluating whole-body lipolysis involves the use of isotopic tracers to measure the glycerol rate of appearance in plasma. Glycerol rates of appearance provide a reasonable index of whole-body lipolytic rates, because glycerol released following hydrolysis of adipose tissue TG stores must enter the bloodstream (58).

It is important to note that data from lipolysis measured in isolated adipocytes did not always agree with data on whole-body lipolytic rates obtained from *in vivo* studies performed on the same subject (57). This could explain part of the discrepancy between studies. Alterations in the local environment of adipocytes, the presence of plasma factors, and the manipulation of tissue could all be explanations of discrepancies in data. In addition, lipolytic activity in specific adipose tissue depots may vary, whereas whole-body kinetics represent a summation of lipolysis from multiple regions. It is important to note that the tracer method cannot appropriately assess visceral adipose tissue lipolysis because the glycerol released enters the portal vein and is cleared by the liver.

The level of FFA in plasma, as an index of lipolysis, represents a balance between FFA delivery into plasma and FFA tissue uptake. However, this estimate may not always provide a reliable measure of whole-body lipolytic activity (47), because fatty acids released during lipolysis can be reesterified within adipocytes. Moreover, a reduction or a mismatch in adipose tissue blood flow during exercise could result in decreased FFA release from adipose tissue, whereas the appearance of glycerol in plasma does not appear to be blood-flow dependent.

Whereas α-adrenergic mechanisms regulate lipolysis at rest, β-adrenergic activity determines the lipolytic rate during exercise (7). An initial decrease of plasma FFA concentration may be due to an immediate increase in the uptake of FFA in muscle (due to capillary dilatation that occurs in the exercising muscle). Because glycerol is not utilized by working muscle, an initial sustained increase of plasma glycerol is seen. At the end of exercise, the increased rate of lipolysis is not immediately normalized, but remains increased for a time. In part, the dilatation of muscle capillaries may dissipate more rapidly then the deactivation of lipolysis, resulting in a pronounced overshoot of plasma FFA (46). This may also relate to exercise intensity or the extent of training of the individual. For example, postexercise lipolytic activity, i.e., lipolytic activity during the recovery period, persisted for at least 90 minutes in untrained subjects (47, 93), but returned to baseline more rapidly in trained individuals (46).

FFA rate of appearance is maximal at 25-40% of maximum oxygen uptake ($\dot{V}O_2$max), and shifts in energy substrate mobilization and utilization occur as exercise intensity increases, particularly at intensities above 70-80% of $\dot{V}O_2$max (77). Above a certain degree of intensity, the muscle preferentially utilizes glycogen stored *in situ*. This progressive shift from lipid to carbohydrate utilization with increasing exercise intensity could be partially regulated by adenosine, which may play a role in stimulating glucose uptake and decreasing fat oxidation without limiting lipolysis during exercise (70). However, these shifts are also associated with the lactate threshold. In spite of conflicting data, it has been suggested that lactate might be a regulatory factor for lipolysis during exercise (14, 38). During strenuous exercise it appears that inhibitory concentrations of circulating lactate are attainable locally because lactate production from anaerobic glycolysis in the muscle is increasing. However, it is not clear that a significant limiting effect on fatty acid kinetics occurred when highly trained individuals exercised above or below the lactate threshold (41).

Adipose Tissue Metabolism in Trained Individuals

Adaptation to physical exercise involves the respiratory and circulatory systems as well as the enzymatic machinery that enables muscles to work more effectively. When an adaptation to endurance exercise has taken place, the muscle is characterized by more oxidative enzyme activity. This particular muscle is now better equipped to work at low intensity for long duration and to use FFAs as the main substrate. There are also changes occurring at the site of lipid mobilization in adipose tissue. These include a more sensitive mechanism for activation of HSL (6), and in untrained individuals an increase in the lipolytic response of adipose tissue to catecholamines after exercise both *in vitro* (81, 86, 92) and *in vivo* (68). However, in endurance-trained individuals, lipolysis has probably already been maximized (47, 77).

Of interest, dexamethasone appears to potentiate the *in vitro* epinephrine-induced lipolytic effect in trained but not untrained individuals (86). However, metyrapone-induced hypocortisolemia (70% suppression) did not alter whole-body substrate utilization despite an attenuated (20-35%) increase in plasma glycerol during exercise (25). This is likely a compensatory response from the other glucoregulatory hormones (insulin, GH, epinephrine, norepinephrine). Overall, these mechanisms, along with adaptations at post-receptor sites, lead to an increased sensitivity of lipid mobilization by the central nervous system (CNS) (6). Adequate lipid mobilization seems to occur at an activity level of the CNS that is lower than before physical training. These adaptations could occur as early as the fifth day into training (68); that is well before maximal increases in muscle mitochondrial enzyme activity have occurred.

Another important adaptation of physical training occurs in adipose tissue. Because fat has advantages over carbohydrate as an energy source, i.e., 37.5 kJ/g versus 16.9 kJ/g, the mass of adipose tissue tends to be lower as training is enhanced. In the physically well-trained individual, adipose tissue mass is reduced with smaller adipocytes. Yet, despite the more limited availability of TG, FFA are still efficiently mobilized by a more sensitive lipolytic response. Indeed, it appears that lipolysis is anything but rate limiting during exercise, and is probably in excess of that required (77). After an overnight fast, lipolysis exceeds fat oxidation by as much as 25-50%, both at rest and during exercise. Therefore, lipolysis clearly does not limit fat oxidation in the fasting state (37, 77, 93).

However, preexercise carbohydrate ingestion (60 g of glucose), with concomitant elevation in insulin levels to 40 μU/ml, partially suppressed lipolysis during one hour of exercise at 45% of $\dot{V}O_2$max and limited fat oxidation (37). A similar effect has been seen at lower insulin concentrations (20 μU/ml) after fructose ingestion (37). In contrast to the fasting state, this decreased lipolytic rate results in a close match between lipolysis and fat oxidation. Glycerol rates of appearance during exercise performed at the same absolute intensity are similar in trained and untrained subjects (46), but higher in trained than in untrained subjects during exercise performed at the same relative intensity (49). Moreover, even if β-blockade reduces the release of FFA from adipocytes, this reduction in energy supply of exercising muscle does not seem to be the cause of impaired exercise performance, at least in untrained individuals (88). Furthermore, there is evidence that suggests that training increases resting adipose TG recycling (68).

As discussed earlier, insulin and perhaps lactate, both of which inhibit lipolysis, are additional factors that relate to lipolysis regulation. In physically trained subjects, plasma insulin is lower (54) and plasma lactate levels increase less during a submaximal workload (68) because the working muscle is now capable of more complete aerobic metabolism with less lactate production. In both cases, the inhibitory action would be diminished in the physically trained state, facilitating the mobilization of FFA from adipose tissue. However, it seems that adipose tissue, like other tissues, becomes more sensitive to insulin with physical training, presumably at the postreceptor level (6).

Adipose Tissue Metabolism and Effect of Exercise in Different Settings

Obesity

There is an inverse relationship between physical activity and body weight. Numerous factors associated with obesity, such as gender, level of obesity, fat distribution, and number and type of fat cells, contribute to the eventual response to exercise. It was shown by Tremblay et al. (87) that following 20 weeks of exercise, men (matched for fat mass) with a high fat-cell size lost six times more fat mass (4.4 kg) than men with low fat-cell size (0.7 kg). Women with either high or low fat-cell size did not lose fat mass. However, there seems to be a weight/fat level threshold above which subjects will lose fat mass following exercise. Results showed that as the level of obesity increases, so does the magnitude of weight loss in men and women (2, 55). Of interest, women seemed to present a higher weight/fat level threshold than men in order to lose fat mass following exercise training (2).

Elderly

Obesity is age dependent, with most individuals increasing their fat stores as they become older. Indeed, aging is associated with a decline in physical activity that contributes to decreased exercise tolerance, decreased lean body mass, increased fat mass, and alterations in glucose and lipoprotein metabolism. Basal and catecholamine-induced lipolysis *in vivo* and *in vitro* decrease in normal-weight older individuals (59). An age-dependent, blunted, lipolytic response to catecholamines, with an associated decreased activation of HSL, contributes to the increase in fat stores in the elderly (59). Elderly persons, compared to young adults, demonstrate a blunted increase in plasma catecholamine concentrations during exercise (82). In addition, insulin-induced antilipolysis and glucose metabolism (78) and LPL (91) are lower in adipose tissue of the elderly when compared to young nonobese subjects. Poehlman et al. (69) suggested that endurance training might increase lipolysis and TG cycling in the elderly. Whole-body lipolysis and plasma FFA rates of appearance are not rate limiting in the elderly (82), and lipolytic rate (glycerol rates of appearance or FFA availability) are not modified by training, at least during the first 60 minutes of exercise (83). However, the reduced rate of fat oxidation (25-35% lower) during exercise in untrained elderly individuals (82) is corrected following endurance training (83).

Gender

Women have less lipolysis during exercise than men (28, 29), perhaps providing an explanation of why men decrease body fat more efficiently with physical training than do women (27, 87). During exercise, the lipolytic rate is higher in abdominal subcutaneous adipose tissue than in gluteal/femoral subcutaneous adipose tissue, especially in women (7). However, there is no difference between the genders at rest. Moreover, in trained men, β-adrenergic-mediated lipolysis is markedly increased without changes in the α_2-adrenergic pathway, whereas in trained women, both an enhancement of the β-adrenergic pathway and a decrease in the α_2-adrenergic pathway efficiency occur (22).

Reduced-Obese Individuals

Weight loss by hypocaloric diet decreases lipolysis and fat oxidation, adaptations that might predispose individuals to weight regain. The blunted utilization of fat as fuel during a 60-minute bout of exercise at 50% $\dot{V}O_2$max, contributes to a positive fat balance and possibly weight gain in formerly obese individuals (72). However, studies of adipose tissue function *in vitro* showed that the addition of exercise training to a hypocaloric diet counteracts the decline in lipolytic responsiveness, fat oxidation, and resting metabolic rate in weight-reduced postmenopausal women (65, 72). Moreover, the lipolytic adaptations were of the same magnitude between subcutaneous abdominal and gluteal/femoral adipose tissue regions (65).

Summary

Triacylglycerols (TG), which are stored in adipocytes in adipose tissue, are the body's major source of energy. Two lipases are involved: (1) lipoprotein lipase (LPL), which is located on the capillary walls of adipose tissue, controls the uptake of circulating TG by adipocytes (storage), and (2) hormone-sensitive lipase (HSL) controls the hydrolysis of adipocyte TG. Insulin and catecholamines have pronounced metabolic effects on human adipose tissue metabolism. Insulin stimulates LPL and inhibits HSL; the opposite is true for catecholamines. There are regional variations in adipocyte TG turnover favoring lipid mobilization in the visceral fat depots and lipid storage in the peripheral subcutaneous sites. The hormonal regulation of adipocyte turnover of TG is altered in obesity, and is most marked in central obesity. There is resistance to insulin stimulation of LPL; however, LPL activity in fasted obese subjects is increased and remains so following weight reduction. Catecholamine-induced lipolysis is enhanced in visceral fat, but decreased in subcutaneous fat.

Numerous adaptations take place with physical training. This results in a more efficient system for oxygen transfer to muscle, which is now able to utilize the unlimited lipid stores instead of the limited carbohydrate that is available. The delivery of lipid substrate to the working muscle also adapts with a higher sensitivity of TG hydrolysis. The system can now operate with less stimulation by the CNS. In addition, less total adipose tissue mass allows the individual an important mechanical advantage for optimal long-term work. Gender differences do exist with respect to how adipose tissue adapts to aerobic exercise training. Physical training also corrects the reduced rate of fat oxidation in the elderly and helps counteract the environment that predisposes reduced-obese subjects to regain weight.

Acknowledgments

Dr. Eckel is supported by a grant from NIDDK (RO1 26356) and from The National Center for Research Resources (RR-00051). Dr. Poirier is supported by a fellowship grant from The R. Samuel McLaughlin Foundation and from l'Association des Cardiologues du Québec.

References

1. Abumrad, N.A., S.A. Melki, and C.M. Harmon. 1990. Transport of fatty acid in the isolated rat adipocyte and in differentiating preadipose cells. *Biochemical Society Transactions* 18: 1130-1132.
2. Andersson, B., X.F. Xu, M. Rebuffe-Scrive, K. Terning, M. Krotkiewski, and P. Bjorntorp. 1991. The effects of exercise, training on body composition

and metabolism in men and women. *International Journal of Obesity* 15: 75-81.

3. Andersson, K., and P. Arner. 1995. Cholinoceptor-mediated effects on glycerol output from human adipose tissue using in situ microdialysis. *British Journal of Pharmacology* 115: 1155-1162.

4. Arner, P. 1988. Control of lipolysis and its relevance to development of obesity in man. *Diabetes Metabolism Reviews* 4: 507-515.

5. Arner, P. 1993. Adenosine, prostaglandins and phosphodiesterase as targets for obesity pharmacotherapy. *International Journal of Obesity and Related Metabolic Disorders* 17: S57-S59.

6. Arner, P. 1995. Impact of exercise on adipose tissue metabolism in humans. *International Journal of Obesity and Related Metabolic Disorders* 19 (Suppl 4): S18-S21.

7. Arner, P., E. Kriegholm, P. Engfeldt, and J. Bolinder. 1990. Adrenergic regulation of lipolysis in situ at rest and during exercise. *Journal of Clinical Investigation* 85: 893-898.

8. Arner, P., H. Lithell, H. Wahrenberg, and M. Bronnegard. 1991. Expression of lipoprotein lipase in different human subcutaneous adipose tissue regions. *Journal of Lipid Research* 32: 423-429.

9. Baillie, A.S., C.T. Coburn, and N.A. Abumrad. 1996. Reversible binding of long-chain fatty acids to purified FAT, the adipose CD36 homolog. *Journal of Membrane Biology* 153: 75-81.

10. Belfiore, F., V. Borzi, E. Napoli, and A.M. Rabuazzo. 1976. Enzymes related to lipogenesis in the adipose tissue of obese subjects. *Metabolism* 25: 483-493.

11. Bessesen, D.H., A.D. Robertson, and R.H. Eckel. 1991. Weight reduction increases adipose but decreases cardiac LPL in reduced-obese Zucker rats. *American Journal of Physiology* 261: E246-E251.

12. Bjorntorp, P., and J. Ostman. 1971. Human adipose tissue dynamics and regulation. *Advances in Metabolic Disorder* 5: 277-327.

13. Blaner, W.S., J.C. Obunike, S.B.Kurlandsky, M. al-Haideri, R. Piantedosi, R.J. Deckelbaum, and I.J. Goldberg. 1994. Lipoprotein lipase hydrolysis of retinyl ester. Possible implications for retinoid uptake by cells. *Journal of Biological Chemistry* 269: 16559-16565.

14. Boyd, A.E., S.R. Giamber, M. Mager, and H.E. Lebovitz. 1974. Lactate inhibition of lipolysis in exercising man. *Metabolism* 23: 531-542.

15. Brouns, F., and G.J. Van der Vusse. 1998. Utilization of lipids during exercise in human subjects: metabolic and dietary constraints. *British Journal of Nutrition* 79: 117-128.

16. Brun, L.D., C. Gagné, P. Julien, A. Tremblay, S. Moorjani, C. Bouchard, and P.J. Lupien. 1989. Familial lipoprotein lipase activity deficiency: study of total body fatness and subcutaneous fat tissue distribution. *Metabolism* 38: 1005-1009.

17. Caro, J.F., L.G. Dohm, W.J. Pories, and M.K. Sinha. 1989. Cellular alterations in liver, skeletal muscle, and adipose tissue responsible for insulin resistance in obesity and type II diabetes. *Diabetes Metabolism Reviews* 5: 665-689.

18. Cianflone, K. 1997. Acylation stimulating protein and the adipocyte. *Journal of Endocrinology* 155: 203-206.
19. Cianflone, K. 1997. The acylation stimulating protein pathway: clinical implications. *Clinical Biochemistry* 30: 301-312.
20. Coppack, S.W., M.D. Jensen, and J.M. Miles. 1994. In vivo regulation of lipolysis in humans. *Journal of Lipid Research* 35: 177-193.
21. Craig, B.W., S.M. Garthwaite, and J.O. Holloszy. 1987. Adipocyte insulin resistance: effects of aging, obesity, exercise, and food restriction. *Journal of Applied Physiology* 62: 95-100.
22. Crampes, F., D. Rivière, M. Beauville, M. Marceron, and M. Garrigues. 1989. Lipolytic response of adipocytes to epinephrine in sedentary and exercise-trained subjects: sex-related differences. *European Journal of Applied Physiology* 59: 249-255.
23. Davies, J.I., and J.E. Souness. 1981. The mechanisms of hormone and drug actions on fatty acid release from adipose tissue. *Review of Pure Applied Pharmacology Science* 2: 1-112.
24. DeFronzo, R.A., E. Jacot, E. Jequier, E. Maeder, J. Wahren, and J.P. Felber. 1981. The effect of insulin on the disposal of intravenous glucose. Results from indirect calorimetry and hepatic and femoral venous catheterization. *Diabetes* 30: 1000-1007.
25. Del Corral, P., E.T. Howley, M. Hartsell, M. Ashraf, and M.S. Younger. 1998. Metabolic effects of low cortisol during exercise in humans. *Journal of Applied Physiology* 84: 939-947.
26. Deshaies, Y., A. Geloen, A. Paulin, A.Marette, and L.J. Bukowiecki. 1993. Tissue-specific alterations in lipoprotein lipase activity in the rat after chronic infusion of isoproterenol. *Hormone and Metabolic Research* 25: 13-16.
27. Després, J.P., and C. Bouchard. 1984. Effects of aerobic training and heredity on body fatness and adipocyte lipolysis in humans. *Journal of Obesity and Weight Regulation* 3: 219-235.
28. Després, J.P., C. Bouchard, R. Savard, A. Tremblay, M. Marcotte, and G. Thériault. 1984. Effects of exercise-training and detraining on fat cell lipolysis in men and women. *European Journal of Applied Physiology* 53: 25-30.
29. Després, J.P., C. Bouchard, R. Savard, A. Tremblay, M. Marcotte, and G. Thériault. 1984. The effect of a 20-week endurance training program on adipose-tissue morphology and lipolysis in men and women. *Metabolism* 33: 235-239.
30. Donahoo, W.T., and R.H. Eckel. 1996. Adipocyte metabolism in obesity. *Current Opinion in Endocrinology and Metabolism* 3: 501-507.
31. Eckel, R.H., 1987. Adipose tissue lipoprotein lipase. In *Lipoprotein lipase*, ed. J. Borensztajn, 79-132. Chicago: Evener.
32. Eckel, R.H., 1989. Lipoprotein lipase. A multifunctional enzyme relevant to common metabolic diseases. *New England Journal of Medicine* 320: 1060-1068.

33. Eckel, R.H., and T.J. Yost. 1987. Weight reduction increases adipose tissue lipoprotein lipase responsiveness in obese women. *Journal of Clinical Investigation* 80: 992-997.

34. Eckel, R.H., T.J. Yost, and D.R. Jensen. 1995. Sustained weight reduction in moderately obese women results in decreased activity of skeletal muscle lipoprotein lipase. *European Journal of Clinical Investigation* 25: 396-402.

35. Fried, S.K., C.D. Russell, N.L. Grauso, and R.E. Brolin. 1993. Lipoprotein lipase regulation by insulin and glucocorticoid in subcutaneous and omental adipose tissues of obese women and men. *Journal of Clinical Investigation* 92: 2191-2198.

36. Galton, D.J., 1968. Lipogenesis in human adipose tissue. *Journal of Lipid Research* 9: 19-26.

37. Horowitz, J.F., R. Mora-Rodriguez, L.O. Byerley, and E.F. Coyle. 1997. Lipolytic suppression following carbohydrate ingestion limits fat oxidation during exercise. *American Journal of Physiology* 273: E768-E775.

38. Issekutz, B.J., W.A. Shaw, and T.B. Issekutz. 1975. Effect of lactate on FFA and glycerol turnover in resting and exercising dogs. *Journal of Applied Physiology* 39: 349-353.

39. Jacobsen, B.K., K. Trygg, I. Hjermann, M.S. Thomassen, C. Real, and K.R. Norum. 1983. Acyl pattern of adipose tissue triglycerides, plasma free fatty acids, and diet of a group of men participating in a primary coronary prevention program (the Oslo Study). *American Journal of Clinical Nutrition* 38: 906-913.

40. Jansson, P.A., A. Larsson, U. Smith, and P. Lonnroth. 1992. Glycerol production in subcutaneous adipose tissue in lean and obese humans. *Journal of Clinical Investigation* 89: 1610-1617.

41. Kanaley, J.A., C.D. Mottram, P.D. Scanlon, and M.D. Jensen. 1995. Fatty acid kinetic responses to running above or below lactate threshold. *Journal of Applied Physiology* 79: 439-447.

42. Kern, P.A., S. Marshall, and R.H. Eckel. 1985. Regulation of lipoprotein lipase in primary cultures of isolated human adipocytes. *Journal of Clinical Investigation* 75: 199-208.

43. Kern, P.A., J.M. Ong, B. Saffari, and J. Carty. 1990. The effects of weight loss on the activity and expression of adipose-tissue lipoprotein lipase in very obese humans. *New England Journal of Medicine* 322: 1053-1059.

44. Kjaer, M., S.F. Pollack, T. Mohr, H. Weiss, G.W. Gleim, F.W. Bach, T. Nicolaisen, H. Galbo, and K.T. Ragnarsson. 1996. Regulation of glucose turnover and hormonal responses during electrical cycling in tetraplegic humans. *American Journal of Physiology* 271: R191-R199.

45. Kjaer, M., N.H. Secher, J. Bangsbo, G. Perko, A. Horn, T. Mohr, and H. Galbo. 1996. Hormonal and metabolic responses to electrically induced cycling during epidural anesthesia in humans. *Journal of Applied Physiology* 80: 2156-2162.

46. Klein, S., E.F. Coyle, and R.R. Wolfe. 1994. Fat metabolism during low-intensity exercise in endurance-trained and untrained men. *American Journal of Physiology* 267: E934-E940.

47. Klein, S., E.F. Coyle, and R.R. Wolfe. 1995. Effect of exercise on lipolytic sensitivity in endurance-trained athletes. *Journal of Applied Physiology* 78: 2201-2206.
48. Klein, S., E.J. Peters, R.E. Shangraw, and R.R. Wolfe. 1991. Lipolytic response to metabolic stress in critically ill patients. *Critical Care Medicine* 19: 776-779.
49. Klein, S., J.M. Weber, E.F. Coyle, and R.R. Wolfe. 1996. Effect of endurance training on glycerol kinetics during strenuous exercise in humans. *Metabolism* 45: 357-361.
50. Klein, S., and R.R. Wolfe. 1992. Carbohydrate restriction regulates the adaptive response to fasting. *American Journal of Physiology* 262: E631-E636.
51. Lafontan, M., and M. Berlan. 1993. Fat cell adrenergic receptors and the control of white and brown fat cell function. *Journal of Lipid Research* 34: 1057-1091.
52. Lam, K., C.L. Carpenter, N.B. Ruderman, J.C. Friel, and K.L. Kelly. 1994. The phosphatidylinositol 3-kinase serine kinase phosphorylates IRS-1. Stimulation by insulin and inhibition by Wortmannin. *Journal of Biological Chemistry* 269: 20648-20652.
53. Lamarche, B., J.P. Després, S. Moorjani, A. Nadeau, P.J. Lupien, A. Tremblay, G. Thériault, and C. Bouchard. 1993. Evidence for a role of insulin in the regulation of abdominal adipose tissue lipoprotein lipase response to exercise training in obese women. *International Journal of Obesity and Related Metabolic Disorders* 17: 255-261.
54. LeBlanc, J., A. Nadeau, M. Boulay, and S. Rousseau-Migneron. 1979. Effects of physical training and adiposity on glucose metabolism and ^{125}I-insulin binding. *Journal of Applied Physiology* 46: 235-239.
55. Lee, L., S. Kumar, and L.C. Leong. 1994. The impact of five-month basic military training on the body weight and body fat of 197 moderately to severely obese Singaporean males aged 17 to 19 years. *International Journal of Obesity and Related Metabolic Disorders* 18: 105-109.
56. Leibel, R.L., N.K. Edens, and S.K. Fried. 1989. Physiologic basis for the control of body fat distribution in humans. *Annual Review of Nutrition* 9: 417-443.
57. Lillioja, S., J. Foley, C. Bogardus, D. Mott, and B.V. Howard. 1986. Free fatty acid metabolism and obesity in man: in vivo in vitro comparisons. *Metabolism* 35: 505-514.
58. Lin, E.C. 1977. Glycerol utilization and its regulation in mammals. *Annual Review of Biochemistry* 46: 765-795.
59. Lonnqvist, F., B. Nyberg, H. Wahrenberg, and P. Arner. 1990. Catecholamine-induced lipolysis in adipose tissue of the elderly. *Journal of Clinical Investigation* 85: 1614-1621.
60. Lonnqvist, F., A. Thome, K. Nilsell, J. Hoffstedt, and P. Arner. 1995. A pathogenic role of visceral fat beta 3-adrenoceptors in obesity. *Journal of Clinical Investigation* 95: 1109-1116.

61. Lonnroth, P., P.A. Jansson, B.B. Fredholm, and U. Smith. 1989. Microdialysis of intercellular adenosine concentration in subcutaneous tissue in humans. *American Journal of Physiology* 256: E250-E255.

62. Marin, P., B. Oden, and P. Bjorntorp. 1995. Assimilation and mobilization of triglycerides in subcutaneous abdominal and femoral adipose tissue in vivo in men: effects of androgens. *Journal of Clinical Endocrinology and Metabolism* 80: 239-243.

63. Mauriège, P., J.P. Després, D. Prud'homme, M.C. Pouliot, M. Marcotte, A. Tremblay, and C. Bouchard. 1991. Regional variation in adipose tissue lipolysis in lean and obese men. *Journal of Lipid Research* 32: 1625-1633.

64. Mauriège, P., A. Marette, C. Atgie, C. Bouchard, G. Thériault, L.K. Bukowiecki, P. Marceau, S. Biron, A. Nadeau, and J.P. Després. 1995. Regional variation in adipose tissue metabolism of severely obese premenopausal women. *Journal of Lipid Research* 36: 672-684.

65. Nicklas, B.J., E.M. Rogus, and A.P. Goldberg. 1997. Exercise blunts declines in lipolysis and fat oxidation after dietary-induced weight loss in obese older women. *American Journal of Physiology* 273: E149-E155.

66. Okada, T., Y. Kawano, T. Sakakibara, O. Hazeki, and M. Ui. 1994. Essential role of phosphatidylinositol 3-kinase in insulin-induced glucose transport and antilipolysis in rat adipocytes. Studies with a selective inhibitor wortmannin. *Journal of Biological Chemistry* 269: 3568-3573.

67. Ong, J.M., B. Saffari, R.B. Simsolo, and P.A. Kern. 1992. Epinephrine inhibits lipoprotein lipase gene expression in rat adipocytes through multiple steps in posttranscriptional processing. *Molecular Endocrinology* 6: 61-69.

68. Phillips, S.M., H.J. Green, M.A. Tarnopolsky, G.F. Heigenhauser, R.E. Hill, and S.M. Grant. 1996. Effects of training duration on substrate turnover and oxidation during exercise. *Journal of Applied Physiology* 81: 2182-2191.

69. Poehlman, E.T., A.W. Gardner, P.J. Arciero, M.I. Goran, and J. Calles-Escandon. 1994. Effects of endurance training on total fat oxidation in elderly persons. *Journal of Applied Physiology* 76: 2281-2287.

70. Raguso, C.A., A.R. Coggan, L.S. Sidossis, A. Gastaldelli, and R.R. Wolfe. 1996. Effect of theophylline on substrate metabolism during exercise. *Metabolism* 45: 1153-1160.

71. Rahn, T., M. Ridderstrale, H. Tornqvist, V. Manganiello, G. Fredrikson, P. Belfrage, and E. Degerman. 1994. Essential role of phosphatidylinositol 3-kinase in insulin-induced activation and phosphorylation of the cGMP-inhibited cAMP phosphodiesterase in rat adipocytes. Studies using the selective inhibitor wortmannin. *FEBS Letters* 350: 314-318.

72. Ranneries, C., J. Bulow, B. Buemann, N.J. Christensen, J. Madsen, and A. Astrup. 1998. Fat metabolism in formerly obese women. *American Journal of Physiology* 274: E155-E161.

73. Raynolds, M.V., P.B. Awald, D.F. Gordon, A. Gutierrez-Hartmann, D.C. Rule, W.M. Wood, and R.H. Eckel. 1990. Lipoprotein lipase gene expression in rat adipocytes is regulated by isoproterenol and

insulin through different mechanisms. *Molecular Endocrinology* 4: 1416-1422.

74. Rebuffe-Scrive, M., L. Enk, N. Crona, P. Lonnroth, L. Abrahamsson, U. Smith, and P. Bjorntorp. 1985. Fat cell metabolism in different regions in women. Effect of menstrual cycle, pregnancy, and lactation. *Journal of Clinical Investigation* 75: 1973-1976.

75. Reynisdottir, S., K. Ellerfeldt, H. Wahrenberg, H. Lithell, and P. Arner. 1994. Multiple lipolysis defects in the insulin resistance (metabolic) syndrome. *Journal of Clinical Investigation* 93: 2590-2599.

76. Reynisdottir, S., H. Wahrenberg, K. Carlstrom, S. Rossner, and P. Arner. 1994. Catecholamine resistance in fat cells of women with upper-body obesity due to decreased expression of beta 2-adrenoceptors. *Diabetologia* 37: 428-435.

77. Romijn, J.A., E.F. Coyle, L.S. Sidossis, A. Gastaldelli, J.F. Horowitz, E. Endert, and R.R. Wolfe. 1993. Regulation of endogenous fat and carbohydrate metabolism in relation to exercise intensity and duration. *American Journal of Physiology* 265: E380-E391.

78. Rowe, J.W., K.L. Minaker, J.A. Pallotta, and J.S. Flier. 1983. Characterization of the insulin resistance of aging. *Journal of Clinical Investigation* 71: 1581-1587.

79. Sadur, C.N., and R.H. Eckel. 1982. Insulin stimulation of adipose tissue lipoprotein lipase. Use of the euglycemic clamp technique. *Journal of Clinical Investigation* 69: 1119-1125.

80. Sadur, C.N., T.J. Yost, and R.H. Eckel. 1984. Insulin responsiveness of adipose tissue lipoprotein lipase is delayed but preserved in obesity. *Journal of Clinical Endocrinology and Metabolism* 59: 1176-1182.

81. Savard, R., J.P. Després, M. Marcotte, G. Thériault, A. Tremblay, and C. Bouchard. 1987. Acute effects of endurance exercise on human adipose tissue metabolism. *Metabolism* 36: 480-485.

82. Sial, S., A.R. Coggan, R. Carroll, J. Goodwin, and S. Klein. 1996. Fat and carbohydrate metabolism during exercise in elderly and young subjects. *American Journal of Physiology* 271: E983-E989.

83. Sial, S., A.R. Coggan, R.C. Hickner, and S. Klein. 1998. Training-induced alterations in fat and carbohydrate metabolism during exercise in elderly subjects. *American Journal of Physiology* 274: E785-E790.

84. Simsolo, R.B., J.M. Ong, and P.A. Kern. 1992. Characterization of lipoprotein lipase activity, secretion, and degradation at different sites of post-translational processing in primary cultures of rat adipocytes. *Journal of Lipid Research* 33: 1777-1784.

85. Sniderman, A.D., K. Cianflone, P. Arner, L.K. Summers, and K.N. Frayn. 1998. The adipocyte, fatty acid trapping, and atherogenesis. *Arteriosclerosis, Thrombosis, and Vascular Biology* 18: 147-151.

86. Toode, K., A. Viru, and A. Eller. 1993. Lipolytic actions of hormones on adipocytes in exercise-trained organisms. *Japanese Journal of Physiology* 43: 253-258.

87. Tremblay, A., J.P. Després, C. Leblanc, and C. Bouchard. 1984. Sex dimorphism in fat loss in response to exercise training. *Journal of Obesity and Weight Regulation* 3: 193-203.
88. van Baak, M.A., J.M. Mooij, and J.A. Wijnen. 1993. Effect of increased plasma non-esterified fatty acid concentrations on endurance performance during beta-adrenoceptor blockade. *International Journal of Sports Medicine* 14: 2-8.
89. van Staveren, W.A., P. Deurenberg, M.B. Katan, J. Burema, L.C. de Groot, and M.D. Hoffmans. 1986. Validity of the fatty acid composition of subcutaneous fat tissue microbiopsies as an estimate of the long-term average fatty acid composition of the diets of separate individuals. *American Journal of Epidemiology* 123: 455-463.
90. Vasseur-Cognet, M., and M.D. Lane. 1993. Trans-acting factors involved in adipogenic differentiation. *Current Opinion in Genetic Development* 3: 238-245.
91. Vessby, B., H. Lithell, J. Boberg, K. Hellsing, and I. Werner. 1976. Gemfibrozil as a lipid lowering compound in hyperlipoproteinaemia. A placebo-controlled cross-over trial. *Proceedings of the Royal Society of Medicine* 69: 32-37.
92. Wahrenberg, H., P. Engfeldt, J. Bolinder, and P. Arner. 1987. Acute adaptation in adrenergic control of lipolysis during physical exercise in humans. *American Journal of Physiology* 253: E383-E390.
93. Wolfe, R.R., S. Klein, F. Carraro, and J.M. Weber. 1990. Role of triglyceride-fatty acid cycle in controlling fat metabolism in humans during and after exercise. *American Journal of Physiology* 258: E382-E389.
94. Yasruel, Z., K. Cianflone, A.D. Sniderman, M. Rosenbloom, M. Walsh, and M.A. Rodriguez. 1991. Effect of acylation stimulating protein on the triacylglycerol synthetic pathway of human adipose tissue. *Lipids* 26: 495-499.
95. Yost, T.J., and R.H. Eckel. 1988. Fat calories may be preferentially stored in reduced-obese women: a permissive pathway for resumption of the obese state. *Journal of Clinical Endocrinology and Metabolism* 67: 259-264.
96. Yost, T.J., and R.H. Eckel. 1992. Regional similarities in the metabolic regulation of adipose tissue lipoprotein lipase. *Metabolism* 41: 33-36.
97. Yost, T.J., D.R. Jensen, and R.H. Eckel. 1993. Tissue-specific lipoprotein lipase: Relationship to body composition and body fat distribution in normal-weight humans. *Obesity Research* 1: 1-4.
98. Yost, T.J., D.R. Jensen, and R.H. Eckel. 1995. Weight regain following sustained weight reduction is predicted by relative insulin sensitivity. *Obesity Research* 3: 583-587.
99. Yukht, A., R.C. Davis, J.M. Ong, G. Ranganathan, and P.A. Kern. 1995. Regulation of lipoprotein lipase translation by epinephrine in 3T3-L1 cells. Importance of the 3' untranslated region. *Journal of Clinical Investigation* 96: 2438-2444.

CHAPTER 10

Skeletal Muscle Metabolism and Obesity

Jean-Aimé Simoneau, PhD[1], and David E. Kelley, MD[2]
[1] Physical Activity Sciences Laboratory, Department of Medicine, Division of
Kinesiology, Department of Social and Preventive Medicine, Faculty of Medicine, Laval University, Ste-Foy, Québec, Canada and [2] Division of Endocrinology and Metabolism, University of Pittsburgh School of Medicine and Department of Veterans Affairs Medical Center, Pittsburgh, Pennsylvania, U.S.A.

Numerous studies have attempted to determine whether tissues of obese individuals
possess metabolic characteristics that could contribute to the excessive storage of
fat that occurs in obesity. An area of potential importance in understanding the
pathogenesis of obesity is to better understand the capacity of muscle to oxidize or
store carbohydrates and lipids during basal (fasting) and insulin-stimulated conditions. Normally, during fasting conditions, skeletal muscle relies predominately
upon lipid oxidation whereas, during insulin-stimulated conditions, muscle switches
predominately to utilization of glucose. If the uptake and oxidation of plasma fatty
acids is reduced within skeletal muscle in obesity, this might have substantial
repercussions for the long-term regulation of fat balance, especially considering the
large contribution of muscle mass (37).

Role of Insulin

Insulin, along with other important hormones, plays a pivotal role in the regulation
of substrate utilization. Blunting of the response to insulin, commonly described as
insulin resistance (38), is caused or exacerbated by obesity. Skeletal muscle is
regarded as a principal site of insulin resistance in obesity (12). Among the potential
causes of insulin resistance in skeletal muscle are a disturbed regulation of insulin
receptor signaling (3), deteriorated insulin-stimulated glucose storage (21, 28, 42,
44, 51), and impaired glucose transport/phosphorylation (27, 40, 41). In addition to
these impairments in the stimulation of glucose metabolism by skeletal muscle, a
more novel concept of abnormal patterns of substrate utilization in obesity concerns
a reduced capacity of skeletal muscle to oxidize fat (10, 28, 29). Indeed, if insulin
resistance can be regarded as a "phenotype," then with respect to skeletal muscle it

has been postulated that this phenotype entails not only defects of glucose metabolism but also abnormal patterns of fatty acid metabolism.

To test the hypothesis that inefficient utilization of fatty acids by skeletal muscle during fasting conditions is a component of obesity-related insulin resistance, we examined fasting patterns of fatty acid metabolism in skeletal muscle of obese subjects and compared them to lean subjects (29). A group of 50 men and women, representing a relatively broad range of values for body mass index, underwent arteriovenous leg balance studies during fasting, as well as insulin-stimulated conditions for measurement of glucose and FFA uptake (based on fractional extraction of oleate) and determination of regional indirect calorimetry [leg respiratory quotient (RQ)]. During fasting conditions, obese subjects had an elevated value for leg RQ relative to lean RQ (0.90 ± 0.01 vs. 0.83 ± 0.02 ; $p < 0.01$), indicative of lower rates of lipid oxidation. However, obese and lean volunteers had similar rates of uptake of plasma FFA across the leg. Thus, a reduced rate of lipid oxidation in skeletal muscle in obesity was not due to lower rates of fatty acid uptake. Moreover, the net value for fat storage (uptake minus rates of oxidation) was greater in obesity, suggesting the mechanism by which lipid content within muscle becomes elevated in obesity. Indicative of the connection between fasting patterns of fatty acid metabolism and the phenotype of insulin resistance, we observed that an elevated value for leg RQ was a significant correlate of insulin-resistant glucose metabolism ($r = -0.57$, $p < 0.001$), the latter measured during an euglycemic hyperinsulinemic clamp. An important observation was that during insulin-stimulated conditions, leg RQ increased significantly in lean subjects but did not change from basal values in obese subjects. Two points should be stressed concerning these findings. First, during insulin-stimulated conditions, the patterns of lipid oxidation in muscle in obesity would appear to reflect higher rates of lipid oxidation than in lean individuals. However, this is not due to a persistent elevation of lipid oxidation in skeletal muscle in obesity; instead, this represents failure to suppress fasting rates of lipid oxidation in obesity. Moreover, during fasting conditions, lipid oxidation is lower in skeletal muscle in obesity. Thus, as the second point, what was observed in lean subjects was that there is an efficient switching between substrates as muscle makes a transition from fasting to insulin-stimulated conditions and in contrast, in obesity, our studies reveal substantial metabolic inflexibility in the regulation of substrate oxidation such that both the response to fasting and to insulin-stimulated conditions are perturbed.

The impaired capacity of skeletal muscle to oxidize fat in obese individuals is not only observed under fasting or insulin-stimulation conditions but also during submaximal physical exercise. Colberg et al. (9) showed that sedentary obese adults rely less on fat oxidation than lean adults during a standardized amount of physical exercise (i.e., total caloric expenditure of obese individuals was matched to that of lean subjects during the exercise test). One interpretation of these data is that muscle triglycerides, which seem to be less utilized during submaximal physical exercise, probably remain at elevated levels within the skeletal muscle cells of obese individuals. These *in vivo* metabolic experiments, both those conducted during resting as well as during exercise studies, have led us to conclude that in obesity, the capacity of skeletal muscle for fatty acid oxidation is reduced and therefore, there is a disposition for lipid accumulation in skeletal muscle of obese individuals.

Storage of Fat in Skeletal Muscle

Patterns of adipose tissue distribution as well as triglyceride deposition within tissues such as skeletal muscle have been related to the expression of insulin resistance. It is now well recognized, based on the work of a number of investigators, that fat deposition within the visceral region, determined from computed tomography scanning, is strongly associated with skeletal muscle insulin resistance and a greater risk of developing cardiovascular diseases (14, 15). Skeletal muscle insulin resistance is a prominent aspect of the insulin-resistance syndrome that is characterized by upper-body adiposity, dyslipidemia, hypertension, and glucose intolerance or frank diabetes mellitus (13, 16, 38). A number of investigations have focused on the associations between abdominal subcutaneous (1, 2) and visceral (17, 24, 31, 39, 43) fat and hyperinsulinemia, glucose intolerance, dyslipidemia, and other cardiovascular risk factors; however, several recent studies also suggest that lipid deposition within skeletal muscle may also be considered as a potentially adverse aspect of regional fat deposition.

Within skeletal muscle, triglyceride can be stored as intramyocellular lipid, in the form of lipid droplets distributed in the cytoplasm of muscle fibers, or interspersed between myocytes as extramyocellular lipid. With the use of ^{1}H magnetic resonance spectroscopy, Boesch et al. (6) has observed that there was a 2- to 2.5-fold variation in the amount of intramyocellular lipid in comparing healthy subjects. During several hours of endurance-type exercise, there appears to be depletion of the intramyocellular triglyceride, consistent with its role as a fuel reserve, and muscle triglyceride stores are then fully restored less than 2 days post-exercise. Based on electron microscope analyses of skeletal muscle of healthy human subjects, the volume occupied by lipid droplets is about 0.5% of the total cell volume (26, 50).

Techniques for the Identification of Skeletal Muscle

By identifying muscle tissue with computed tomography (CT) scans, Kelley et al. (30) reported that obese men had an increased amount of muscle tissue, yet the attenuation values determined by CT, which are commonly expressed as Houndsfield values, are lower in obese compared to lean individuals. Attenuation values are the biophysical characteristics of tissue identified by CT imaging and are the basis of the various shading from black to gray to white in tissue contrast seen on CT images. By convention, attenuation values are based on a relation to that of water (set to 0), and adipose tissue has negative values, while bone has markedly high values and various lean tissues have intermediate values. Using skeletal muscle in lean, healthy individuals as a reference, the range of attenuation values in obesity is shifted into a lower range. The chemical change that is probably causing this shift is an increased content of triglyceride within the muscle in obesity, due to the fact that adipose tissue has a negative attenuation value. Increased fat deposition within muscle and a lower range of attenuation values is a phenomenon previously reported in other types of skeletal muscle diseases (4, 8, 11). One of our studies demonstrated that both abdominal visceral adipose tissue and skeletal muscle fat accounted for almost 60% of the variance in insulin-stimulated leg glucose storage. However, skeletal muscle fat content had the strongest predictive value for insulin resistance in an analysis restricted to obese individuals, with visceral fat content adding independent signifi-cance (45). These findings were not only confirmed by Pan et al. (34) and Dugas et al. (18), but also in our recent study that involved a larger group of subjects (25).

More invasive techniques, such as the use of muscle biopsies, have clearly supported the concept that skeletal muscle in obese men contains a greater amount of lipid (determined from whole muscle homogenates) than that of lean men (32). Additional research indicates that individuals with a high triglyceride content in their skeletal muscle generally manifest insulin resistance (34, 35, 47). Moreover, obese patients who have reached the clinical stage of type 2 diabetes mellitus, have a four- to fivefold increase in skeletal muscle triglyceride concentrations compared to healthy control subjects (5). It is important to mention that, similar to the study of Landin et al. (32), these studies were unable to distinguish the localization of fat (intra- versus extramuscular fat). As mentioned above, fat within skeletal muscle can be found either within connective tissue between muscle fibers or stored directly within muscle cells. Measuring total triglycerides from whole-muscle homogenates does not tell us whether fat is localized inside of or between muscle fibers. How obesity affects these two different compartments remained unclear until the publi-cation of recent studies. In addition to whole-muscle homogenate triglyceride measurements, Phillips et al. (35) showed that the intramyocellular triglyceride content of muscle, determined from muscle sections stained for fat with the oil red O histochemical staining technique, was inversely related to insulin sensitivity.

A similar approach has been used by Thériault et al. (48) to analyze skeletal muscle of lean individuals, obese individuals, and obese individuals with type 2 diabetes. The area occupied by lipid in at least 80 muscle fibers per subject was quantified with image analysis software (NIH Image). The results revealed that lipid accumulation was about twice as high in muscles of obese individuals and individuals with obese type 2 diabetes compared to lean individuals. We recently conducted a similar investigation in obese and lean individuals (Malenfant et al., unpublished results), and in addition to oil red O staining, muscle fibers were histochemically stained to reveal their myosin heavy chain content for the categorization of the different muscle fiber types (I, IIA, and IIB). Analyses of lipid droplets within individual muscle fibers revealed that their number (but not their size) was almost twice as high in all three muscle fibers of obese compared to lean subjects (figure 10.1). Thus, an increased content of fat appears to be a key structural abnormality of skeletal muscle in obese individuals.

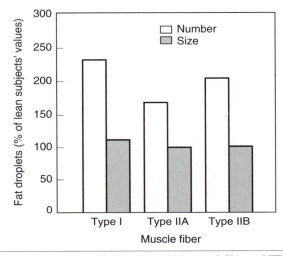

Figure 10.1   Number and size of fat droplets within type I, IIA, and IIB fibers in skeletal muscle of obese individuals expressed as a percentage of lean subjects' values.

The identification of metabolic defects that cause an imbalance between the uptake of fatty acids and the oxidation of fatty acids within skeletal muscle is a topic of active research. In research conducted in the laboratory of Simoneau and colleagues, a goal was to identify whether there were metabolic markers that could signify the respective role of entry of fatty acids through the membrane of the muscle cell, intracellular transport, import into mitochondria, or oxidation within mitochondria. To achieve these aims, a group of lean and obese individuals with a normal glucose tolerance underwent percutaneous biopsy of vastus lateralis skeletal muscle to determine the activity or content of heparin-releasable lipoprotein lipase (LPL), cytosolic fatty acid binding protein (FABPc), carnitine palmitoyl transferase (CPT), citrate synthase (CS), and uncoupling protein 2 (UCP2) content. The results, expressed as percentages of the lean subjects' values, are summarized in figure 10.2. Abnormalities of LPL expression in skeletal muscle could have metabolic consequences, including effects on lipid and lipoprotein concentration, energy homeostasis, body weight, and body composition. However, as reflected by the activity of heparin-releasable LPL, the capacity of skeletal muscle for the uptake of triglycerides did not differ in obese compared to lean subjects. These results are in agreement with prior observations in nondiabetic obese and nonobese individuals (19), although some conflicting data have also been reported (20, 21, 36).

Although our current understanding of its precise physiological roles has not advanced as far as knowledge of its structure (49), the cytosolic isoform of the fatty acid binding protein can be viewed as a molecule that provides niches for hydrophobic binding of fatty acids and long-chain acyl CoA esters within cells. The content of FABPc was not different between the two groups. Conversely, muscle CPT (i.e., the enzyme complex regarded as a rate-limiting step for the entrance and oxidation

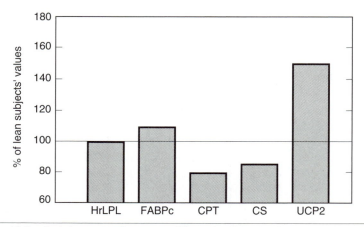

Figure 10.2   Activity or protein content of heparin releasable lipoprotein lipase (HrLPL), cytosolic fatty acid binding protein (FABPc), carnitine palmitoyl transferase (CPT), citrate synthase (CS), and uncoupling protein 2 content (UCP2) in obese individuals expressed as a percentage of lean subjects' values.

of long-chain fatty acid CoA esters in mitochondria) and CS activities (a marker of the mitochondrial volume density of muscle) were lower in obese subjects. The discovery within the past few years of at least two UCP homologues (UCP2 and UCP3) has led to the supposition that these proteins may have functional roles in regulating energy expenditure (7, 22) and that the content of UCP2 and UCP3 might be altered in obesity. While obesity-related differences in UCP2 expression in skeletal muscle seem to be a logical postulate (23), Millet et al. (33) have shown that UCP2 mRNA levels in human skeletal muscle were neither related to body mass index nor to resting metabolic rate. Using an antibody to determine the protein content of UCP2 in the skeletal muscle of lean and obese individuals, we have recently reported that the content of UCP2 is increased in obesity and that this increase is correlated with the reduced reliance on fat oxidation during fasting conditions (46). These cross-sectional studies challenge the hypothesis that the functional role of UCP2 is to regulate thermogenic capacity of skeletal muscle; however, further investigations using interventions known to affect patterns of energy expenditure would be very useful in examining the functional role of UCP2. Thus, these metabolic data are strongly suggestive that in obesity skeletal muscle is disposed toward fat esterification rather than oxidation.

Conclusion

An increased fat content in skeletal muscle appears to be a key structural abnormality in obese individuals. Our recent observations, which entailed a cross-sectional comparison of lean and obese subjects, reaffirm this finding while addressing a more novel concept that the levels of metabolic markers of skeletal muscle for the utilization of fatty acids are altered in obesity in a manner that favors lipid accumulation. The task that clearly remains is to continue to more precisely define the nature of the defects within the pathways of fat metabolism, to discern the contribution of environmental and genetic influences, and to make use of these insights to develop effective treatment strategies for obese individuals.

Acknowledgments

The projects described in this paper were supported by funding from National Institutes of Health R01 DK49200-02 and 5M01RR00056 (General Clinical Research Center).

References

1. Abate, N., A. Garg, R. M. Peshock, J. Stray-Gundersen, B. Adams-Huet, and S. M. Grundy. 1996. Relationship of generalized and regional adiposity to insulin sensitivity in men with NIDDM. *Diabetes* 45: 1684-1693.
2. Abate, N., A. Garg, R. M. Peshock, J. Stray-Gundersen, and S. M. Grundy. 1995. Relationships of generalized and regional adiposity to insulin sensitivity in men. *Journal of Clinical Investigation* 96: 88-98.
3. Ahmad, F., J. L. Azevedo Jr, R. Cortright, G. L. Dohm, and B. J. Goldstein. 1997. Alterations in skeletal muscle protein-tyrosine phosphatase activity and expression in insulin-resistant human obesity and diabetes. *Journal of Clinical Investigation* 100: 449-458.
4. Alanen, A. M., B. Falck, H. Kalimo, M. E. Komu, and V. H. Sonninen. 1994. Ultrasound, computed tomography and magnetic resonance imaging in myopathies: correlations with electromyography and histopathology. *Acta Neurologica Scandinavica* 89: 336-346.
5. Beck-Nielsen, H., and L. C. Groop. 1994. Metabolic and genetic characterization of prediabetic states. Sequence of events leading to non-insulin-dependent diabetes mellitus. *Journal of Clinical Investigation* 94: 1714-1721.
6. Boesch, C., J. Slotboom, H. Hoppeler, and R. Kreis. 1997. In vivo determination of intra-myocellular lipids in human muscle by means of localized ^{1}H-MR-spectroscopy. *Magnetic Resonance Medicine* 37: 484-493.
7. Boss, O., S. Samec, F. Kühne, P. Biglenga, F. Assimacopoulos-Jeannet, J. Seydoux, J.-P. Giacobino, and P. Muzzin. 1998. Uncoupling protein-3 expression in rodent skeletal muscle is modulated by food intake but not by changes in environmental temperature. *Journal of Biological Chemistry* 273: 5-8.
8. Bulcke, J., D. Crolla, J. L. Termotte, A. Baert, Y. Palmer, and R. van den Bergh. 1981. Computed tomography of muscle. *Muscle & Nerve* 4: 67-72.

9. Colberg, S. R., J. M. Hagberg, S. D. McCole, J. M. Zmuda, P. D. Thompson, and D. E. Kelley. 1996. Utilization of glycogen but not plasma glucose is reduced in individuals with NIDDM during mild-intensity exercise. *Journal of Applied Physiology* 81: 2027-2033.

10. Colberg, S. R., J.-A. Simoneau, F. L. Thaete, and D. E. Kelley. 1995. Skeletal muscle utilization of free fatty acids in women with visceral obesity. *Journal of Clinical Investigation* 95: 1846-1853.

11. De Kerviler, E., A. Leroy-Willig, D. Duboc, B. Eymard, and A. Syrota. 1996. MR quantification of muscle fatty replacement in McArdle's disease. *Magnetic Resonance Imaging* 10: 1137-1141.

12. DeFronzo, R. A., R. C. Bonadonna, and E. Ferrannini. 1992. Pathogenesis of NIDDM. A balanced overview. *Diabetes Care* 15: 318-368.

13. DeFronzo, R. A., and E. Ferrannini. 1991. A multifaceted syndrome responsible for NIDDM, obesity, hypertension, dyslipidemia, and atherosclerotic cardiovascular disease. *Diabetes Care* 14: 173-194.

14. Després, J.-P. 1998. The insulin resistance-dyslipidemic syndrome of visceral obesity: Effect on patients' risk. *Obesity Research* 6 (Suppl 1): 8S-17S.

15. Després, J.-P., and B. Lamarche. 1993. Effects of diet and physical activity on adiposity and body fat distribution: implications for the prevention of cardiovascular disease. *Nutrition Research Reviews* 6: 137-159.

16. Després, J.-P., S. Moorjani, P. J. Lupien, A. Tremblay, A. Nadeau, and C. Bouchard. 1990. The regional distribution of body fat, plasma lipoproteins and cardiovascular disease. *Arteriosclerosis* 10: 497-511.

17. Després, J.-P., A. Nadeau, A. Tremblay, M. Ferland, S. Moorjani, P. J. Lupien, G. Thériault, S. Pinault, and C. Bouchard. 1989. Role of deep abdominal fat in the association between regional adipose tissue distribution and glucose tolerance in obese women. *Diabetes* 38: 304-309.

18. Dugas, S., M. Brochu, M. Dumont, A. Nadeau, D. Prud'homme, and J.-P. Després. 1997. Skeletal muscle fat content assessed by CT: Associations with the insulin resistance syndrome in women. *International Journal of Obesity* 21: S35.

19. Eckel, R. H., T. J. Yost, and D. R. Jensen. 1995. Alterations in lipoprotein lipase in insulin resistance. *International Journal of Obesity* 19: S16-S21.

20. Eckel, R. H., T. J. Yost, and D. R. Jensen. 1995. Sustained weight reduction in moderately obese women results in decreased activity of skeletal muscle lipoprotein lipase. *European Journal of Clinical Investigation* 25: 396-402.

21. Evans, D. J., M. Murray, and A. H. Kissebah. 1984. Relationship between skeletal muscle insulin resistance, insulin-mediated glucose disposal, and insulin binding: effects of obesity and body fat topography. *Journal of Clinical Investigation* 74: 1515-1525.

22. Fleury, C., M. Neverova, S. Collins, S. Raimbault, O. Champigny, C. Levi-Mayrueis, F. Bouillard, M. F. Seldin, R. S. Surwit, D. Ricquier, and C. H. Warden. 1997. Uncoupling protein-2: a novel gene linked to obesity and hyperinsulinemia. *Nature Genetics* 15: 269-272.

23. Flier, J. S., and B. B. Lowell. 1997. Obesity research springs a proton leak. *Nature Genetics* 15: 223-224.
24. Fujioka, S., Y. Matsuzawa, K. Tokunaga, and S. Tarui. 1987. Contribution of intra-abdominal fat accumulation to the impairment of glucose and lipid metabolism in human obesity. *Metabolism* 36: 54-59.
25. Goodpaster, B. H., F. L. Thaete, J.-A. Simoneau, and D. E. Kelley. 1997. Subcutaneous abdominal fat and thigh muscle composition predict insulin sensitivity independently of visceral fat. *Diabetes* 46: 1579-1585.
26. Hoppeler, H. 1986. Exercise-induced ultrastructural changes in skeletal muscle. *International Journal of Sports Medicine* 7: 187-204.
27. Kelley, D. E., M. A. Mintun, S. C. Watkins, J.-A. Simoneau, F. Jadali, A. Fredrickson, J. Beattie, and R. Thériault. 1996. The effect of NIDDM and obesity on glucose transport and phosphorylation in skeletal muscle. *Journal of Clinical Investigation* 97: 2705-2713.
28. Kelley, D. E., and J.-A. Simoneau. 1997. Mechanisms of insulin resistance in obesity. In *Clinical research in diabetes*, eds B. Draznin, 57-72. Totowa, NJ: Humana Press.
29. Kelley, D. E., J.-A. Simoneau, B. Goodpaster, and F. Troost. 1997. Defects of skeletal muscle fatty acid metabolism in obesity. *Obesity Research* 5: 21S.
30. Kelley, D. E., S. Slasky, and J. Janosky. 1991. Skeletal muscle density: Effects of obesity and non-insulin-dependent diabetes mellitus. *American Journal of Clinical Nutrition* 54: 509-515.
31. Kissebah, A. H., and G. R. Krakower. 1994. Regional adiposity and morbidity. *Physiological Review* 74: 761-811.
32. Landin, K., F. Lindgärde, B. Saltin, and L. Wilhelmsen. 1988. Decreased skeletal muscle potassium in obesity. *Acta Medica Scandinavica* 223: 507-513.
33. Millet, L., H. Vidal, F. Andreelli, D. Larrouy, J.-P. Riou, D. Ricquier, M. Laville, and D. Langin. 1997. Increased uncoupling protein-2 and -3 mRNA expression during fasting in obese and lean humans. *Journal of Clinical Investigation* 100: 2665-2670.
34. Pan, D. A., S. Lillioja, A. D. Kriketos, M. R. Milner, L. A. Baur, C. Bogardus, A. B. Jenkins, and L. H. Storlien. 1997. Skeletal muscle triglyceride levels are inversely related to insulin action. *Diabetes* 46: 983-988.
35. Phillips, D. I. W., S. Caddy, V. Ilic, B. A. Fielding, K. N. Frayn, A. C. Borthwick, and R. Taylor. 1996. Intramuscular triglyceride and muscle insulin sensitivity: Evidence for a relationship in nondiabetic subjects. *Metabolism* 8: 947-950.
36. Pollare, T., B. Vessby, and H. Lithell. 1991. Lipoprotein lipase activity in skeletal muscle is related to insulin sensitivity. *Arteriosclerosis & Thrombosis* 11: 1192-1203.
37. Ravussin, E., and B. A. Swinburn. 1992. Pathophysiology of obesity. *Lancet* 340: 404-408.
38. Reaven, G. M. 1995. Pathophysiology of insulin resistance in human disease. *Physiological Review* 75: 473-486.

39. Ross, R., L. Fortier, and R. Hudson. 1996. Separate associations between visceral and subcutaneous adipose tissue distribution, insulin and glucose levels in obese women. *Diabetes Care* 19: 1404-1411.
40. Rothman, D. L., I. Magnusson, G. W. Cline, G. I. Shulman, and R. G. Shulman. 1995. Decreased muscle glucose transport/phosphorylation is an early defect in the pathogenesis of non-insulin-dependent diabetes mellitus. *Journal of Clinical Investigation* 92: 983-987.
41. Rothman, D. L., R. G. Shulman, and G. I. Shulman. 1992. ^{31}P nuclear magnetic resonance measurements of muscle glucose-6-phosphate: evidence for a reduced insulin-dependent diabetes mellitus. *Journal of Clinical Investigation* 89: 1069-1075.
42. Schalin-Jäntti, C., E. Laurila, M. Löfman, and L. C. Groop. 1995. Determinants of insulin-stimulated skeletal muscle glycogen metabolism in man. *European Journal of Clinical Investigation* 25: 693-698.
43. Seidell, J. C., P. Bjorntorp, L. Sjostrom, R. Sannerstedt, M. Krotkiewski, and H. Kvist. 1988. Regional distribution of muscle and fat mass in men—new insight into the risk of abdominal obesity using computed tomography. *International Journal of Obesity* 13: 289-303.
44. Shulman, G. I., D. L. Rothman, T. Jue, P. Stein, R. A. DeFronzo, and R. G. Shulman. 1990. Quantification of muscle glycogen synthesis in normal subjects and subjects with non-insulin-dependent diabetes by ^{13}C nuclear magnetic resonance spectroscopy. *New England Journal of Medicine* 322: 223-228.
45. Simoneau, J.-A., S. R. Colberg, F. L. Thaete, and D. E. Kelley. 1995. Skeletal muscle glycolytic and oxidative enzyme capacities are determinants of insulin sensitivity and muscle composition in obese women. *Federation of American Societies of Experimental Biology Journal* 9: 273-278.
46. Simoneau, J.-A., D. E. Kelley, M. Neverova, and C. H. Warden. 1998. Overexpression of muscle uncoupling protein-2 content in human obesity associates with reduced skeletal muscle lipid utilization. *Federation of American Societies of Experimental Biology Journal* 12: 1739-1745.
47. Storlien, L., A. Jenkins, D. Chisholm, W. Pascoe, S. Khouri, and E. Kraegen. 1991. Influence of dietary fat composition on development of insulin resistance in rats: relationship to muscle triglyceride and ω-3 fatty acids in muscle phospholipids. *Diabetes* 40: 280-289.
48. Thériault, R., B. Goodpaster, and D. E. Kelley. 1998. Intramuscular lipid content quantified by histochemistry is increased in obesity and type 2 diabetes mellitus. *Diabetes* 47: A315.
49. Veerkamp, J. H., and R. J. A. Paulussen. 1987. Fatty acid transport in muscle: the role of fatty acid-binding proteins. *Biochemical Society Transactions* 15: 331-336.
50. Wang, N., R. S. Hikida, R. S. Staron, and J.-A. Simoneau. 1993. Muscle fiber types of women after resistance training—quantitative ultrastructure and enzyme activity. *Pflügers Archives* 424: 494-502.
51. Yki-Järvinen, H., and V. A. Koivisto. 1983. Effects of body composition on insulin sensitivity. *Diabetes* 32: 765-969.

PART III

Physical Activity in the Prevention and Treatment of Obesity

CHAPTER 11

Physical Activity and Body Composition in Children and Adolescents

Bernard Gutin, PhD and Paule Barbeau, PhD

Georgia Prevention Institute, Medical College of Georgia, Augusta, Georgia, U.S.A.

We will begin this chapter by reviewing what is known about the determinants of physical activity and the role of exercise and diet in the etiology of juvenile obesity. Abbreviations are listed in table 11.1. We will then summarize what is known from nonexperimental studies about the relationship of exercise and body composition to health in children and adolescents, with a focus on preclinical markers for future coronary artery disease (CAD) and non-insulin-dependent diabetes mellitus (NIDDM), because they are the health problems most closely associated with obesity. Next, we will review experimental studies of exercise and physical training, which provide the clearest evidence of causal relationships between exercise, body composition, and health.

Wherever possible, we will focus on research that used youths as subjects. However, with respect to many of the specific topics involved, little or nothing is known from studies of youths; indeed, many important questions have not yet been fully answered in adults. Where data on juveniles are limited, the reader is referred to other chapters in this book that deal with these topics in adults. Because the growth and maturation of youths may interact with other factors to influence

Table 11.1 List of Abbreviations

CAD	coronary artery disease
HDLC	high density lipoprotein cholesterol
LDLC	low density lipoprotein cholesterol
MRI	magnetic resonance imaging
NIDDM	non-insulin dependent diabetes mellitus
PAI-1	plasminogen activator inhibitor-1
RMR	resting metabolic rate

body composition and health, extrapolations of findings from adults to children need to be treated as hypotheses to be tested directly in subsequent studies with children.

We will give special emphasis to recent studies, especially those that have used new techniques to assess body composition and energy expenditure. For example, dual x-ray absorptiometry is easily administered to children (33) and is valid (139) and reliable (60). With respect to regional deposition of fat, important new information has been provided by studies that have used computed tomography scanning or magnetic resonance imaging (MRI) to distinguish visceral adipose tissue from subcutaneous abdominal adipose tissue (14, 16, 105, 155). MRI is especially appropriate for repeated measurements on children, because it does not involve any radiation exposure.

With respect to measurement of energy expenditure, our understanding of some components of 24-hour energy expenditure has improved in recent years due to increased use of metabolic chambers. However, the restriction of subjects to the chamber precludes investigation of energy expenditure during free-living physical activity. The development of the doubly labeled water procedure has improved assessment of free-living energy expenditure and physical activity, as described elsewhere in this volume (91). Although a distinction can be drawn between the terms "physical activity" (i.e., all movement that elevates energy expenditure) and "exercise" (more structured and intense activity), for purposes of this chapter they will be used interchangeably to refer to movement associated with various degrees of energy expenditure.

Determinants of Physical Activity in Youths

Research dealing with free-living activity as a behavioral outcome is limited by lack of satisfactory measurement instruments (127). As more objective methods of activity assessment are developed, such as doubly labeled water, our understanding of this topic may expand. Because cardiovascular fitness is related to habitual physical activity (108), and because systematic increases in activity (i.e., physical training) lead to increases in cardiovascular fitness (131, 156), fitness information provides an indirect index of physical activity.

A complete review of the determinants of physical activity is that by Sallis et al. (127); only a brief summary of the literature is possible in this chapter. Based on their review and other recent papers, the following conclusions may be drawn. (1) Boys are more active than girls (71, 144); this is supported by data showing that even as young as 5 to 6 years of age boys are more fit than girls (63). (2) Whites seem to be more active in aerobic activities than African- or Mexican-Americans, which is consistent with recent data (71, 112, 145) showing that even after controlling for body fatness, African-American youths were lower in cardiovascular fitness than were whites. (3) Physical activity tends to decline during the teen years (71), especially in girls (108). (4) Confidence in one's abilities to perform exercise (i.e., physical self-efficacy) is a strong predictor of future physical activity in adolescents (118). (5) Peers, siblings, and parents are influences on physical activity, probably through some combination of modeling, prompting, and reinforcement. (6) Youths are more active in the winter than other seasons, on weekends compared to weekdays, and outdoors compared to indoors. (7) Children are more likely to engage in intermittent, rather than sustained, bouts of high-intensity exercise (128). In addition, time spent sitting or watching television is related to low physical activity levels (15, 107), which may account for why some studies have reported that large amounts of television viewing are associated with obesity (47).

Etiology of Juvenile Obesity

Although there is clear evidence that genetic factors play a role in obesity, it is also apparent that nongenetic factors are important, especially in technologically advanced societies (12). Regardless of whether the underlying impetus is genetic or environmental, variations in body energy stores are necessarily due to variations in one or more components of energy expenditure or intake. As described elsewhere in this volume (see part II), because it is difficult to measure many of the involved variables precisely, little definitive information is available from prospective studies on the role of energy expenditure and diet in the natural history of obesity in people of any age, especially juveniles.

In light of the recent discovery in mice that the hormone leptin decreases energy intake and increases energy expenditure, some investigations have been undertaken in children. Along with Caprio et al. (17), we (55) found that leptin was closely correlated with subcutaneous abdominal adipose tissue and total fat mass in children, and less closely correlated with visceral adipose tissue. Lahlou et al. (79) found that serum leptin levels were positively correlated with fasting insulin levels, adiposity, and weight gain the previous year, but were not associated with resting energy expenditure or lower energy intake; indeed, the obese children ingested 2 to 3 times more energy (measured with self-reports) than the lean children. Salbe et al. (126) examined cross-sectional relationships among these factors in 5-year-old Pima Indian children. They found that leptin concentrations, which were closely correlated to percent body fat, were also correlated with physical activity level (the ratio of total to resting energy expenditure, measured with doubly labeled water) after adjustment for body fat ($r = .26$, $p < .01$). Nagy et al. (99) found African-American and Caucasian girls to have higher leptin levels than African-American and Caucasian boys; however, these differences were no longer significant after controlling for total body composition, visceral adipose tissue, and subcutaneous abdominal adipose tissue.

These results indicate that in children leptin is a marker of total body adiposity, but does not appear to suppress energy intake or halt fat deposition, suggesting some type of leptin resistance. Given the small amount of information available in youths, very few definite conclusions can yet be made.

As noted later in this chapter, visceral adipose tissue may play an especially important role in the development of CAD/NIDDM risk factors starting in child-hood. Unfortunately, no information is yet available about the etiology of this fat depot in juveniles.

Resting Metabolism

The components of total energy expenditure and the extent to which these are predictive of future obesity in adults are described elsewhere in this volume (91, 124). Because the thermic effect of food is the component of energy expenditure that is least reproducible (115) and most tedious to measure, it has not been extensively studied and its role in the etiology of juvenile obesity is unclear. Cross-sectional studies do not provide a consistent picture concerning differences in the thermic effect of food between lean and obese youths (5, 86), and no prospective studies in children have been reported concerning whether low levels of this component of energy expenditure predispose youths to accretion of fat.

Resting metabolic rate (RMR) has been more extensively studied. Some cross-ethnic comparisons support the idea that RMR can account for obesity in some populations. For example, one possible reason for the greater prevalence of obesity among African-Americans and their greater difficulty in losing weight (76) is an inherent tendency to have a lower RMR, as has been shown in some child studies (72, 96). On the other hand, a study comparing white and Pima Indian 5-year-olds found no difference between these two ethnic groups in RMR, physical activity energy expenditure, or total energy expenditure, suggesting that excess food intake may be responsible for the high incidence of obesity in Pima Indians (125).

One cross-sectional study found that children of obese parents had relatively low RMR (49), suggesting that they were at increased risk for the development of obesity. However, when these children were followed up 12 years later (50), the children of the obese parents were not found to be more obese than the children of nonobese parents. Two recent reviews of the doubly labeled water literature (26, 42), along with other recent studies (154), concluded that low levels of resting energy expenditure are probably not responsible for obesity in children.

Physical Activity

Although physical activity constitutes on average a relatively small portion of total energy expenditure, it has potential importance in influencing obesity development for several reasons. First, it is largely volitional. Second, because it varies so much from individual to individual, it is the component of energy expenditure that accounts for variability in total energy expenditure. Third, activity can increase fat-free mass, the main determinant of RMR (104), with long-term consequences for energy balance. Fourth, exercise training can influence substrate utilization, thereby playing a role in how ingested nutrients are partitioned into fat and fat-free mass.

It is problematic to determine from nonexperimental studies whether exercise and body fatness are related because of the difficulty in knowing how to express physical activity. If it is expressed as activity energy expenditure, then some have found no association between activity and fatness (83), and others have found the paradoxical result that obese youths are *more* active than nonobese youths (5, 85). Furthermore, one study has suggested that whereas physical inactivity was associated with fat mass in 6- to 12-year-old boys, the same was not true for girls, suggesting that the factors contributing to obesity may be sex specific (51).

When interpreting data on energy expenditure during activity, it is necessary to take into account that a heavier youth uses more energy to move the body a given distance. Thus, if a lean and an obese youth display the same free-living activity energy expenditure, it represents less movement in the obese child. Consequently, it is necessary to adjust activity energy expenditure for body weight or composition to determine if variations in movement are associated with fatness. The problem concerns the exponent to use in making this adjustment. If energy expenditure is simply divided by weight—i.e., an exponent of one is used—then an overcorrection may result, automatically creating a negative correlation between adjusted energy expenditure and fatness (113). Unfortunately, the correction factor varies for different children, depending on how much of their activity involves carrying their body weight (e.g., walking/running) and how much involves activities in which the body weight is supported and most of the work is external (e.g., cycling). Although the variety of procedures used to correct for body weight may account for the discrepant findings of cross-sectional studies (23, 24, 45), a recent analysis of doubly labeled water studies by DeLany (26) concluded that low levels of physical activity were associated with higher levels of body fatness.

Some of the difficulties of cross-sectional studies also afflict prospective cohort (i.e., non-experimental) studies, with the result that such studies have not provided a consistent picture either (43, 119, 150). The statistically sophisticated study by Goran et al. (43) illustrates the complexity of the matter. These authors estimated activity to be the difference between total 24-hour energy expenditure and resting energy expenditure, expressed either in absolute terms or adjusted (by regression) for fat-free mass. Therefore, the energy cost of transporting the fat mass was included in the activity energy expenditure, but fat mass was not included in the adjustment. This presumably led to an inflated baseline activity energy expenditure in the fatter children. The children who were fattest at baseline were the ones whose rate of increase in fatness was the greatest over time. Thus, an inflated baseline activity level for them may have obscured the potential contribution of low levels of actual movement to accretion of fat. As a result, interpretation of the results of such studies is controversial (29).

Time-motion analysis may provide a more direct index of how much exercise the child does. Cross-sectional studies of this nature show that inactive children are fatter (45, 83) even while ingesting less energy (25); however, in cross-sectional studies it is impossible to tell whether the inactivity caused more fatness or whether higher fatness caused the inactivity. A clearer picture emerges from recent epidemiologic studies in which the exercise levels of children were estimated in other ways. When exercise level was estimated by their parents in relation to other children of the same age, it was found that exercise and family history of obesity were principal risk factors for later development of childhood obesity (97) or higher body mass index (75). A study that used the Caltrac movement sensor to measure activity found that preschoolers who were classified as inactive were 3.8 times as likely as active children to have an increasing triceps skinfold slope during the average 2.5-year follow-up (93).

To the degree that cardiovascular fitness can be accepted as a proxy for physical activity, the results of a 3-year longitudinal study of 7- to 12-year-olds (70) are pertinent. It was found that those children who increased the most in maximal oxygen consumption were those who increased the most in fat-free mass and left ventricular mass, but who increased the least in skinfold fatness. Although adult studies suggest that another factor to consider is the intensity of the exercise, little is known about the role of different exercise intensities in the etiology of childhood fatness.

Exercise may influence accumulation of fat by improving the use of lipids as a substrate for energy (88, 129). Consistent with this hypothesis are recently reported data showing that obese children, during the dynamic phase of fat deposition, had a decreased degree of lipolysis in response to epinephrine infusion (13). However, Le Stunff and Bougnères (80) found *increased* lipid oxidation in obese children, even in the relatively early stages of obesity. Maffeis et al. (84) also observed greater lipid oxidation in obese children, perhaps as a compensatory mechanism for their higher levels of fat intake. The ability of lipid oxidation to predict future changes in the fatness of children has not yet been elucidated in prospective studies. Perhaps reduced lipid oxidation plays a role only prior to the onset of obesity, especially in those whose fat intake is not especially high, after which lipid oxidation increases as part of a feedback loop to prevent further accretion of fat.

Diet

It is evident that accurate and reliable quantitative information about energy and macronutrient intake would be valuable, but such data are difficult to obtain, especially in children. Cross-sectional studies using diet recall methods have generally not found a relationship between body fatness and energy intake (93). However, using the doubly labeled water procedure to validate self-reported measures of energy intake, it was found that obese adolescents (5) tended to underreport energy intake to a greater degree than lean adolescents. Consequently, methodological considerations make it difficult to draw any clear conclusions about the role of total free-living energy intake in the etiology of obesity.

With respect to fat intake, most (32, 41, 84, 101, 146), but not all juvenile studies (22, 98), support the idea that diets high in fat are associated with body fatness or gain in weight. Prospective studies testing the hypothesis of a synergistic effect of a high-fat diet and sedentary behavior have not yet been reported in juveniles.

Summary

Because of the difficulty of measuring activity and diet, compounded by the uncertainty of how to express "physical activity," few definitive conclusions are warranted about the etiology of juvenile obesity. Perhaps the most reasonable conclusion is similar to one reached in a study of 10-year weight changes in a national cohort of adults (152), i.e., that low physical activity leads to weight gain, while weight gain leads to further diminution of activity, forming a vicious cycle. This would imply that interventions that either decrease fatness or increase activity might turn the cycle in the other, more favorable, direction.

Exercise, Body Composition, and Health: Nonexperimental Designs

Many of the health problems associated with obesity, such as CAD and NIDDM, manifest themselves in the form of morbidity and mortality during the adult years. However, the origins of these problems can be traced to childhood. For example, an autopsy study of individuals 2 to 39 years old who died from noncardiac causes, on whom prior information concerning risk factors was available, found that individuals who had increased numbers of risk factors also displayed more severe asymptomatic coronary and aortic atherosclerosis (8). These studies suggest that some of the damage produced by these risk factors begins in youths and that lifestyle interventions to reduce risk factors early in life may be warranted.

Because lifestyle habits are to some degree formed early in life and track into adulthood (74), effective early intervention may have a lifelong impact. In the United Kingdom, it was shown that active adults tended to have been active as children, and sedentary adults who had been active as children were more likely to be persuaded to become active (69). In addition, body weight lost in weight control intervention programs is more likely to be sustained in children than in adults (34).

The road to obesity-related lifestyle habits may be related to temperament and caregiver practices during infancy. Wells et al. (149) investigated the relationship between infant temperament at 12 weeks, and fatness and physical activity patterns at 2 to 3.5 years of age. They found that infants who were easily soothed had lesser skinfold thicknesses in childhood than did infants who were fussy or cried a lot and those who displayed distress to limitations (i.e., showing distress while waiting for food, being in a confined position, being dressed, etc.). Easily soothed infants were also more likely to be awake and active during childhood, rather than upset or watching television. Interestingly, fussiness was strongly associated with carbohydrate consumption, but not fat consumption, suggesting that caretakers may tend to feed difficult infants more sugary foods to appease them. Thus, temperament in infancy, and coping methods invoked by caretakers to deal with this temperament, may play a role in forming dietary likes and dislikes, as well as lifestyle habits.

The links among various risk factors for CAD and NIDDM (i.e., hypertension, dyslipidemia, and hyperinsulinemia) indicate the presence of an underlying metabolic syndrome sometimes named Syndrome X or the insulin resistance syndrome (27, 116). It is clear that obesity, particularly abdominal obesity, and physical inactivity are associated with this syndrome (27, 151). Children who are overweight and have excess fat at 9 years of age are more likely to be overweight, have excess fat, and have more cardiovascular risk factors at age 11 than their leaner peers (31). Moreover, recent studies have shown that poor cardiovascular fitness, excess body fatness, and elevated CAD/NIDDM risk factors are all interrelated (11, 19, 62, 78, 114, 117, 148), even in children as young as 5 to 6 years of age (63). Autopsy studies of youths who died from noncardiac causes showed that obesity and elevated glycohemoglobin were associated with coronary lesions (89); these data illustrate the deleterious effects of obesity early in life, as well as supporting the idea that a prediabetic state plays a role in the early development of atherosclerosis.

Because child obesity and poorer fitness have increased in prevalence over the last 10 to 15 years (77, 143), the implications for the future public health are worrisome. African-American females are at especially high risk for obesity and associated health problems (111); even in childhood, African-American girls are fatter than Caucasian girls (102). However, even after adjusting for ethnicity, obesity contributes to disease risk (48). One illustration of the seriousness of the secular trend to increased obesity is the parallel increase in the incidence of NIDDM (i.e., what is called "adult-onset diabetes") and obesity among youngsters in Cincinnati over a 10-year period (110). Support for the positive influence of exercise in this scenario is provided by the results of prospective studies that found favorable changes in CAD risk in those children who improved the most in physical fitness (67, 134).

Although there is some evidence that activity and adiposity can exert independent influences on CAD/NIDDM risk factors (21, 134, 138), there is also evidence that activity and fitness tend to exert much of their favorable effect on CAD/NIDDM risk factors through their influence on adiposity. For example, in a sample of 1092 third-grade children, obesity was associated with higher systolic blood pressure and total cholesterol, whereas self-reported physical activity levels were not related to the risk factors (90). Another study found that endurance running performance was associated with a more favorable risk profile, and that body mass seemed to be the most important factor explaining the difference in lipid and insulin values in adolescents of differing physical fitness (9). We found that both low cardiovascular fitness and high body fatness were correlated with unfavorable CAD/NIDDM risk factor levels in 7- to 10-year-olds (62); however, when multiple regression was applied, only percent fat explained a significant independent proportion of the variance in the risk factors. We followed this study with one in which weight was controlled experimentally rather than statistically; that is, cardiovascular fitness was expressed as submaximal heart rate during the weight-supported task of supine cycling (58), thus dissociating fitness from weight (and fatness). We found that percent fat was significantly related to unfavorable levels of systolic blood pressure, triacylglycerol, HDL cholesterol, the ratio of total to HDL cholesterol, and insulin, whereas cardiovascular fitness was not significantly correlated with any of these risk factors.

There is some evidence that elevated left ventricular mass, which is predictive of future cardiovascular morbidity (28), is associated with higher levels of physical activity in youths (141). However, it may be that different components of left ventricular mass (i.e., wall thickness and cavity size) are influenced differently by different stimuli. That is, hypertension imposes an afterload on the heart, which results in greater wall thickness, whereas aerobic exercise produces a greater venous return (preload), thereby increasing cavity size. When we compared left ventricular dimensions in 8- to 12-year-old distance runners and nonrunners, the data were consistent with this hypothesis, i.e., the runners had significantly larger internal dimensions, but somewhat (not significantly) thinner walls (64). However, Rowland et al. (123) failed to find differences in cardiac dimensions in prepubertal runners and nonrunners. Thus, more work is needed with larger numbers of subjects to clarify this matter.

We (56) recently investigated the extent to which body composition of children explained variation in left ventricular geometry and function. As an index of function we used mid-wall fractional shortening, an aspect of left ventricular systolic function that identifies individuals at elevated risk for future cardiovascular mortality who are otherwise undetected by conventional endocardial shortening indices (136). We found that greater percent fat was associated with unfavorable levels of left ventricular mass, relative wall thickness, and mid-wall fractional shortening. These results suggest that already in youths, excess fatness is associated with unfavorable left ventricular geometry and function.

Endothelial dysfunction, a relatively early event in atherogenesis (92), is associated with various manifestations of cardiovascular disease in children (18), leading us to investigate the influence of cardiovascular fitness and body fatness in 7- to 13-year-olds (140). Endothelium-dependent arterial dilation was determined by measuring with an echocardiograph the difference in diameter of the superficial femoral artery at baseline and after release of a tourniquet applied to the leg for 5 minutes. Better endothelial function, as represented by a greater amount of dilation in response to the increased blood flow that occurs when the tourniquet is released, was associated with lower percent fat and better cardiovascular fitness.

With respect to regional deposition of fat, our early studies of 5- to 11-year-olds (62, 63) detected no association between anthropometric indices of fat patterning and risk factors. Asayama et al. (4) studied 6- to 12-year-old Japanese children and found that indices of general adiposity and fat patterning (i.e., waist/hip and waist/thigh circumferences) were correlated with CAD risk factors in boys; however, in girls only the indices of fat patterning were correlated with the risk factors. The Bogalusa Study, which can detect weak relationships among variables as statistically significant because large numbers of children are studied, found a tendency for central, but not peripheral, skinfold thickness to be related to CAD/NIDDM risk factors (40).

An important limitation of child studies that used anthropometry or dual x-ray absorptiometry is that they could not distinguish subcutaneous abdominal adipose tissue from the possibly more deleterious visceral adipose tissue (27). Another factor to consider is that in children, very little of the abdominal fat is in the visceral adipose tissue compartment relative to the amount in subcutaneous abdominal adipose tissue (14, 38, 46, 105). Therefore, at an early stage in the development of the CAD/NIDDM syndrome (i.e., in childhood), it may be necessary to measure visceral adipose tissue directly in order to uncover its relationship to other risk factors.

A study that used MRI in obese children (14) found that visceral adipose tissue, but not subcutaneous abdominal adipose tissue, was significantly related to LDL cholesterol and triacylglycerol. Furthermore, a study involving obese and non-obese adolescent girls found that visceral adipose tissue was significantly correlated with triacylglycerol, HDL cholesterol, and insulin in the obese girls only (16). On the other hand, a study of non-obese 7- to 10-year-old girls found no significant relationship between visceral adipose tissue and triacylglycerol, HDL cholesterol, or insulin (155). Therefore, the visceral adipose tissue–risk factor relationship appears to be more pronounced in obese than in lean children.

We recently reported the relationships among these factors in 7- to 11-year-old obese children (105), using MRI to measure visceral adipose tissue and subcutaneous abdominal adipose tissue. Table 11.2 shows the bivariate correlations between the body composition variables and the risk factors. Subsequent to the bivariate correlational analyses, we applied multiple regressions to these data. This approach revealed that for most of the lipid and lipoprotein parameters (triacylglycerol, HDL cholesterol, the ratio of total to HDL cholesterol, LDL particle size, and apolipoprotein B), visceral adipose tissue was the only predictor variable retained in the final models. The negative relationship between LDL particle size and visceral adipose tissue is noteworthy because of the evidence that small, dense LDL particles are relatively more atherogenic than larger, more buoyant particles (137). For nonlipid risk factors (insulin and systolic blood pressure), total fat mass rather than visceral adipose tissue was retained as a significant predictor in the regression models. Neither subcutaneous abdominal adipose tissue, gender, or ethnicity was retained in the final regression models. Because of interrelations among indices of adiposity, no definite conclusions about which aspect of adiposity is most deleterious are yet warranted. In our sample of obese children, visceral adipose tissue and fat mass had shared variance of 45% (i.e., $r^2 = 0.45$); thus, although these aspects of body composition are related, to some degree they provide information about different aspects of adiposity.

Table 11.2   Correlations Between Adiposity Measures and Cardiovascular Risk Factors in Obese 7- to 11-Year-Old Children.

Cardiovascular risk factor	Number of children	Measures of adiposity			
		VAT	SAAT	%fat	TFM
Triacylglycerol	60	0.46[b]	0.22	0.06	0.18
Total cholesterol	62	0.08	-0.03	0.01	-0.06
HDLC	62	-0.40[b]	-0.32[a]	-0.08	-0.28[a]
TC/HDLC	62	0.45[b]	0.35[b]	0.12	0.29[a]
LDLC	60	0.02	-0.01	0.04	-0.03
LDL size	41	-0.32[a]	-0.21	-0.01	-0.13
Apo A-I	62	-0.05	-0.09	-0.09	-0.09
Apo B	62	0.32[a]	0.32[a]	0.15	0.29[a]
Insulin	56	0.34[a]	0.40[b]	0.35[b]	0.49[a]
Glucose	60	0.14	0.14	0.06	0.12
Systolic BP	64	0.21	0.28[a]	0.15	0.33[b]
Diastolic BP	64	-0.19	-0.11	-0.04	-0.07

VAT = visceral adipose tissue, SAAT = subcutaneous abdominal adipose tissue, %fat = percent body fat, TFM = total fat mass, HDLC = high density lipoprotein cholesterol, TC = total cholesterol, LDLC = low density lipoprotein cholesterol, Apo A-I = apolipoprotein A subfraction I, Apo B = apolopoprotein B, BP = blood pressure.

[a] $p < .05$;   [b] $p < .01$.

Reprinted, by permission, from S. Owens, B. Gutin, M. Ferguson, J. Allison, W. Karp, and N.-A. Le, 1998, "Visceral adipose tissue and cardiovascular risk factors in obese children," *Journal of Pediatrics* 133: 41-45.

With respect to the role of ethnicity in fat patterning, Goran et al. (44) found in prepubertal children that, after controlling for subcutaneous abdominal adipose tissue, white children, especially the girls, had significantly higher levels of visceral adipose tissue. This is consistent with findings in Caucasian and African-American women (20). Thus, to some degree the greater prevalence of obesity in African-American females might be mitigated by their having lower levels of fatness in the most atherogenic region.

Some hemostatic indices have emerged as risk factors for cardiovascular disease and stroke, including fibrinogen, plasminogen activator inhibitor-1 (PAI-1), and D-Dimer (72, 106, 135). Because little is known about the relationship of body composition to these indices in youths, we examined these relationships in a group of 7- to 11-year-old obese children who varied from 27 to 61 percent fat (37); insulin was included in the analysis because it is one possible mechanism through which fatness may influence hemostatic activity (132). As shown in table 11.3, various adiposity measurements were associated with all three hemostatic indices, suggesting that clotting/fibrinolysis is another pathway through which body composition influences the early etiology of cardiovascular disease.

The evidence reviewed is consistent with the notions that several major "adult" health problems actually have their origins in childhood, and that excess body fat, especially visceral fat, is associated with preclinical markers for some of these diseases. However, causal inferences drawn from the results of nonexperimental designs must be considered very tentative until supported by experimental studies that determine the extent to which changes in fatness are followed by changes in the fatness-related preclinical markers.

Table 11.3   Correlations Between Plasma Hemostatic Factors, Adiposity Measures, and Insulin[a]

Variable	Fib	PAI-1	D-Dimer	%fat	VAT	SAAT	TFM	FFM	BMI
%fat	0.42[c]	0.08	0.40[c]	—	0.40[c]	0.78[c,d]	0.85[c,d]	0.25	0.78[c,d]
VAT	0.21	0.49[c]	0.16	0.40[c]	—	0.55[c]	0.51[c]	0.47[c]	0.48[c]
SAAT	0.40[c]	0.32[b]	0.37[b]	0.78[c,d]	0.55[c]	—	0.94[c]	0.65[c]	0.93[c]
TFM	0.42[c]	0.28	0.40[c]	0.85[c,d]	0.51[c]	0.94[c]	—	0.71[c]	0.93[c]
FFM	0.23	0.50[c]	0.27	0.25	0.47[c]	0.65[c]	0.71[c]	—	0.62[c]
BMI	0.41[c]	0.24	0.43[c]	0.78[c,d]	0.48[c]	0.93[c]	0.93[c]	0.62[c]	—
Insulin	0.11	0.61[c,d]	0.13	0.42[c,d]	0.58[c,d]	0.59[c,d]	0.63[c,d]	0.68[c,d]	0.55[c,d]

Fib = fibrinogen, PAI-1 = plasminogen activator inhibitor-1, %fat = percent body fat, VAT = visceral adipose tissue, SAAT = subcutaneous abdominal adipose tissue, TFM = total fat mass, FFM = fat-free mass, BMI = body mass index.

[a]Log transformation used for VAT, PAI-1, and D-Dimer;   [b]$p < .05$;   [c]$p < .01$;
[d]Spearman correlation coefficient.

Reprinted, by permission, from M.A. Ferguson, B. Gutin, S. Owens, M. Litaker, R. Tracy, and J. Allison, 1998, "Fat distribution and hemostatic measures in obese children," *American Journal of Clinical Nutrition* 67: 1136-1140. © Am. J. Clin. Nutr. American Society for Clinical Nutrition.

Experimental Studies of Exercise and Physical Training

Although cross-sectional and longitudinal cohort studies can provide a foundation for hypotheses concerning the relationships among exercise, body composition, and health, only controlled experiments can apply rigorous tests of these hypotheses. For example, knowing that in free-living youths there is a cyclic relationship between inactivity, weight gain, more inactivity, more weight gain, and so on may be less illuminating than seeing what happens when youths increase or decrease their activity level.

Indirect support for the value of exercise early in life comes from experimental studies on animals that have shown physical training to prevent obesity (103) and to greatly attenuate development of the CAD/NIDDM risk syndrome (116). Although it is reasonable to formulate hypotheses from animal studies, the complexity of the growing human requires that the hypotheses be tested directly in children at different stages of maturity.

A simple model of what might occur when a youth begins an exercise program is that the energy expenditure of the exercise is added to the youth's customary daily energy expenditure, leading to an increase in 24-hour energy expenditure. If energy intake is uninfluenced by the physical training, an energy deficit will result, body energy stores will be reduced, and favorable changes will take place in all the CAD/NIDDM risk factors related to obesity. But what if children who participate in a vigorous exercise program decrease their normal daily physical activities (e.g., walking or cycling to school) or increase time spent in sedentary activity (e.g., watching television), with the result that their 24-hour energy expenditure is similar to that of children who do not participate in the physical training? Furthermore, what if the interaction of training with different types of diets sometimes causes energy intake to increase more than energy expenditure, thereby leading to a positive rather than a negative energy balance? What if the links between fatness and risk factors in children noted in cross-sectional research are partly or largely coincidental rather than causal, such that improvements in the former do not necessarily lead to improvements in the latter? Finally, what if different types or intensities of physical training influence these factors in different ways in youths of various types (e.g., those with positive or negative family histories of obesity; or those with varying maturity levels, genders, or ethnicities)?

It is clear that this matter is far from simple. As a result, definitive answers to these questions are not available in people of any age, especially children. We will attempt to piece together what is known in order to arrive at some hypotheses and conclusions concerning the effects of exercise.

Energy Expenditure, Diet, and Body Composition

Restriction of energy intake (i.e., dieting) allows a youth to immediately produce a relatively large daily energy deficit compared to what would be appropriate for exercise alone (6). For example, many obese youths can reduce energy intake by 1000 kcal/d, while it would be difficult and perhaps dangerous for an unfit youth to try to increase physical activity by that amount. Thus, dieting is likely to be more effective for short-term weight loss. However, dieting leads to a reduction of resting metabolism in proportion to decreases in fat-free mass (87), thus setting the stage for the dieter to regain the lost weight when the diet stops (133). Little is known about whether growing children who achieve an energy deficit through exercise maintain or even increase fat-free mass and resting metabolism as they decline in fat mass. To our knowledge, no direct experimental information is available concerning the effect of physical training on the resting metabolic rate and the thermic effect of food in children.

Another aspect of 24-hour energy expenditure that may be influenced by physical training is energy expenditure during the hours when the person is not engaged in the training itself. When exercise was added to dieting in children, an increase in energy expenditure (measured by doubly labeled water) occurred, only half of which could be explained by the energy expended during the exercise sessions; the other half seemed to be due to an increase in energy expenditure outside the formal exercise sessions (10). Thus, the training did not lead to a compensatory reduction in spontaneous physical activity.

Several studies of exercise interventions that did not try to influence diet showed favorable effects on body composition in 7- to 10-year-olds (61, 95, 104, 131), suggesting that the youths did not fully compensate for the energy expended in the physical training sessions by altering their energy intake or spontaneous physical activity. The last study also measured visceral and subcutaneous abdominal adipose tissue. Figure 11.1 shows that the increases in both visceral adipose tissue and subcutaneous abdominal adipose tissue of the control group were significantly greater than those of the physical training group (104). To our knowledge, this is the first study of the effects of training on visceral adipose tissue in children or adolescents to be reported.

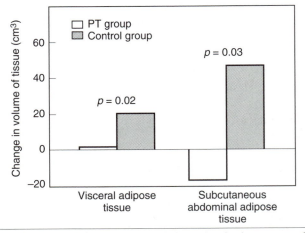

Figure 11.1   Histogram of volume change of visceral and subcutaneous abdominal adipose tissue over a 4-month period in physical training (PT) and control groups. The volume of both types of fat deposits increase significantly more for the control group than for the PT group.

Reprinted, by permission, from B. Gutin and S. Owens, 1999, "Role of exercise intervention in improving body fat distribution and risk profile in children," *American Journal of Human Biology* 11: 237-247. Copyright © 1999, Wiley-Liss, Inc., a subsidiary of John Wiley & Sons, Inc.

We also conducted a study that involved a modified crossover design that allowed us to see the effects of training and detraining on body composition in obese 7- to 10-year-olds (54). The youths were randomly assigned to participate in the physical training during the first or second 4-month periods of the 8-month experimental period. Measurements were made at baseline, after 4 months, and after 8 months, allowing us to view the pattern of changes in body composition over an 8-month period during which the growing youths participated in more or less regular exercise. The three-point data were analyzed using time by group analyses of variance to determine the effects of the training and detraining. Figure 11.2 shows that for percent fat (the primary outcome variable), the time by group interaction was significant and the pattern was as predicted, i.e., group 1 declined in percent fat during the period of training and increased somewhat during the next 4 months, while group 2 remained stable during the first 4 months and declined during the 4 months when they engaged in the training. It is noteworthy that the declines in percent fat were due to significant declines in fat mass, along with (nonsignificant) increases in fat-free mass during the periods of training. This is in contrast to the reductions in fat-free mass that occur when dieting alone is used for weight control (133).

To explore one possible underlying mechanism for the changes in body mass and composition, we determined the influence of the training on leptin levels in a subset

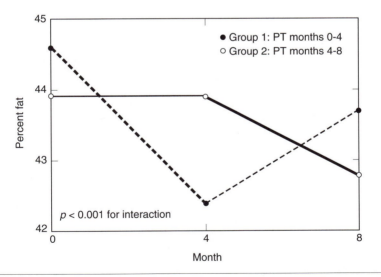

Figure 11.2   Percent fat plotted against time in months. Data are means for percent fat over the three time points of the study. Group 1 engaged in physical training (PT) for the first 4 months of the intervention period and group 2 engaged in physical training for the second 4 months of the intervention period; the thickened lines denote the periods of training for each group. The p-value is for the group by time interaction in the ANOVA.

Reprinted, by permission, from B. Gutin, S. Owens, T. Okuyama, S. Riggs, M. Ferguson, and M. Litaker, 1999, "Effect of physical training and its cessation on percent fat and bone density of children with obesity," *Obesity Research* 7: 208-214.

of 34 subjects of this study (55). Figure 11.3 shows that the pattern of change over the three time points was significantly different for the groups, with leptin levels declining during periods of training in both groups, and increasing rather sharply in group 1 in the 4-month period after cessation of physical training. It is noteworthy that when the data were corrected for changes in fat mass (or total mass) the pattern was similar and remained significant. This result is consistent with the idea that leptin serves as an "alarm" hormone, i.e., in the face of any kind of energy deficit, whether from increased energy expenditure or reduced energy intake (81), plasma levels are reduced to help defend the body against weight loss. Perhaps this mechanism played a role in the accumulation of fat exhibited by the subjects in group 1 over the 4-month period following cessation of training. Thus, youths that participate in a vigorous training program may have a tendency to gain fat if they cease the training, emphasizing the importance of maintaining a high level of physical activity throughout life.

Almost no information is available concerning the optimal dose of exercise for treatment of childhood obesity (6). With respect to exercise intensity, if total energy used during the exercise sessions is the critical factor in producing an energy deficit, then simply lengthening low-intensity sessions would allow a youth to use the same amount of energy as would be used in a shorter session of higher intensity. Savage et al. (131) used this approach with prepubertal boys and found skinfold fat to decline similarly in the low- and high-intensity groups, even though the high-intensity training resulted in a clearer improvement in cardiovascular fitness.

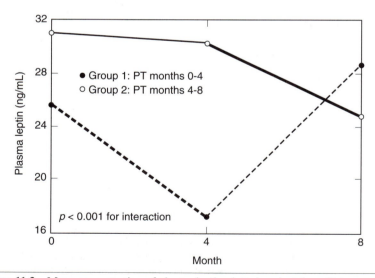

Figure 11.3   Mean concentration of plasma leptin plotted against time. Data are means for leptin over the three time points of the study.

Reprinted, by permission, from B. Gutin, L. Ramsey, P. Barbeau, W. Cannady, M. Ferguson, M. Litaker, and S. Owens, 1999, "Plasma leptin concentrations in obese children: Changes during 4-mo. periods with and without physical training," *American Journal of Clinical Nutrition* 69: 388-394. © Am. J. Clin. Nutr. American Society for Clinical Nutrition.

However, even when the energy expenditure during the exercise sessions themselves is controlled, there are reasons to suspect that higher intensity exercise may be more efficacious in reducing fatness if the intervention continues for a longer period than the typical 2 to 4 months. First, the post-exercise metabolism may be greater following higher intensity exercise; the small amount of extra energy used following each session may gradually accumulate to meaningful levels over long periods of time. Second, if high-intensity training increases cardiovascular fitness more effectively (131), then the youth would be progressively able to use up more energy in a given amount of training time, eventually leading to greater fat loss.

To our knowledge, no information from experimental studies is available concerning the volume of exercise that is optimal for prevention and treatment of childhood obesity. The National Association for Sport and Physical Education (NASPE) recommends that children obtain at least 1 to 2 hours of exercise each day (100). A theoretical analysis conducted for the World Health Organization (153, p. 125), although done for a hypothetical adult male rather than for a hypothetical child, suggests that 1 to 2 hours of walking each day would be needed to raise the average activity level into a range that would be protective against excessive weight gain. Some quantitative estimates can be made from our studies, in which each child's heart rate is calibrated against his/her energy expenditure in a multi-stage test in the laboratory, permitting an estimate of energy expenditure during the exercise sessions to be made from the average heart rate during the sessions. In our study of 7- to 10-year-olds (104), the average heart rate was 158 beats per minute (bpm), the exercise duration was 40 minutes, the energy expended was approximately 925 kJ (220 kcal) per session, and the attendance at the exercise classes was 4.4 days/week; the children reduced their percent fat by 2.2 units over the 4-month period. Thus, it seems reasonable to hypothesize that obese youths who incorporated 4 to 6 hours/week of moderate-to-vigorous exercise into their lifestyles might expect to reduce their percent fat by several units per year, a clinically meaningful amount. Although no experimental data are available concerning how much exercise is needed for prevention of juvenile obesity, it is reasonable to hypothesize that a similar level of activity would help nonobese youths, including those with a propensity to become obese, to maintain a healthy body composition.

In addition to encouraging greater amounts of moderate and vigorous exercise, it may be wise to attack the problem from the other end of the activity spectrum, i.e., sedentary behavior, as exemplified by TV watching, which has been linked to obesity (30, 47). In a clever use of alternative reinforcement approaches, Epstein et al. (35) showed that reinforcing children for reducing sedentary behaviors such as TV watching was more effective for reduction of percent fat than reinforcing them for increased physical activity or a combination of the two.

CAD/NIDDM Risk Factors

Studies of physical training and CAD/NIDDM risk factors in youths have sometimes failed to provide clear evidence that the training was effective (3, 122), leading some to conclude that regular exercise does not improve these aspects of cardiovascular health in children (121). However, a conclusive judgment on this matter may not yet be warranted (53) because of the following limitations of these studies: (1) The training dose may have been insufficient in magnitude or inadequately controlled; (2) the number of subjects may have been too small; and (3) the subjects may have been so healthy that no improvement in risk factors would be expected. For example, Alpert and Wilmore (2) concluded that training studies that used hypertensive youths were more likely to show reduced blood pressure than studies using normotensive youths.

Two studies that were carried out for relatively long periods help to clarify the actual long-term effect of regular exercise. Hansen et al. (65) reported the effect on blood pressure of adding 3 days/week of physical education to a normal 2 days/week program in children. They found that after 3 months of the intervention the groups did not differ significantly in fitness or blood pressure, whereas after 8 months the exercise groups had increased significantly in fitness and declined significantly in systolic and diastolic blood pressure. A 2-year study of obese Japanese children, which involved 7 days of aerobic exercise per week, found substantial decreases in skinfold fatness and increased HDL cholesterol after 1 and 2 years (130). Others have shown that obese adolescents who engaged in a diet plus exercise treatment reduced their fatness and other CAD/NIDDM risk factor levels more than the group that used diet alone (7, 120).

We conducted a small study of the effects of a 10-week intervention period of supervised physical training or lifestyle education in obese 7- to 11-year-old African-American girls (59). The physical training group showed a significant increase in cardiovascular fitness and a significant decrease in percent body fat, while the lifestyle education group declined significantly more in dietary energy and percent of energy intake from fat. The interventions were similarly effective in improving some CAD/ NIDDM risk factors, perhaps through different pathways, i.e., the physical training improved fitness and fatness, while the lifestyle education improved diet. Another study that compared exercise alone to diet alone in obese 9- to 12-year-olds with elevated fasting cholesterol levels found that after 6 weeks of intervention, both groups lowered their triacylglycerol, ApoA, and insulin levels, despite having had no weight loss or change in body mass index (66). Furthermore, the diet group had a large attrition rate (40%) whereas the exercise group had 0% attrition, suggesting that exercise, rather than diet, might be the better treatment for some CAD/NIDDM risk factors in children.

Figure 11.4 shows the significant three-point pattern for insulin in our study of 7- to 10-year-olds; a similar and significant pattern was seen for triacylglycerol, but not for other lipids and lipoproteins. Thus, insulin and triacylglycerol seem to be the components of the insulin-resistance syndrome that are most sensitive to changes in adiposity caused by exercise.

We also measured some other components of cardiovascular health, including cardiac autonomic traffic. Higher levels of parasympathetic activity, which are indicative of better cardiovascular and metabolic health, are associated with greater amounts of beat-to-beat variability, i.e., heart period variability (also called heart rate variability) (73) in electrocardiogram intervals. A preliminary report from the first half of the subjects who completed the first 4-month period (57) showed that the training led to increased parasympathetic activity. Figure 11.5 shows the significant three-point pattern for all the subjects of the root mean square of successive

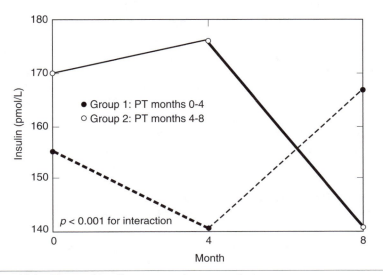

Figure 11.4 Means of insulin concentration versus time.

Reprinted, by permission, from M.A. Ferguson, B. Gutin, N.-A. Le, W. Karp, M. Litaker, M. Humphries, T. Okuyama, S. Riggs, and S. Owens, 1999, "Effects of exercise training and its cessation on components of the insulin resistance syndrome in obese children," *International Journal of Obesity* 22: 889-895.

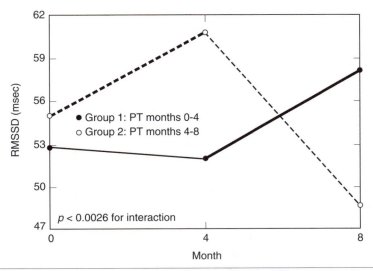

Figure 11.5 Means for the root mean square of successive differences (RMSSD) of heart period variability plotted against time.

Reprinted, by permission, from B. Gutin and S. Owens, 1999, "Role of exercise intervention in improving body fat distribution and risk profile in children," *American Journal of Human Biology* 11: 237-247. Copyright © 1999, Wiley-Liss, Inc., a subsidiary of John Wiley & Sons, Inc.

differences (RMSSD), an index of parasympathetic activity; values increased for both groups during the periods of training compared to the periods of no training (52).

Thus, the physical training, without any dietary intervention, had the favorable effect of improving total body composition while preventing the natural increase in visceral adipose tissue that occurs as children mature, as well as improving some other obesity-associated aspects of cardiovascular health. However, some other obesity-associated factors (e.g. hemodynamic and left ventricular parameters) were not significantly influenced by the training, perhaps because the changes in body composition produced in 4 months were relatively modest in magnitude.

Directions for Future Research

The determination of the optimal exercise dose for body composition and other aspects of health is of primary importance. Relatively little is known about how much, what modality, or what intensity of exercise is most efficacious in altering cardiovascular health in people of any age, and especially little is known about this matter in children. The same can be said about visceral adipose tissue. It is conceivable that adult guidelines for exercise prescription are inadequate for children. For example, 30 minutes per day of moderate exercise (147) may not be enough to compensate for the large reductions in activity and fitness brought about by factors such as the reduction of school physical education, the danger of spending time outdoors during nonschool hours, and the evolving dominance of sedentary activities such as television, video games, and computers.

In deciding how to alter lifestyle to improve body composition, it is necessary to consider both exercise and diet together. Therefore, efficacy studies of interactions between different training doses and different types of diets are needed to provide a strong scientific foundation for school or community interventions. There are several lines of evidence supporting the hypothesis that a combination of exercise and a low-fat (high-carbohydrate) diet may be optimal for reduction of fatness.

An issue related to the concept of exercise intensity, about which very little is known, is the role of resistance training in weight control programs. It does seem clear that resistance training can help youths to improve their muscular strength (142) and endurance (109), thereby helping their performance and future participation in sports and dance activities. Because strength training usually involves taking the muscle group close to, or all the way to, momentary failure, it is high in intensity even though the overall energy expended may not be as high as in aerobic training sessions. The high intensity of the exercise, and the resulting increases in muscle (i.e., lean) tissue, may lead to favorable changes in body composition. Moreover, the intense muscular activity may enhance insulin action and improve CAD/NIDDM risk status, independent of body fatness or cardiovascular fitness (68).

Relatively little is known about how gender, ethnicity, and pubertal status interact with exercise and diet to influence body composition and associated CAD/NIDDM risk factors. Such information would help in the design of interventions targeted to the appropriate groups and outcome variables.

Summary

With respect to the etiology and treatment of obesity, it is likely that exercise acts synergistically with low-fat diets to prevent obesity and to treat it effectively. Controlled interventions that have included both of these factors have generally been more successful than either alone. This may be because the low-fat and low-energy diets immediately lead to a substantial energy deficit, while the exercise adds somewhat to the energy deficit, preserves fat-free mass and facilitates long-term retention of the weight loss. Although it is clear that visceral adipose tissue is associated with the CAD/NIDDM risk syndrome starting in childhood, almost nothing is known about its etiology or the degree to which various combinations of exercise or diet can influence it during childhood.

A recent large-scale study in Thailand (97) casts an interesting light on the roles of activity and diet in preventing childhood obesity. It was found that for young children a family history of obesity and a low exercise level were the main risk factors for later development of childhood obesity. The authors concluded that increasing exercise was clearly the most appropriate intervention to prevent obesity in Thailand, because the risk of nutritional deficiency is still high and a mass intervention program focusing on reduction of energy intake might make the situation worse. The World Health Organization (153) has found that this condition of increasing obesity coexisting with undernutrition is seen in many countries.

It is becoming clear that already in young children, excess total body and visceral adiposity are linked to unfavorable levels of preclinical markers for "adult" diseases such as CAD and NIDDM. Data from cross-sectional and nonexperimental longitudinal studies suggest that exercise and fitness during childhood have favorable effects on these preclinical markers, with much of the effect mediated by the influence of exercise on body composition. However, randomized trials of controlled physical training over periods of up to 4 months have produced mixed results, with some failing to provide evidence of favorable changes in these preclinical markers.

One way to reconcile these somewhat discrepant results is to postulate that the nonexperimental studies allow the uncovering of causal relations that emerge after several years of exposure to the detrimental effects of inactivity or excess fatness, while short-term experimental studies do not produce an increase in activity over a long enough period for favorable changes in the preclinical markers to take place. This concept is supported by the few controlled intervention studies that were carried out over relatively long periods of 8 months to 2 years (65, 130) and the studies from our laboratory mentioned above that clearly documented the substantial dose of training imparted to the youths over several-month periods. The other possibility is that exercise has a limited or negligible causal effect on preclinical markers during childhood (121), and the relationships found in the nonexperimental studies are due to the common influence of some other factor, such as a genetic predisposition to be active/fit/lean on the one hand, and to have favorable levels of the risk markers on the other. Longer-term experimental studies are needed to clarify the true impact of exercise.

In considering the relationship of exercise to health in youths, it is necessary to note that exercise poses some risks in the form of sports injuries. In fact, Aaron and LaPorte (1) suggest that many of the purported benefits of regular exercise in youth are myths ("an active child is a healthy child"; "an active child will be an active adult"), and that on balance the risk of injury and related medical costs outweigh the potential benefits of exercise for most youth. However, as we have seen in this chapter, the evidence of a beneficial effect of exercise on the body composition and health of children is rapidly accumulating. In addition, longitudinal projects such as the Amsterdam Study (74) have shown tracking from adolescence into adulthood of variables such as physical activity, body composition, and cardiovascular risk factors.

The secular trend toward greater child and adolescent obesity over the last two decades (39, 143) suggests that past public health efforts to improve activity and diet behaviors have been overcome by societal trends toward less activity and/or greater intake of high-fat foods. To date, programs that have taken a public health approach in school and community settings have had only modest success in altering diet, exercise, and preclinical markers of adult diseases (82). However, the recent Surgeon General's report (147) shows that the scientific evidence linking physical activity to health is convincing. Perhaps this increasing recognition of the importance of physical activity will energize funding agencies to sponsor more research on these topics, providing an improved foundation for the development of more effective interventions. In the meantime, there is enough known to encourage parents, health professionals, teachers, and legislatures to implement measures to increase physical activity of our children, thereby contributing to their present and future health.

References

1. Aaron, D.J., and R.E. Laporte. 1997. Physical activity, adolescence, and health: an epidemiological perspective. In *Exercise and Sport Sciences Reviews,* ed. J. Holloszy, 25:391-405. Baltimore: Williams & Wilkins.
2. Alpert, B.S., and J.H. Wilmore. 1994. Physical activity and blood pressure in adolescents. *Pediatric Exercise Science* 6:361-380.
3. Armstrong, N., and B. Simons-Morton. 1994. Physical activity and blood lipids in adolescents. *Pediatric Exercise Science* 6:381-405.
4. Asayama K., H. Hayashibe, K. Dobashi, Y. Kawada, and S. Nakazawa. 1995. Relationships between biochemical abnormalities and anthropometric indi-

ces of overweight, adiposity and body fat distribution in Japanese elementary school children. *International Journal of Obesity* 19:253-259.

5. Bandini, L., D. Schoeller, H. Cyr, and W. Dietz. 1990. Validity of reported energy intake in obese and nonobese adolescents. *American Journal of Clinical Nutrition* 52:421-425.

6. Bar-Or, O., J. Foreyt, C. Bouchard, K.D. Brownell, W.H. Dietz, E. Ravussin, A.D. Salbe, S. Schwenger, S. St. Jeor, and B. Torun. 1998. Physical activity, genetic, and nutritional considerations in childhood weight management. *Medicine and Science in Sports and Exercise* 30:2-10.

7. Becque, M., V. Katch, A. Rocchini, C. Marks, and C. Moorehead. 1988. Coronary risk incidence of obese adolescents: reduction by exercise plus diet interventions. *Pediatrics* 81:605-612.

8. Berenson, G.S., S.R. Srinivasan, W. Bao, W.P. Newman III, R.E. Tracy, and W.A. Wattigney. 1998. Association between multiple cardiovascular risk factors and atherosclerosis in children and young adults. *New England Journal of Medicine* 338:1650-1656.

9. Bergstrom, E., O. Hernell, and L.A. Persson. 1997. Endurance running performance in relation to cardiovascular risk indicators in adolescents. *International Journal of Sports Medicine* 18:300-307.

10. Blaak, E.E., K.R. Westerterp, O. Bar-Or, L.J.M. Wouters, and W.H.M. Saris. 1992. Total energy expenditure and spontaneous activity in relation to training in obese boys. *American Journal of Clinical Nutrition* 55:777-782.

11. Boreham, C.A., J. Twisk, M.J. Savage, G.W. Cran, and J.J. Strain. 1997. Physical activity, sports participation, and risk factors in adolescents. *Medicine and Science in Sports and Exercise* 29:788-793.

12. Bouchard, C. 1994. Genetics of obesity: overview and research directions. In *The Genetics of Obesity*, ed. C. Bouchard, 223-266. Boca Raton:CRC Press.

13. Bougnères, P., C. Le Stunff, C. Pecqueur, E. Pinglier, P. Adnot, and D. Ricquier. 1997. In vivo resistance of lipolysis to epinephrine. *Journal of Clinical Investigation* 99:2568-2573.

14. Brambilla, P., P. Manzoni, S. Sironi, P. Simone, A. Del Maschio, B. di Natale, and G. Chiumello. 1994. Peripheral and abdominal adiposity in childhood obesity. *International Journal of Obesity* 18:795-800.

15. Bratteby, L.A., B. Sandhagen, M. Lotborn, and G. Samuelson. 1997. Daily energy-expenditure and physical-activity assessed by an activity diary in 374 randomly selected 15-year-old adolescents. *European Journal of Clinical Nutrition* 51:592-600.

16. Caprio, S., L.D. Hyman, S. McCarthy, R. Lange, M. Bronson, and W.V. Tamborlane. 1996. Fat distribution and cardiovascular risk factors in obese adolescent girls: importance of the intraabdominal fat depot. *American Journal of Clinical Nutrition* 64:12-17.

17. Caprio, S., W.V. Tamborlane, D. Silver, C. Robinson, R. Leibel, S. McCarthy, A. Grozman, A. Belous, D. Maggs, and R.S. Sherwin. 1996. Hyperleptinemia: an early sign of juvenile obesity. Relations to body fat depots and insulin concentrations. *American Journal of Physiology* 271:E626-E630.

18. Celermajer, D.S. 1997. Endothelial dysfunction: Does it matter? Is it reversible? *Journal of the American College of Cardiology* 30:325-333.
19. Chu, N.F., E.B. Rimm, D.J. Wang, H.S. Liou, and S.M. Shieh. 1998. Clustering of cardiovascular disease risk factors among obese schoolchildren: the Taipei Children Heart Study. *American Journal of Clinical Nutrition* 67:1141-1146.
20. Conway, J.M., S.Z. Yanovski, N.A. Avila, and V.S. Hubbard. 1995. Visceral adipose tissue differences in black and white women. *American Journal of Clinical Nutrition* 61:765-771.
21. Craig, S.B., L.G. Bandini, A.H. Lichtenstein, E.J. Schaefer, and W.H. Dietz. 1996. The impact of physical activity on lipids, lipoproteins, and blood pressure in preadolescent girls. *Pediatrics* 98:389-395.
22. Davies, P.S. 1997. Diet composition and body mass index in preschool children. *European Journal of Clinical Nutrition* 51:443-448.
23. Davies, P., J. Gregory, and A. White. 1995. Physical activity and body fatness in pre-school children. *International Journal of Obesity* 19:6-11.
24. Davies, P.S., J.C. Wells, C.A. Fieldhouse, J.M. Day, and A. Lucas. 1995. Parental body composition and infant energy expenditure. *American Journal of Clinical Nutrition* 61:1026-1029.
25. Deheeger, M., M.F. Rolland-Cachera, and A.M. Fontvieille. 1997. Physical activity and body composition in 10-year-old French children: linkages with nutritional intake. *International Journal of Obesity* 21:372-379.
26. DeLany, J.P. 1998. Role of energy expenditure in the development of pediatric obesity. *American Journal of Clinical Nutrition* 68(suppl.):950S-955S.
27. Després, J-P. 1997. Visceral obesity, insulin resistance, and dyslipidemia: contribution of endurance exercise training to the treatment of the plurimetabolic syndrome. In *Exercise and Sports Sciences Reviews*, ed. J. Holloszy, 25:271-300. Baltimore: Williams & Wilkins.
28. Devereux, R., F. de Simone, A. Ganau, and M. Roman. 1994. Left ventricular hypertrophy and geometric remodeling in hypertension: stimuli, functional consequences and prognostic implications. *Journal of Hypertension* 12(suppl.):S117-S127.
29. Dietz, W.H. 1998. Does energy expenditure affect changes in body fat in children? *American Journal of Clinical Nutrition* 67:190-191.
30. Dietz, W.H., and S.L. Gortmaker. 1985. Do we fatten our children at the television set? Obesity and television viewing in children and adolescents. *Pediatrics* 75:807-812.
31. Dwyer, J.T., E.J. Stone, M. Yang, H. Feldman, L.S. Webber, A. Must, C.L. Perry, P.R. Nader, and G.S. Parcel. 1998. Predictors of overweight and overfatness in a multiethnic pediatric population. *American Journal of Clinical Nutrition* 67:602-610.
32. Eck, L.H., R.C. Klesges, C.L. Hanson, and D. Slawson. 1992. Children at familial risk for obesity: an examination of dietary intake, physical activity and weight status. *International Journal of Obesity* 16:71-78.

33. Ellis, K., R. Shypailo, J. Pratt, and W. Pond. 1994. Accuracy of dual-energy x-ray absorptiometry for body-composition measurements in children. *American Journal of Clinical Nutrition* 60:660-665.
34. Epstein, L.H., A.M. Valoski, M.A. Kalarchian, and J. McCurley. 1995. Do children lose and maintain weight easier than adults: A comparison of child and parent weight changes from six months to ten years. *Obesity Research* 3:411-417.
35. Epstein, L.H., A.M. Valoski, L.S. Vara, J. McCurley, L. Wisniewski, M.A. Kalarchian, K.R. Klein, and L.R. Shrager. 1995. Effects of decreasing sedentary behavior and increasing activity on weight change in obese children. *Health Psychology* 2:109-115.
36. Ferguson, M.A., B. Gutin, N.-A. Le, W. Karp, M. Litaker, M. Humphries, T. Okuyama, S. Riggs, and S. Owens. 1999. Effects of exercise training and its cessation on components of the insulin resistance syndrome in obese children. *International Journal of Obesity* 22:889-895.
37. Ferguson, M.A., B. Gutin, S. Owens, M. Litaker, R. Tracy, and J. Allison. 1998. Fat distribution and hemostatic measures in obese children. *American Journal of Clinical Nutrition* 67:1136-1140.
38. Fox, K., D. Peters, N. Armstrong, P. Sharpe, and M. Bell. 1993. Abdominal fat deposition in 11-year-old children. *International Journal of Obesity* 17:11-16.
39. Freedman, D.S., S.R. Srinivasan, R.A. Valdez, D.F. Williamson, and G.S. Berenson. 1997. Secular increases in relative weight and adiposity among children over two decades: The Bogalusa Study. *Pediatrics* 99:420-426.
40. Freedman, D.S. 1995. The importance of body fat distribution in early life. *American Journal of the Medical Sciences* 310(suppl.):S72-S76.
41. Gazzaniga, J.M., and T.L. Burns. 1993. Relationship between diet composition and body fatness, with adjustment for resting energy expenditure and physical activity, in prepubescent children. *American Journal of Clinical Nutrition* 58:21-28.
42. Goran, M.I., and M. Sun. 1998. Total energy expenditure and physical activity in prepubertal children: recent advances based on the application of the doubly labeled water method. *American Journal of Clinical Nutrition* 68(suppl.):944S-949S.
43. Goran, M.I., R. Shewchuk, B.A. Gower, T.R. Nagy, W.H. Carpenter, and R.K. Johnson. 1998. Longitudinal changes in fatness in white children: no effect of childhood expenditure. *American Journal of Clinical Nutrition* 67:309-316.
44. Goran, M.I., T.R. Nagy, M.S. Treuth, C. Trowbridge, C. Dezenberg, A. McGloin, and B.A. Gower. 1997. Visceral fat in white and African-American prepubertal children. *American Journal of Clinical Nutrition* 65:1703-1708.
45. Goran, M.I., G. Hunter, and R. Johnson. 1996. Physical activity related energy expenditure and fat mass in young children. *International Journal of Obesity* 20:1-8.

46. Goran, M.I., M. Kaskoun, and W.P. Shuman. 1995. Intra-abdominal adipose tissue in young children. *International Journal of Obesity* 19:279-283.

47. Gortmaker, S.L., A. Must, A.M. Sobol, K. Peterson, G.A. Colditz, and W.H. Dietz. 1996. Television viewing as a cause of increasing obesity among children in the United States, 1986-1990. *Archives of Pediatric and Adolescent Medicine* 150:356-362.

48. Gower, B.A., T.R. Nagy, C.A. Trowbridge, C. Dezenberg, and M.I. Goran. 1998. Fat distribution and insulin response in prepubertal African American and white children. *American Journal of Clinical Nutrition* 67:821-827.

49. Griffiths, M., and P.R. Payne. 1976. Energy expenditure in small children of obese and non-obese parents. *Nature* 260:698-700.

50. Griffiths, M., P.R. Payne, A.J. Stunkard, J.P.W. Rivers, and M. Cox. 1990. Metabolic rate and physical development in children at risk of obesity. *Lancet* 336:76-78.

51. Guillaume, M., L. Lapidus, P. Bjorntorp, and A. Lambert. 1997. Physical activity, obesity, and cardiovascular risk factors in children. The Belgian Luxembourg Child Study II. *Obesity Research* 5:549-556.

52. Gutin, B., and S. Owens. 1999. Role of exercise intervention in improving body fat distribution and risk profile in children. *American Journal of Human Biology* 11:237-247.

53. Gutin, B., and S. Owens. 1996. Is there a scientific rationale supporting the value of exercise for the present and future cardiovascular health of children? *Pediatric Exercise Science* 8:294-302.

54. Gutin, B., S. Owens, T. Okuyama, S. Riggs, M. Ferguson, and M. Litaker. 1999. Effect of physical training and its cessation upon percent fat and bone density of obese children. *Obesity Research* 7:208-214.

55. Gutin B., L. Ramsey, P. Barbeau, W. Cannady, M. Ferguson, M. Litaker, and S. Owens. 1999. Plasma leptin concentrations in obese children: changes during 4-mo periods with and without physical training. *American Journal of Clinical Nutrition* 69:388-394.

56. Gutin, B., F. Treiber, S. Owens, and G. Mensah. 1998. Relations of body composition to left ventricular geometry and function in children. *Journal of Pediatrics* 132:1023-1027.

57. Gutin, B., S. Owens, G. Slavens, S. Riggs, and F. Treiber. 1997. Effect of physical training on heart period variability in obese children. *Journal of Pediatrics* 130:938-943.

58. Gutin, B., S. Owens, F. Treiber, S. Islam, W. Karp, and G. Slavens. 1997. Weight-independent cardiovascular fitness and coronary risk factors. *Archives of Pediatric and Adolescent Medicine* 151:462-465.

59. Gutin, B., N. Cucuzzo, S. Islam, C. Smith, and M.E. Stachura. 1996. Physical training, lifestyle education, and coronary risk factors in obese girls. *Medicine and Science in Sports and Exercise* 28:19-23.

60. Gutin, B., M. Litaker, S. Islam, T. Manos, C. Smith, and F. Treiber. 1996. Body composition measurement in 9-11-y-old children by dual x-ray

absorptiometry, skinfold thickness measurements, and bioimpedance analysis. *American Journal of Clinical Nutrition* 25:287-292.

61. Gutin, B., N. Cucuzzo, S. Islam, C. Smith, and M.E. Stachura. 1995. Physical training improves body composition of black obese 7- to 11-year-old girls. *Obesity Research* 3:305-312.

62. Gutin, B., S. Islam, T. Manos, N. Cucuzzo, C. Smith, and M. Stachura. 1994. Relation of percentage of body fat and maximal aerobic capacity to risk factors for atherosclerosis and diabetes in black and white seven- to eleven-year-old children. *Journal of Pediatrics* 125:847-852.

63. Gutin, B., C. Basch, S. Shea, I. Contento, M. DeLozier, J. Rips, M. Irgoyen, and P. Zybert. 1990. Blood pressure, fitness, and fatness in 5- and 6-year-old children. *Journal of the American Medical Association* 264:1123-1127.

64. Gutin, B., N. Mayers, J. Levy, and M. Herman. 1988. Physiologic and echocardiographic studies of age-group runners. In *Competitive Sports for Children and Youths,* eds. E. Brown and C. Branta, 117-128. Champaign, IL: Human Kinetics.

65. Hansen, H., K. Froberg, N. Hyldebrandt, and J. Nielson. 1991. A controlled study of eight months of physical training and reduction of blood pressure in children: the Odense schoolchild study. *International Journal of Obesity* 18:795-800.

66. Hardin, D.S., J.D. Hebert, T. Bayden, M. Dehart, and L. Mazur. 1997. Treatment of childhood syndrome X. *Pediatrics* 100:E51-E54.

67. Hofman, A., and H.J. Walter. 1989. The association between physical fitness and cardiovascular risk factors in children in a five-year follow-up study. *International Journal of Epidemiology* 18:830-835.

68. Hurley, B.F., J.M. Hagberg, A.P. Goldberg, D.R. Seals, A.A. Ehsani, R.E. Brennan, and J.O. Holloszy. 1988. Resistive training can reduce coronary risk factors without altering VO_2max or percent body fat. *Medicine and Science in Sports and Exercise* 20:150-154.

69. James, W.P.T. 1996. Chapter discussion. In *The Origins and Consequences of Obesity,* eds. D.J. Chadwick and G. Cardew, 252-253. Chichester, NY: Wiley.

70. Janz, K.F., and L.T. Mahoney. 1997. Three-year follow-up of changes in aerobic fitness during puberty: the Muscatine study. *Research Quarterly for Exercise and Sport* 68:1-9.

71. Kann, L., S.A. Kinchen, B.I. Williams, J.G. Ross, R. Lowry, C.V. Hill, J.A. Grunbaum, P.S. Blumson, J.L. Collins, and L.J. Kolbe. 1998. Youth risk behavior surveillance—United States, 1997. In *CDC Surveillance Summaries.* Morbidity and Mortality Weekly Report. 47:23-24.

72. Kaplan, A.S., B.S. Zemel, and V.A. Stallings. 1996. Differences in resting energy expenditure in prepubertal black children and white children. *Journal of Pediatrics* 129:643-647.

73. Kautzner, J., and A.J. Camm. 1997. Clinical relevance of heart rate variability. *Clinical Cardiology* 20:162-168.

74. Kemper, H., and W. van Mechelen. 1995. Physical fitness and the relationship to physical activity. In *The Amsterdam Growth Study*, ed. H. Kemper, 174-188. Champaign, IL: Human Kinetics.

75. Klesges, R.C., L.M. Klesges, L.H. Eck, and M.L. Shelton. 1995. A longitudinal analysis of accelerated weight gain in preschool children. *Pediatrics* 95:126-130.

76. Kumanyika, S. 1993. Ethnicity and obesity development in children. *Annals of the New York Academy of Sciences* 699:81-92.

77. Kuntzleman, C.T. 1993. Childhood fitness: what is happening? What needs to be done? *Preventive Medicine* 22:520-532.

78. Kwiterovich, P.O. 1993. Prevention of coronary disease starting in childhood: what risk factors should be identified and treated? *Coronary Artery Disease* 4:611-630.

79. Lahlou, N., P. Landais, D. De Boissieu, and P.F. Bougnères. 1997. Circulating leptin in normal children and during the dynamic phase of juvenile obesity: relation to body fatness, energy metabolism, caloric intake, and sexual dimorphism. *Diabetes* 46:989-993.

80. Le Stunff, C., and P.F. Bougnères. 1993. Time course of increased lipid and decreased glucose oxidation during early phase of childhood obesity. *Diabetes* 42:1010-1016.

81. Levine, A.S., and C.J. Billington. 1998. Do circulating leptin concentrations reflect body adiposity or energy flux? *American Journal of Clinical Nutrition* 68:761-762.

82. Luepker, R.V., C.L. Perry, S.M. McKinlay, P.R. Nader, G.S. Parcel, E.J. Stone, L.S. Webber, J.P. Elder, H.A. Feldman, C.C. Johnson, S.H. Kelder, and M.W. Wu. 1996. Outcomes of a field trial to improve children's dietary patterns and physical activity. *Journal of the American Medical Association* 275:768-776.

83. Maffeis, C., M. Zaffanello, and Y. Schutz. 1997. Relationship between physical inactivity and adiposity in prepubertal boys. *Journal of Pediatrics* 131:288-292.

84. Maffeis, C., L. Pinelli, and Y. Schutz. 1996. Fat intake and adiposity in 8- to 11-year-old children. *International Journal of Obesity* 20:170-174.

85. Maffeis, C., M. Zaffanello, L. Pinelli, and Y. Schutz. 1996. Total energy expenditure and patterns of activity in 8–10-year-old obese and nonobese children. *Journal of Pediatric Gastroenterology and Nutrition* 23:256-261.

86. Maffeis, C., Y. Schutz, R. Micciolo, L. Zoccante, and L. Pinelli. 1993. Resting metabolic rate in six- to ten-year-old obese and nonobese children. *Journal of Pediatrics* 122:556-562.

87. Maffeis, C., Y. Schutz, and L. Pinelli. 1992. Effect of weight loss on resting energy expenditure in obese prepubertal children. *International Journal of Obesity* 16:41-47.

88. Mayers, N., and B. Gutin. 1979. Physiological characteristics of elite prepubertal cross country runners. *Medicine and Science in Sports* 11:172-176.

89. McGill, H.C. Jr., J.P. Strong, R.E. Tracy, C.A. McMahan, and M.C. Oalmann. 1995. Relation of a postmortem renal index of hypertension to atherosclerosis in youth. The Pathobiological Determinants of Atherosclerosis in Youth (PDAY) Research Group. *Arteriosclerosis, Thrombosis & Vascular Biology* 15:2222-2228.

90. McMurray, R.G., J.S. Harrell, A.A. Levine, and S.A. Gansky. 1995. Childhood obesity elevates blood pressure and total cholesterol independent of physical activity. *International Journal of Obesity* 19:881-886.

91. Melby, C.L., R.C. Ho, and J.O. Hill. 2000. Assessment of energy expenditure. In *Physical activity and obesity,* ed. C. Bouchard, chapter in this volume, Champaign, IL: Human Kinetics.

92. Meredith, I.T., A.C. Yeung, and F.F. Weidinger. 1993. Role of impaired endothelium-dependent vasodilation in ischemic manifestations of coronary artery disease. *Circulation* 87:56-66.

93. Miller, W.C., A.K. Linderman, J. Wallace, and M. Niederpruem. 1990. Diet composition, energy intake, and exercise in relation to body fat in men and women. *American Journal of Clinical Nutrition* 52:426-430.

94. Moore, L.L., U.D.T. Nguyen, K.J. Rothman, L.A. Cupples, and R.C. Ellison. 1995. Preschool physical activity level and change in body fatness in young children. *American Journal of Epidemiology* 142:982-988.

95. Morris, F.L., G.A. Naughton, J.L. Gibbs, J.S. Carlson, and J.D. Wark. 1997. Prospective ten-month exercise intervention in premenarcheal girls: positive effects on bone and lean mass. *Journal of Bone Mineral Research* 12:1453-1462.

96. Morrison, J.A., M.P. Alfaro, P. Khoury, B.B. Thornton, and S.R. Daniels. 1996. Determinants of resting energy expenditure in young black girls and young white girls. *Journal of Pediatrics* 129:637-642.

97. Mo-suwan, L., and A.F. Geater. 1996. Risk factors for childhood obesity in a transitional society in Thailand. *International Journal of Obesity* 20:697-703.

98. Muecke, L., B. Simons-Morton, I.W. Huang, and G. Parcel. 1992. Is childhood obesity associated with high-fat foods and low physical activity? *Journal of School Health* 62:19-23.

99. Nagy, T.R., B.A. Gower, C.A. Trowbridge, C. Dezenberg, R. M. Shewchuk, and M.I. Goran. 1997. Effects of gender, ethnicity, body composition, and fat distribution on serum leptin concentrations in children. *Journal of Clinical Endocrinology and Metabolism* 82:2148-2152.

100. National Association for Sport and Physical Education (NASPE). 1998. Physical activity guidelines for pre-adolescent children.

101. Nguyen, V.T., D.E. Larson, R.K. Johnson, and M.I. Goran. 1996. Fat intake and adiposity in children of lean and obese parents. *American Journal of Clinical Nutrition* 63:507-513.

102. Obarzanek, E., G. Schrieber, P. Crawford, S. Goldman, P. Barrier, M. Frederick, and E. Lakatos. 1994. Energy intake and physical activity in

relation to indexes of body fat: the National Heart, Lung, and Blood Institute Growth and Health Study. *American Journal of Clinical Nutrition* 60:15-22.

103. Oscai, L. 1989. Exercise and obesity: emphasis on animal models. In *Perspectives in Exercise Science and Sports Medicine: Youth, Exercise and Sport,* eds. C. Gisolfi and D. Lamb, 2:273-292. Indianapolis: Benchmark Press.

104. Owens, S., B. Gutin, J. Allison, S. Riggs, M. Ferguson, M. Litaker, and W. Thompson. 1999. Effect of physical training on total and visceral fat in obese children. *Medicine and Science in Sports and Exercise* 31:143-148.

105. Owens, S., B. Gutin, M. Ferguson, J. Allison, W. Karp, and N.-A. Le. 1998. Visceral adipose tissue and cardiovascular risk factors in obese children. *Journal of Pediatrics* 133:41-45.

106. Panchenko, E., A. Dobrovolsky, K. Davletov, E. Titaeva, A. Kravets, J. Podinovskaya, and Y. Karpov. 1995. D-Dimer and fibrinolysis in patients with various degrees of atherosclerosis. *European Heart Journal* 16:38-42.

107. Pate, R.R., S.G. Trost, G.M. Felton, D.S. Ward, M. Dowda, and R. Saunders. 1997. Correlates of physical-activity behavior in rural youth. *Research Quarterly for Exercise and Sport* 68:241-248.

108. Pate, R.R., B.J. Long, and G. Heath. 1994. Descriptive epidemiology of physical activity in adolescents. *Pediatric Exercise Science* 6:434-447.

109. Payne, V.G., J.R. Morrow, L. Johnson, and S.N. Dalton. 1997. Resistance training in children and youth: a meta-analysis. *Research Quarterly in Exercise and Sport* 68:80-88.

110. Pinhas-Hamiel, O., L.M. Dolan, S.R. Daniels, D. Standford, P.R. Khoury, and P. Zeitler. 1996. Increased incidence of non-insulin-dependent diabetes mellitus among adolescents. *Journal of Pediatrics* 128:608-615.

111. Pi-Sunyer, F. 1991. Health implications of obesity. *American Journal of Clinical Nutrition* 53(suppl.):1595S-1603S.

112. Pivarnik, J.M., M.S. Bray, A.C. Hergenroeder, R.B. Hill, and W.W. Wong. 1995. Ethnicity affects aerobic fitness in U.S. adolescent girls. *Medicine and Science in Sports and Exercise* 27:1635-1638.

113. Prentice, A.M., G.R. Goldberg, P.R. Murgatroyd, and T.J. Cole. 1996. Physical activity and obesity: problems in correcting expenditure for body size. *International Journal of Obesity* 20:688-691.

114. Raitakari, O.T., S. Taimela, K.V. Porkka, R. Telama, I. Valimaki, H.K. Akerblom, and J.S. Viikari. 1997. Associations between physical-activity and risk-factors for coronary heart-disease—The cardiovascular risk in young Finns study. *Medicine and Science in Sports and Exercise* 29:1055-1061.

115. Ravussin, E., and B.A. Swinburn. 1992. Pathophysiology of obesity. *Lancet* 340:404-408.

116. Reaven, G. 1988. Role of insulin resistance in human disease. *Diabetes* 37:1595-1607.

117. Reaven P., P.R. Nader, C. Berry, and T. Holy. 1998. Cardiovascular disease insulin risk in Mexican-American and Anglo-American children and mothers. *Pediatrics* 101:E121-E127.

118. Reynolds, K., J. Killen, S. Beyson, D. Maron, C. Taylor, N. Maccoby, and J. Farquhar. 1990. Psychosocial predictors of physical activity in adolescents. *Preventive Medicine* 19:541-551.
119. Roberts, S., J. Savage, W. Coward, B. Chew, and A. Lucus. 1988. Energy expenditure and intake in infants born to lean and overweight mothers. *New England Journal of Medicine* 318:461-466.
120. Rocchini, A., V. Katch, A. Schork, and R. Kelch. 1987. Insulin and blood pressure during weight loss in obese adolescents. *Hypertension* 10:267-273.
121. Rowland, T. 1996. Is there a scientific rationale supporting the value of exercise for the present and future cardiovascular health of children? The con argument. *Pediatric Exercise Science* 8:303-309.
122. Rowland, T.W., L. Martel, P. Vanderburgh, T. Marros, and N. Charkardian. 1996. The influence of short-term aerobic training on blood lipids in healthy 10-12 year old children. *International Journal of Sports Medicine* 17:487-492.
123. Rowland, T.W., V.B. Unnithan, N.G. MacFarlane, N.G. Gibson, and J.Y. Paton. 1994. Clinical manifestations of the 'athlete's heart' in prepubertal male runners. *International Journal of Sports Medicine* 15:515-519.
124. Salbe, A.D., and E. Ravussin. 2000. The determinants of obesity. In *Physical activity and obesity*, ed. C. Bouchard, chapter in this volume, Champaign, IL: Human Kinetics.
125. Salbe, A.D., A.M. Fontvieille, I.T. Harper, and E. Ravussin. 1997. Low levels of physical activity in 5-year-old children. *Journal of Pediatrics* 131:423-429.
126. Salbe, A.D., M. Nicolson, and E. Ravussin. 1997. Total energy expenditure and the level of physical activity correlate with plasma leptin concentrations in five-year-old children. *Journal of Clinical Investigation* 99:592-595.
127. Sallis, J.F., B.G. Simons-Morton, E.J. Stone, C.B. Corbin, L.H. Epstein, N. Faucette, R.J. Iannotti, J.D. Killen, R.C. Klesges, C.K. Petray, T.W. Rowland, and W.C. Taylor. 1996. Determinants of physical activity and interventions in youth. *Medicine and Science in Sports and Exercise* 24(suppl.):S248-S257.
128. Sallo, M., and R. Silla. 1997. Physical activity with moderate to vigorous intensity in preschool and first-grade school children. *Pediatric Exercise Science* 9:44-54.
129. Saris, W.H. 1995. Effects of energy restriction and exercise on the sympathetic nervous system. *International Journal of Obesity* 19(suppl.):S17-S23.
130. Sasaki, J., M. Shindo, H. Tanaka, M. Ando, and K. Arakawa. 1987. A long-term aerobic exercise program decreases the obesity index and increases the high density lipoprotein cholesterol concentration in obese children. *International Journal of Obesity* 11:339-345.
131. Savage, M.P., M.M. Petratis, W.H. Thomson, K. Berg, J.L. Smith, and S.P. Sady. 1986. Exercise training effects on serum lipids of prepubescent boys and adult men. *Medicine and Science in Sports and Exercise* 18:197-204.

132. Schneider, D.J., and B.E. Sobel. 1991. Augmentation of synthesis of plasminogen activator inhibitor type-1 by insulin and insulin-like growth factor type-1: implications for vascular disease in hyperinsulinemic states. *Proceedings of the National Academy of Sciences USA* 88:9959-9963.

133. Schwingshandl, J., and M. Borkenstein. 1995. Changes in lean body mass in obese children during a weight reduction program: effect on short term and long term outcome. *International Journal of Obesity* 19:752-755.

134. Shea, S., C. Basch, B. Gutin, A. Stein, I. Contento, M. Irigoyen, and P. Zybert. 1994. The rate of increase in blood pressure in children 5 years of age is related to changes in aerobic fitness and body mass index. *Pediatrics* 94:456-470.

135. Shimomura I., T. Funahashi, M. Takahashi, K. Maeda, K. Kotani, T. Nakamura, S. Yamashita, M. Miura, Y. Fukuda, K. Takemura, K. Tokunaga, and Y. Matsuzawa. 1996. Enhanced expression of PAI-1 in visceral fat: possible contributor to vascular disease in obesity. *Nature Medicine* 2:800-803.

136. Simone, G. de, R.B. Devereux, G.F. Mureddu, M.J. Roman, A. Ganau, M.H. Alderman, F. Contaldo, and J.H. Laragh. 1996. Influence of obesity on left ventricular midwall mechanics in arterial hypertension. *Hypertension* 28:276-283.

137. Stampfer, M.J., R.M. Krauss, J. Ma, P.J. Blanche, L.G. Holl, F.M. Sacks, and C.H. Hennekens. 1996. A prospective study of triglyceride level, low-density lipoprotein particle diameter, and risk of myocardial infarction. *Journal of the American Medical Association* 276:882-888.

138. Suter, E., and Hawes, M. 1993. Relationship of physical activity, body fat, diet, and blood lipid profile in youths 10-15 yr. *Medicine and Science in Sports and Exercise* 25:748-754.

139. Svendsen, O.L., J. Haarbo, C. Hassager, and C. Christiansen. 1993. Accuracy of measurements of body composition by dual-energy x-ray absorptiometry in vivo. *American Journal of Clinical Nutrition* 57:605-608.

140. Treiber, F., D. Papavassiliou, B. Gutin, D. Malpass, W. Yi, S. Islam, H. Davis, and W. Strong. 1997. Determinants of endothelium-dependent femoral artery vasodilation in youth. *Psychosomatic Medicine* 59:376-381.

141. Treiber, F., F. McCaffrey, K. Pflieger, W.R. Strong, and H. Davis. 1993. Determinants of left ventricular mass in children. *American Journal of Hypertension* 6:505-513.

142. Treuth, M.S., G.R. Hunter, C. Pichon, R. Figueroa-Colon, and M.I. Goran. 1998. Fitness and energy expenditure after strength training in obese prepubertal girls. *Medicine and Science in Sports and Exercise* 30:1130-1136.

143. Troiano, R.P., K.M. Flegal, R.J. Kuczmarski, S.M. Campbell, and C.L. Johnson. 1995. Overweight prevalence and trends for children and adolescents—The National Health and Nutrition Examination Surveys, 1963 to 1991. *Archives of Pediatric and Adolescent Medicine* 149:1085-1091.

144. Trost, S.G., R.R. Pate, M. Dowda, R. Saunders, D.S. Ward, and G. Felton. 1996. Gender differences in physical activity and determinants of physical activity in rural fifth grade children. *Journal of School Health* 66:145-150.

145. Trowbridge, C.A., B.A. Gower, T.R. Nagy, G.R. Hunter, M.S. Treuth, and M.I. Goran. 1997. Maximal aerobic capacity in African-American and Caucasian children. *American Journal of Physiology* 273(4 pt 1):E809-E814.

146. Tucker, L.A., G.T. Seljaas, and R.L. Hager. 1997. Body fat percentage of children varies according to their diet composition. *Journal of the American Dietetic Association* 97:981-986.

147. U.S. Department of Health and Human Services. 1996. *Physical activity and health: a report from the Surgeon General.* Atlanta, GA: U.S. Department of Health and Human Services, Centers for Disease Control and Prevention, National Center for Chronic Disease Prevention and Health Promotion.

148. Webber, L.S., W.A. Wattigney, S.R. Srinivasan, and G.S. Berenson. 1995. Obesity studies in Bogalusa. *American Journal of Medical Science* 310(suppl.):S53-S61.

149. Wells, J.C., M. Stanley, A.S. Laidlaw, J.M.E. Day, M. Stafford, and P.S.W. Davies. 1997. Investigation of the relationship between infant temperament and later body composition. *International Journal of Obesity* 21:400-406.

150. Wells, J.C.K., M. Stanley, A.S. Laidlaw, J.M. Day, and P.S. Davies. 1996. The relationship between components of infant energy expenditure and child-hood body fatness. *International Journal of Obesity* 20:848-853.

151. Williams, B. 1994. Insulin resistance: the shape of things to come. *Lancet* 344:521-524.

152. Williamson, D.F., J. Madans, R.F. Anda, J.C. Kleinman, H.S. Kahn, and T. Byers. 1993. Recreational physical activity and ten-year weight change in a US national cohort. *International Journal of Obesity* 17:279-286.

153. World Health Organization. 1998. *Obesity: Preventing and managing the global epidemic. Report of a WHO consultation on obesity.*

154. Wurmser, H., R. Laessle, K. Jacob, S. Langhard, H. Uhl, A. Angst, A. Muller, and K.M. Pirke. 1998. Resting metabolic rate in preadolescent girls at high risk of obesity. *International Journal of Obesity* 22:793-799.

155. Yanovski, J., S. Yanovski, K. Filmer, V. Hubbard, N. Avila, and B. Lewis. 1996. Differences in body composition of black and white girls. *American Journal of Clinical Nutrition* 64:833-839.

156. Yoshizawa, S., H. Honda, N. Nakamura, K. Itoh, and N. Watanabe. 1997. Effects of an 18-month endurance run training program on maximal aerobic power in 4- to 6-year-old girls. *Pediatric Exercise Science* 9:33-43.

CHAPTER 12

Physical Activity Level and Weight Control in Adults

Andrew M. Prentice, PhD [1], and Susan A. Jebb, SRD, PhD [2]

[1] MRC International Nutrition Group, Public Health Nutrition Unit, London School of Hygiene and Tropical Medicine, London, UK and [2] MRC Human Nutrition Research, Cambridge, UK

Most people would readily accept that low levels of physical activity are one of the leading risk factors for the development of obesity and its attendant co-morbidities. However, the concrete evidence in support of this contention is surprisingly fragile, and there continues to be a considerable research effort in this direction. Some lay observers have wondered why scientists spend so much effort attempting to prove what is perfectly obvious. This criticism is only partly justified; it is important to refine our understanding of the relationship between physical activity and obesity in order to formulate the most appropriate public health measures and to generate intervention policies.

If we start by accepting the overall assumption that excessively sedentary behavior poses a risk for weight gain in adults, and that the technological revolution of the late 20th century is one of the prime causes of epidemic obesity in affluent nations (21), then the subsidiary questions we need to address include these: Where does physical inactivity rank in the hierarchy of risk factors? Are all people equally vulnerable to weight gain when inactive? Are there particular characteristics that predispose some groups to inactivity and its associated weight gain (e.g., lower social class and education level, city dwellers, females in adolescence)? Are there genes that predispose individuals or segments of society to slothfulness, or to a liking for vigorous activity (e.g., by modulating endorphin action)? Is there a threshold level of activity that will protect against weight gain and, if so, is this similar in all individuals and against all dietary backgrounds? In terms of weight gain (as distinct from coronary health) do we need to distinguish between different intensities of activity, or is it simply the cumulative energy expenditure that matters? What is the influence of physical activity in modulating the link between obesity and its recognized co-morbidites?

Some of these questions are still unanswerable using existing research findings, but the remainder will be addressed in this chapter. We start with a discussion of some of the serious limitations of the epidemiological and experimental evidence that exists so far.

Problems in Assessing the Relationship Between Physical Activity and Obesity

There are so many problems associated with the existing literature on physical activity and obesity that most rigorous systematic reviews would be forced to exclude almost all studies. We believe that this would be a self-defeating exercise and that many of the studies give useful insights as long as they are interpreted with due caution and with a clear understanding of their limitations.

The chief problem with trying to perform any secular trend analysis to identify the main etiologic determinants of the current epidemic of obesity is that physical activity has only recently emerged as a variable of interest in health research. With the exception of work on the relationship between *exercise* and coronary heart disease, which dates back to the 1950s, research on *habitual activity patterns* as possible determinants of metabolic fitness and obesity has a short history. This creates a paucity of baseline data and forces the somewhat unsatisfactory use of proxy measures of inactivity, such as car ownership and TV viewing (25).

Recent awareness of the pivotal role of activity in maintaining metabolic and psychological fitness has led to major efforts to devise better methods for the quantitative assessment of total energy expenditure (TEE) and activity patterns (see the chapters in part II of this volume). These methods suffer from a common problem in epidemiological research, namely that their precision and reliability tends to be reciprocally related to their feasibility in large-scale studies. Thus, self-reported measures based on brief (sometimes single-factor) questionnaires are bound to be of limited validity, especially because they are also vulnerable to overreporting of activity, particularly by overweight and obese subjects (16). The fact that, in spite of their crudeness, such procedures have frequently revealed quite strong inverse correlations between activity and ill health indicates to us that the true underlying associations must be very strong. At the other extreme is the doubly labeled water ($^2H_2$$^{18}O$) method that we have used extensively in our laboratory (3, 22, 23). This method gives an integrated measure of TEE over 10-14 days and is very robust to observer effects. When combined with a measurement of basal metabolic rate (BMR) by indirect calorimetry, it allows estimation of "activity-plus-thermogenesis," computed as TEE minus BMR (from the expression TEE = BMR + thermogenesis + activity). Because thermogenesis is generally a small and rather constant fraction of expenditure, the variable TEE minus BMR provides an excellent measure of habitual activity. The problem is that the technique is rather complex, labor intensive, expensive, and thus unfeasible for most large-scale work. There are also interpretative difficulties concerning the correct way to adjust activity energy expenditure for differences in body size and fatness (24). Between these two extremes in research technique—large-scale and low resolution or small-scale and high resolution—lies a range of measures based on different methods for monitoring movement or heart rate (19).

Wareham (31) has recently demonstrated that cumbersome procedures, such as 4-day ambulatory heart-rate monitoring combined with individual calibration against indirect calorimetry, can be achieved in large study cohorts (ranging to several thousand subjects). He further recommends that the combined use of several methods with uncorrelated errors (preferably validated against doubly labeled water), and their synthesis into a composite activity index or a multi-factorial analysis, will theoretically provide the best way of assessing habitual physical activity (31, 32). Apart from Wareham's work, which is still in progress, we are unaware of any studies that have taken such care in defining the exposure variable, and the literature needs to be read with this limitation in mind.

Another major problem with secular and prospective cohort studies of obesity relates to the difficulty of matching the appropriate time to assess exposures (such as diet and activity) to an outcome variable (obesity). Exposures show profound societal and individual variations; the outcome variable represents the cumulative effects of years or decades of energy imbalance. The importance of this difficulty will become apparent later when we discuss some of the prospective studies that find quite different associations between activity and weight gain depending on whether they use baseline activity, activity at follow-up, or a combination of the two.

Interpretation of the research is also confused by the fact that "activity" is such a complex synthesis of minor physical movements [often termed "fidgeting" (27), or more recently "non-exercise activity thermogenesis" (NEAT) (15)], the movements of everyday life, and more extreme motor activity associated with manual work or sport. We are at the very early stages of understanding the relative importance of these components to the maintenance of metabolic health. Dietz (7) has introduced a further dimension to the problem by suggesting that inactivity should not simply be viewed as the opposite of activity. It is possible, for instance, that excessive TV viewing or computer usage can lead to venous stasis, which could have profound impact on metabolic control and variables such as vascular tone. Furthermore, such sedentary behaviors may be associated with other negative behaviors, such as excess snacking and alcohol consumption. The success of weight-loss interventions in adolescents, which target avoidance of inactivity rather than promotion of exercise, underscores the importance of this dimension to the problem (9).

Another issue that needs to be considered is the distinction between risk factors that may tend to cause a medical condition and those that protect against it. In the case of obesity, our view is that activity is certainly protective, but inactivity is not necessarily a causal factor. There are many very sedentary people who do not gain weight, but there are far fewer physically active people who are overweight, although it is certainly possible if the activity is associated with excess consumption (often of alcohol). Obesity is particularly difficult to analyze in this respect, because in contrast to other physiological disturbances such as NIDDM, it is visible to the sufferer and carries a social stigma. Thus, many people resort to other behavioral strategies (especially restrained eating) that prevent inactivity from being expressed as an obesity phenotype. This cognitive control generates a large degree of biobehavioral "noise" that must certainly dilute the true associations between inactivity and obesity. Such dilution may be more or less pronounced in different social and racial groups depending upon the external pressures to avoid obesity.

These methodological difficulties have previously been examined in greater detail by Williamson (33), who concluded that many of the contradictory findings in the field "result from the complex nature of the mechanisms underlying weight change and from the limited techniques available for assessing diet and physical activity among the general population." All of these difficulties should be borne in mind as we now discuss the existing evidence relating physical activity to weight control in adults. There are few studies that have considered many of these factors, and there are none that have adequately accounted for all of them.

Evidence From Secular and Age Trend Association Studies

In 1995 we published an ecological analysis of the causes of obesity in Britain that drew on a variety of different data sources ranging from experimental to large-scale nation wide statistics (25). The purpose of the paper was to examine the relative roles of modern diets and lifestyles in contributing to the rapidly escalating prevalence of obesity; in other words, to quantify the effects of gluttony versus sloth. The U.K. is exceptional in having maintained a National Food Survey (NFS) for over 50 years (18). This allows annual estimates of food and nutrient purchase and consumption patterns in a large randomly selected sample of British families. Analysis of this data indicates that *per capita* energy intake has actually been declining quite substantially in the past 30 years at exactly the time that obesity rates have started to rise so fast. Although there are certain limitations to the NFS data, the fact that energy intake has declined is corroborated by other studies even though the extent of the decline is uncertain (26). The resultant paradox of escalating obesity against a background of declining energy intake indicates that energy expenditure must be declining at an even faster rate. This contention is shown in figure 12.1, which plots dietary and activity variables against the secular trend in obesity. The figure indicates a much closer concordance between obesity and the proxy measures of inactivity than between obesity and total energy or fat intake.

Similar trends are shown for an analysis (using separate data sources) of the social class gradient in obesity (figure 12.2). On the basis of the analysis summarized in figures 12.1 and 12.2, and on a broad range of other statistics summarizing changes in activity and work patterns, we concluded that physical inactivity is at least as

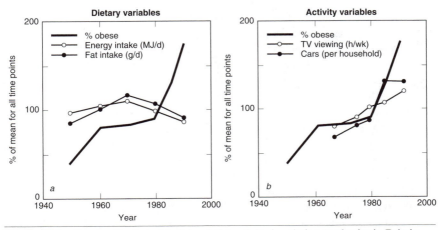

Figure 12.1 Secular trends in (a) diet and (b) activity in relation to obesity in Britain. Each plotted point is the percent of the mean value of all data from 1950 to 1990. Data for diet from National Food Survey (18); data for obesity from Office of Population Censuses and Surveys (2) and historical surveys; data for television viewing and car ownership from Central Statistical Office (5).

This figure was first published in the *BMJ* [A.M. Prentice, S.A. Jebb, "Obesity in Britain: Gluttony or Sloth?" *British Medical Journal,* 311: 437-439.] and is reproduced by permission of the *BMJ*.

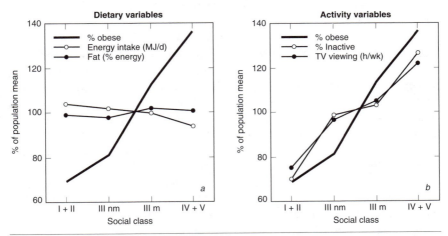

Figure 12.2 Social class trends in (a) diet and (b) activity in relation to obesity in British women. Social class I + II is highest, social class IV + V is lowest; nm = non-manual, m = manual. Data for diet from Gregory et al. (12); data for obesity from Office of Population Censuses and Surveys (2); data for television viewing and car ownership from Central Statistical Office (5).

This figure was first published in the *BMJ* [A.M. Prentice, S.A. Jebb, "Obesity in Britain: Gluttony or Sloth?" *British Medical Journal,* 311: 437-439.] and is reproduced by permission of the *BMJ*.

important as high-fat, energy-dense diets in the etiology of obesity, and possibly represents the dominant factor (25).

The ecological data on which this analysis was performed are frankly weak, but the paper has received little criticism and has been frequently cited, suggesting that it agrees with most people's intuitive understanding of the importance of sedentary lifestyle. To our knowledge this remains the only detailed ecological study of this type. It should be noted that it makes no attempt to analyze the relative importance of gluttony or sloth at an individual level, but rather demonstrates that sedentary behaviors represent a key risk factor for a group or population level; and these behaviors will increase the likelihood of the full obesity phenotype being expressed in genetically prone individuals.

Elsewhere, we have argued that the decrease in habitual levels of physical activity brought on by the electronic/technological revolution, increased affluence, and the popularity of home entertainment may represent the most rapid evolutionary challenge that *homo sapiens* has ever been forced to confront, and that the development of mass obesity represents one of the grade-shifts in species morphology that occur from time to time in response to overwhelming change in the external environment (21). In this context the epidemic of obesity is a perfectly understandable consequence of modern behavioral choices and its avoidance requires the individual to recognize the power of these external environmental influences and to create a protective microenvironment that includes plenty of discretionary physical activity.

Evidence From Cross-Sectional Analyses

This chapter makes no attempt to provide an exhaustive review of all available studies. Such summaries are available elsewhere (33), and systematic reviews are in preparation for publication in the near future. Instead we will draw on a few of the key publications to illustrate the relationships that have emerged. We openly acknowledge selection bias in favor of those studies showing that activity protects against the development of obesity. For the sake of balance, the following publications are examples of studies that have revealed only weak associations between physical activity and weight regulation: Lissner et al. found links to be present only among high fat consumers (17); the China Health and Nutrition Survey found associations only in women (20); and in the Amsterdam Growth and Health Study, physical activity was related to the sum of skinfolds but not to BMI (30).

Figure 12.3 illustrates some of the most frequently cited results in this field (28). Figure 12.3a shows that the prevalence of obesity in 25- to 65-year-old Finns was strongly related to self-reported levels of leisure-time physical activity (determined by a 13-factor questionnaire), with a greater than twofold gradient across the thirds of activity ($p < 0.001$ for trends). This type of association is vulnerable to the possibility of *post hoc* changes in activity as a result of weight gain (i.e., the issue of reverse causality). The data plotted in figure 12.3b are less vulnerable to this. The figure illustrates the relative risk of gaining more than 5 kg over the preceding

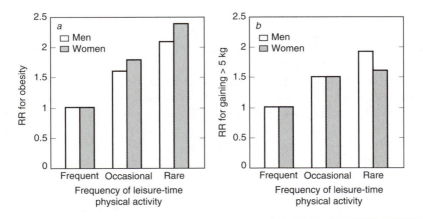

Figure 12.3   Cross-sectional associations between leisure-time physical activity, obesity, and excess weight gain in adult Finns. (a) Relative risk (RR) of obesity in 2726 men and 2915 women referenced to the frequent activity group. (b) Relative risk of gaining more than 5 kg over the previous decade in 6484 men and 6143 women. Data from Rissanen et al. (28).

decade (as opposed to the prevalence of obesity itself), so there will be a mixture of lean and obese in the various weight gain categories. In this case people reporting low levels of leisure-time physical activity were almost twice as likely to have gained more than 5 kg.

Figure 12.4 shows a similar analysis of the risk of inappropriate weight gain from a separate group of Finns (13). Again the analysis has looked cross-sectionally at the

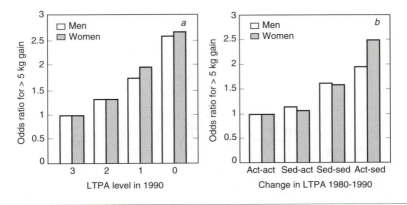

Figure 12.4   Associations between leisure-time physical activity (LTPA) and excess weight gain in adult Finns. (a) Cross-sectional analysis of weight gain in relation to self-reported activity at follow-up (using the highest level as referent). (b) Retrospective analysis of the effects of changing level of activity over the course of the study (using people who remained active throughout the referent category). Data are from 2564 men and 2695 women aged 19 to 63 y (13).

risk of antecedent weight gain compared to a single estimate of activity from a 23-item questionnaire, and as in the Rissanen data the odds ratio for excess gain was about 2.5 in the most inactive group (p < 0.001 in both genders). This study will be recited in the section on prospective studies.

The Health Surveys of England conducted annually over the past 8 years have consistently revealed a highly significant inverse association between the prevalence of obesity and sports participation. Time spent walking is also inversely correlated with obesity, but the associations are less strong (2). The difference between participants and nonparticipants in sports is about twofold, but is likely to be diluted by classification error and once again reflects the complexity of the causal influences on obesity. An important aspect of this data is that it shows that the greatest public health dividend would be achieved if people who are completely sedentary become more active. In terms of obesity, there is little benefit to be achieved if people who are already active become even more active. Other studies replicate this finding (e.g., 1, 4).

In the United States, the National Health and Nutrition Examination Survey (NHANES-1) data have also been used to investigate the effects of recreational physical activity on 10-year weight change between 1971-1975 and 1982-1984 (34). It had previously been noted that nonrecreational activity had small and inconsistent associations with body weight. Figure 12.5 shows the results of cross-sectional analysis of self-reported activity at follow-up against antecedent weight change for over 3500 men and 5800 women. Inactivity is especially related to extreme weight gain. In this analysis the activity ratings at baseline were not predictive of subsequent weight gain. Once again, this could be because the follow-up measures reflect *post hoc* changes in activity as a consequence of weight gain,

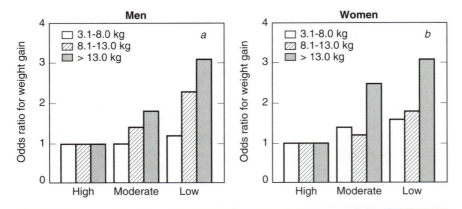

Figure 12.5   Odds ratios for weight gain in (a) men and (b) women in the NHANES-1 Epidemiologic Follow-Up Study. Odds ratios are for 10-year weight gain in relation to self-reported recreational physical activity at follow-up. High, moderate, and low refer to activity levels. Ages 25 to 74 y; men, n = 3515; women, n = 5810. Data from Williamson et al. (34).

although the authors favor other explanations. They speculate that the population as a whole may have become more sensitized to physical activity as a determinant of health at the second round of measurements, and thus might be less prone to misclassification. They also speculate that intentional weight-modulating exercise such as jogging would be more prevalent in the 1980s than in the 1970s, thus giving better discrimination between exercisers and nonexercisers.

These associations, and analogous ones from similar studies (e.g., 14, 30), have emerged in spite of methodological limitations and the possible presence of *post hoc* effects, some of which would tend to generate reverse associations; for example, the observation that overweight people may be more prone to overestimate their activity level (16) or may have taken up exercise as a weight-reducing strategy.

Evidence From Prospective Studies

Prospective studies (i.e., those that look forward from initial baseline measurements) have both advantages and disadvantages. The chief advantage is that the measures of exposure and outcome are contemporaneous, although there are no published studies that make activity measurements over more than one or two discrete time points. The main disadvantage is that, in the context of a slowly developing syndrome like obesity, the follow-up periods so far available are rather short. Some studies have looked over intervals as short as 2 years (e.g., 11). Studies of a 10-year duration are the maximum currently available for which there are large sample sizes (13, 28, 34). Some longer studies have rather limited sample sizes (30). The discriminatory power is limited over short periods because mean body-mass gain of affluent populations tends to be about 1 g/d (i.e., less than half a kilogram per year), and the standard deviations are similarly small. These problems are compounded by a day-to-day fluctuation in body weight of about ±1 kg or more, especially when measurements are made under poorly standardized conditions (nonfasted, lightly clothed, sometimes self-reported, and therefore using nonstandardized scales) as in some large studies. Therefore, once again, the fact that some rather strong predictive power can be attributed to physical activity suggests that its true significance is great.

The recent report on predictors of weight change in the U.S. Health Professionals Follow-Up Study summarizes 4-year weight trends in almost 20,000 men studied in 1988 and 1992 (6). Activity was assessed at the beginning and end of the 4-year period using a 7-factor questionnaire that focused on vigorous activity and TV/VCR viewing habits. Results of the regression model of weight change against patterns of both activity and inactivity are shown in figure 12.6. The trends are not particularly large, but are highly significant and indicate that decreasing activity below a nominal cutoff of 1.5 hours/week would predict an average 0.6 kg weight gain [95% confidence intervals (CI) 0.3 to 0.8 kg], while an increase in activity to above 1.5 hours/week would predict a weight loss of 0.9 kg (95% CI-1.2 to-0.6 kg).

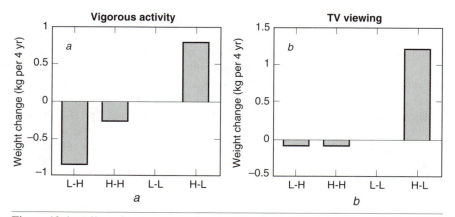

Figure 12.6   Effect of changes in activity habits on average 4-year weight changes in the Health Professionals Follow-Up Study. Weight changes were derived from regression analysis on 19,478 men aged 40 to 75 y measured 1988-1992. H = high, L = low, H-L indicates a change from high to low, H-H indicates activity remained high, and so forth. Data from Coakley et al. (6).

An increase in TV viewing to above 14 hours/week predicted a weight gain averaging 1.2 kg (95% CI 0.4 to 2.0 kg) (figure 12.7).

An alternative way of representing the effects of changing activity is shown in figure 12.7 in plots of the cross-sectional prevalence rates of obesity in 45- to 54-year-olds at four measurement periods that span 6 years from 1986-1992. The left-hand bar from each set gives an indication of the baseline situation and shows that

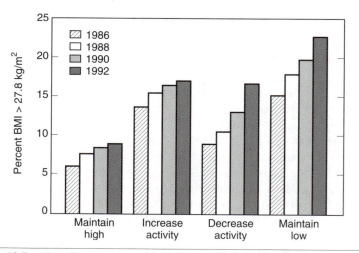

Figure 12.7   Changing prevalence of obesity over time for different patterns of recreational vigorous physical activity in the Health Professionals Follow-Up Study. Data based on 3666 men aged < 55 y from Coakley et al. (6).

those who started with activity levels above the 1.5 hours/week cutoff (i.e., the "maintain high" and "decrease activity" groups) had lower initial prevalence of obesity, averaging 6 and 9% compared to 13 and 15% in those who started with activity levels below 1.5 hours/week. The subsequent change in prevalence over time was lower in the groups who maintained high activity or increased their activity (~3% rise in prevalence) compared to those who maintained low activity or decreased their activity (~8% rise in prevalence).

The prospective analysis of the Haapanen study of adult Finns is shown in figure 12.4b, together with the cross-sectional analysis of this study (13). In both men and women, the odds ratio for a > 5 kg weight gain was significantly raised in those who remained sedentary throughout (odds ratio ~1.6) and in those who decreased activity (odds ratio ~2.2).

The strongest associations between obesity and activity levels come from Williamson's analyses of the effects of recreational physical activity in the NHANES-1 study (34). In men the odds ratios of 2.3 for those who maintained low activity or who decreased their activity were only marginally significant, but in women the equivalent odds ratios of 7.1 and 6.2 were highly significant.

The study by DiPietro et al. (8) provides some of the most objective data on changes in activity patterns by measuring the change in participants' endurance times on a treadmill test. In the Aerobics Center Longitudinal Study they showed that improvements in cardiorespiratory fitness attenuate age-related weight gain in healthy men and women. The strong relationship between change in endurance and weight change is illustrated in figure 12.8. As in many of these studies, the possibility of reverse causality (i.e., weight gain leading to decreased fitness) cannot be excluded, and this study is particularly vulnerable to the issue of noncontiguous

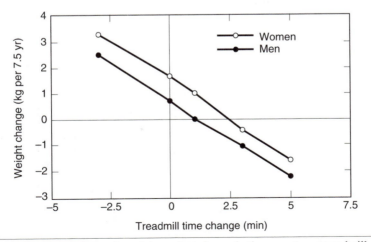

Figure 12.8   Plot of change in weight against change in time spent on a treadmill, showing the relationship between change in fitness and weight gain in the Aerobics Center Longitudinal Study. Results are derived from multiple linear regression coefficients adjusted for possible confounding variables. Data from DiPietro et al. (8).

data collection, because the mean interval of 1.8 years between fitness tests was compared to an interval of 7.5 years between weight assessments.

Is There a Protective Level of Activity?

Results from the Healthy Worker Project of 20,000 people in 32 worksites in the U.S. are suggestive of a threshold effect (11). The baseline analysis showed that high-intensity activity was a strong inverse predictor of body mass index (BMI) in both men and women. Walking was a significant predictor in men, but not in women. Group sports and job activity were not significant discriminators. Both walking and high-intensity activity were significant negative predictors of weight gain at follow-up, but group sports and job activity were not. It should be noted, however, that these differences may reflect more a statistical threshold than a real biological threshold. Because correlation dilution attenuates all of the associations, it may simply be that it is only the strongest associations that exceed conventional statistical cutoffs.

Although based on a small sample, Schoeller's study of weight regain is one of the most powerful, because it used doubly labeled water to objectively assess the energy expended on physical activity (29). Figure 12.9a shows physical activity levels (PAL = total energy expenditure/BMR) for the groups classified as active, moderately active, and sedentary. Figure 12.9b shows that post-diet weight rebound was much lower in the active group. By reference to the scatter plot of individual PAL values versus weight regain, the authors concluded that there was a threshold at about 47 kJ/kg/d extra activity needed to prevent weight gain. This would be equivalent to about 80 min/d moderate activity or 35 min/d vigorous activity. While

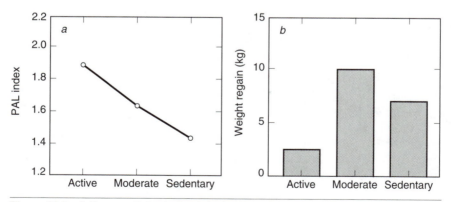

Figure 12.9   Influence of physical activity level on post-dieting weight regain. (a) Physical activity level (PAL) measured by doubly labeled water in 32 women immediately after profound weight loss (23 ± 9 kg). (b) 12-month weight regain in relation to physical activity. Data from Schoeller et al. (29).

providing useful guidelines, these figures should not be overinterpreted; the scatter plot does not indicate a discrete cutoff, and indeed it would be surprising if it did given the multitude of influences on body weight.

Ferro-Luzzi and Martino (10) have also suggested that an activity threshold exists at about $1.8 \times$ BMR above which obesity is eightfold less common in men, whereas analysis using $1.7 \times$ BMR as the threshold shows little difference in the prevalence of obesity above and below this level. Their data, however, are based on cross-country analyses with all the possible confounders that may exist.

Summary

In spite of all of the caveats that we have mentioned in this paper and the relatively poor quality of data available, there is still an overwhelming case in favor of the conclusion that physical activity (especially discretionary leisure-time and recreational physical activity) is strongly related to successful long-term weight maintenance. The fact that such strong associations emerge from such crude measures of exposure makes this conclusion all the more robust. Many people have empirically proved this fact to their own satisfaction and have built activity into their lifestyle as a weight-regulatory strategy. The challenge for public health professionals is to spread the physical-activity message to the majority of the population, who have not yet understood the damage that the modern sedentary lifestyle is doing to their health.

References

1. Barlow, C.E., H.W. Kohl, L.W. Gibbons, and S.N. Blair. 1995. Physical fitness, mortality and obesity. *International Journal of Obesity* 19: (suppl 4): S41-S44.
2. Bennett, N., T. Dodd, J. Flatley, S. Freeth, and K. Bolling. 1995. *Health Survey for England 1993*. London: HMSO.
3. Black, A.E., W.A. Coward, T.J. Cole, and A.M. Prentice. 1996. Human energy expenditure in affluent societies: an analysis of 574 doubly-labelled water measurements. *European Journal of Clinical Nutrition* 50: 72-92.
4. Blair, S.N., E. Horton, A.S. Leon, I.M. Lee, B.L. Drinkwater, R.K. Dishman, M. Mackey, and M.L. Kienholz. 1996. Physical activity, nutrition, and chronic disease. *Medicine and Science in Sports and Exercise* 28: 335-349.
5. Central Statistical Office. 1994. *Social Trends 24*. London: HMSO.
6. Coakley, E.H., E.B. Rimm, G. Colditz, I. Kawachi, and W. Willett. 1998. Predictors of weight change in men: Results from The Health Professionals Follow-Up Study. *International Journal of Obesity* 22: 89-96.

7. Dietz, W.H. 1996. The role of lifestyle in health: the epidemiology and consequences of inactivity. *Proceedings of the Nutrition Society* 55: 829-840.

8. DiPietro, L., H.W. Kohl, C.E. Barlow, and S.N. Blair. 1998. Improvements in cardiovascular fitness attenuate age-related weight gain in healthy men and women: the Aerobics Centre Longitudinal Study. *International Journal of Obesity* 22: 55-62.

9. Epstein, L.H. 1996. Family-based behavioural intervention for obese children. *International Journal of Obesity* 20: (suppl 1): S14-S21.

10. Ferro-Luzzi, A. and L. Martino. 1996. Obesity and physical activity. In *The origins and consequences of obesity,* eds. D. Chadwick and G. Cardew, 207-227. Chichester, NY: John Wiley & Sons.

11. French, S.A., R.W. Jeffery, J.L. Forster, P.G. McGovern, S.H. Kelder, and J.E. Baxter. 1994. Predictors of weight change over two years among a population of working adults: the Healthy Worker project. *International Journal of Obesity* 18: 145-154.

12. Gregory, J., K. Foster, H. Tyler, and M. Wiseman. 1990. *The Dietary and Nutritional Survey of British Adults.* London: HMSO.

13. Haapanen, N., S. Miilunpalo, M. Pasanen, P. Oja, and I. Vuori. 1997. Association between leisure time physical activity and 10-year body mass change among working-aged men and women. *International Journal of Obesity* 21: 288-296.

14. Klesges, R., L. Klesges, C. Haddock, and L. Eck. 1992. A longitudinal analysis of the impact of dietary intake and physical activity on weight change in adults. *American Journal of Clinical Nutrition* 55: 818-822.

15. Levine, J.A., N.L. Eberhardt, and M.D. Jensen. 1999. Role of nonexercise activity thermogenesis in resistance to fat gain in humans. *Science* 283: 212-214.

16. Lichtman, S.W., K. Pisarska, E. Berman, M. Perstone, H. Dowling, E. Offenbacher, H. Weisel, S. Heshka, D.E. Matthews, and S.B. Heymsfield. 1993. Discrepancy between self-reported and actual caloric intake and exercise in obese subjects. *New England Journal of Medicine* 327: 1893-1898.

17. Lissner, L., B.L. Heitmann, and C. Bengtsson. 1997. Low-fat diets may prevent weight gain in sedentary women: Prospective observations from the Population Study of Women in Gothenburg, Sweden. *Obesity Research* 5: 43-48.

18. Ministry of Agriculture Fisheries and Foods. 1940-98. *Household Food Consumption and Expenditure.* London: HMSO.

19. Murgatroyd, P.R., P.S. Shetty, and A.M. Prentice. 1993. Techniques for the measurement of human energy expenditure. *International Journal of Obesity* 17: 549-568.

20. Paeratukul, S., B.M. Popkin, G. Keyou, L.S. Adair, and J. Stevens. 1998. Changes in diet and physical activity affect the body mass index of Chinese adults. *International Journal of Obesity* 22: 424-431.

21. Prentice, A.M. 1997. Obesity—the inevitable penalty of civilisation? *British Medical Bulletin* 53: 229-237.
22. Prentice, A.M., A.E. Black, W.A. Coward, and T.J. Cole. 1996. Energy expenditure in affluent societies: an analysis of 319 doubly-labelled water measurements. *European Journal of Clinical Nutrition* 50: 93-97.
23. Prentice, A.M., A.E. Black, W.A. Coward, H.L. Davies, G.R. Goldberg, P.R. Murgatroyd, J. Ashford, M. Sawyer, and R.G. Whitehead. 1986. High levels of energy expenditure in obese women. *British Medical Journal* 292: 983-987.
24. Prentice, A.M., G.R. Goldberg, P.R. Murgatroyd, and T.J. Cole. 1996. Physical activity and obesity: problems in correcting expenditure for body size. *International Journal of Obesity* 20: 688-691.
25. Prentice, A.M., and S.A. Jebb. 1995. Obesity in Britain: Gluttony or sloth? *British Medical Journal* 311: 437-439.
26. Prentice, A.M., and S.A. Jebb. 1995. Obesity in Britain. *British Medical Journal* 311: 1568-1569.
27. Ravussin, E., S. Lillioja, W.C. Knowler, L. Christin, D. Freymond, W.G. Abbott, V. Boyce, B.V. Howard, and C. Bogardus. 1988. Reduced rate of energy expenditure as a risk factor for body weight gain. *New England Journal of Medicine* 318: 467-472.
28. Rissanen, A.M., M. Heliovaara, P. Knekt, A. Reunanen, and A. Aroma. 1991. Determinants of weight gain and overweight in adult Finns. *European Journal of Clinical Nutrition* 45: 419-430.
29. Schoeller, D.A., K. Shay, and R.F. Kushner. 1997. How much physical activity is needed to minimize weight gain in previously obese women? *American Journal of Clinical Nutrition* 66: 551-556.
30. Twisk, J.W.R., H.C.G. Kemper, W.V. Mechelen, G.B. Post, and F.J.V. Lenthe. 1998. Body fatness: Longitudinal relationship of body mass index and the sum of skinfolds with other risk factors for coronary heart disease. *International Journal of Obesity* 22: 915-922.
31. Wareham, N.J., S.H.J. Hennings, N.E. Day, and A.M. Prentice. 1997. Feasibility of heart-rate monitoring to estimate total level and pattern of energy expenditure in a population-based epidemiological study: the Ely young cohort feasibility study. *British Journal of Nutrition* 78: 889-900.
32. Wareham, N.J., S.H.J. Hennings, C.D. Byrne, C.N. Hales, A.M. Prentice, and N.E. Day. 1998. A quantitative analysis of the relationship between habitual energy expenditure, fitness and the metabolic cardiovascular syndrome. *British Journal of Nutrition* 80: 235-241.
33. Williamson, D.F. 1996. Dietary intake and physical activity as predictors of weight gain in observational, prospective studies of adults. *Nutrition Reviews* 54: S101-S109.
34. Williamson, D.F., J. Madans, R.F. Anda, J.C. Kleinman, H.S. Kahn, and T. Byers. 1993. Recreational physical activity and ten-year weight change in a US national cohort. *International Journal of Obesity* 17: 279-286.

CHAPTER 13

Physical Activity Level and Weight Control in Older Citizens

Eric T. Poehlman, PhD

Department of Medicine, University of Vermont, Burlington, Vermont, U.S.A.

Exercise is important in the prevention and management of cardiovascular disease, peripheral vascular disease, and obesity (1). Low levels of physical fitness have been shown to be a greater predictor of mortality than smoking, obesity, hypertension, or hypercholesterolemia (7). Previously sedentary men who adopted regular exercise or physical activity during a 4-year period experienced risk reductions of 44% for all-cause mortality and 52% for cardiovascular mortality during a subsequent 5-year period. As many as 250,000 deaths per year in the United States are attributable to physical inactivity (27). If just 30 percent of the adult population were to adopt regular physical activity, 30,000 lives per year would be saved through reductions in coronary disease and diabetes alone (35).

The decline in physical activity with advancing age likely contributes to the increase in obesity. Older individuals are overweight or obese when energy intake chronically exceeds energy expenditure. The prevalence of obesity is increasing in older men and women, generally because the age of the population in the U.S. has increased. In 1900, only 4% of the total population was over 65 years of age. Today, this rapidly growing segment of the American population comprises about 12% of the total population, encompassing about 30 million people. Further projections into the 21st century indicate that the elderly will account for 17 to 20% of the population by the year 2025.

Obesity in younger individuals is generally characterized by elevated levels of fat mass and fat-free mass, whereas obesity in the elderly is reflective of high levels of body fat, but low quantities of fat-free mass. The erosion of fat-free mass is a reproducible finding with advancing age and contributes to loss of muscular strength, higher rates of disability, and a general decline in quality of life (18). Thus, pharmacologic and/or lifestyle strategies that seek to reduce body fat and preserve fat-free mass are important therapeutic considerations in the elderly.

The U.S. Surgeon General's report (47) on physical activity and health states that 60 percent of the adult United States population is not active enough to achieve health benefits. Data from nationally representative survey studies have shown that African-American adults, older adults, and those of lower socioeconomic status are even less likely to engage in regular exercise or physical activity (19). These levels of participation in physical activity are disappointing, given that even modest participation in physical activity, such as low-intensity walking, are associated with a lower overall mortality rate (21). Thus, encouraging older people even to moderately increase their physical activity levels would likely improve their overall health and functional independence. In this chapter, we will specifically examine the effects of increasing physical activity on energy expenditure, body-fat distribution, and fat metabolism in older individuals.

Physical Activity and Energy Expenditure

Significant changes in body composition occur in adults with aging. The amount of fat in the body increases and is stored preferentially in abdominal rather than peripheral tissues; the lean component, including body water, skeletal muscle, and organ mass and bone mineral content decrease (5, 18, 22, 46). It is generally thought that these deleterious changes in body composition with advancing age are due to a small positive energy imbalance between energy intake and energy expenditure, due to an increasing sedentary lifestyle. The effects of aging on the components of energy expenditure have previously been reviewed (28, 29). Thus, we will not consider the effects of aging on energy intake and energy expenditure. We will instead focus on the role of physical activity as a therapeutic intervention to offset deleterious changes in physical activity energy expenditure and body composition in aging individuals.

The most variable component of the daily energy expenditure is physical activity. Physical activity energy expenditure includes the energy expended above the resting metabolic rate and the thermic effect of feeding; it also includes the energy expended through voluntary exercise and involuntary activity such as shivering, fidgeting, and postural control. In sedentary individuals, the thermic effect of activity may comprise as little as 100 kilocalories per day; in highly active older individuals, it may approach 1500 to 2000 kilocalories per day. Thus, physical activity represents a significant factor governing the balance of daily energy expenditure and food intake in older humans because it is extremely variable and subject to voluntary control. Older males, in general, tend to have a greater caloric expenditure associated with physical activity than older females, due partially to the greater energy cost of moving a larger body mass.

Measurement Issues

The measurement of the energy expenditure that results from physical activity has traditionally presented several methodological challenges. Indirect calorimetry, using a mouthpiece or face mask, has been used to assess oxygen consumption and carbon dioxide production. This method generally yields reliable and accurate measurements of the energy cost of physical activity in a laboratory setting, but provides no information about the energy cost of physical activity under free-living conditions because of the stationary nature of the equipment. In an attempt to avoid some of the problems associated with the measurement of free-living physical activity, several less complicated (and less accurate) methods have been devised. These methods employ physiological measurements, observation, and records of physical activity, as well as activity diaries or recall. Heart-rate recording, used to measure energy expenditure, is based on the correlation between heart rate and oxygen consumption during moderate to heavy exercise. The correlation, however, is much poorer at lower levels of physical activity, and a subject's heart rate may be altered by such events as anxiety or change in posture without significant changes in oxygen consumption. Physical activity diaries and physical activity recall instruments have been used to quantify the energy costs of different activities over a representative period of time. Record keeping is often inaccurate and may interfere with the subject's normal activities. Furthermore, the subject's recall of physical activity depends on the subject's memory, which may not always be reliable. Measuring motion using a device such as a pedometer or an accelerometer may provide an index of physical activity (i.e., counts), but does not accurately measure daily energy expenditure over extended periods of time.

Room calorimeters are available for conducting short-term studies (several days) of energy expenditure in humans when the object is to measure the combined effects of resting metabolic rate, the thermic effect of a meal, and the energy expenditure of physical activity. Physical activity level is quantified by a radar system that is activated by movement of the subject within the chamber. Free-living physical activity, however, is blunted in the room calorimeter because of its confining nature. Thus, room calorimeters do not offer the best model for accurately quantifying free-living physical activity and total energy expenditure in older individuals.

The doubly labeled water technique is a method used to determine energy requirements in free-living populations and in subjects in whom traditional measures of energy expenditure, using indirect calorimetry, have proven impractical and difficult. The basis of this technique is that following the administration of two stable isotopes of water (^2H_2O and $H_2{}^{18}O$), ^2H_2O is lost from the body in water alone, whereas $H_2{}^{18}O$ is lost not only in water but also as $C^{18}O_2$ via the carbonic anhydrase system. The difference in the two turnover rates is therefore related to the carbon dioxide production rate, and with a knowledge of the fuel mixture oxidized (from the composition of the diet), energy expenditure can be calculated.

Advantages of the doubly labeled water technique are as follows: (1) It measures total daily energy expenditure, which includes an integrated measure of resting metabolic rate, the thermic response to feeding, and the energy expenditure of physical activity; (2) it permits an unbiased measurement of free-living energy expenditure; and (3) measurements are conducted over extended periods of time (1 to 3 weeks). Daily free-living physical activity is calculated from the difference between the total daily energy expenditure and the combined energy expenditures of the resting metabolic rate and the thermic effect of meals. We will examine studies that examined the effects of physical activity on total and physical activity energy expenditure in older persons.

Effects of Exercise Training on Physical Activity

The reader is referred to a previous review (42) that summarized the effects of exercise training on daily energy expenditure in younger populations using doubly labeled water and room calorimeters. There is sparse data, however, regarding the effects of endurance or resistance training on physical activity energy expenditure in older individuals using doubly labeled water methodology. It is generally assumed that regular participation in endurance exercise results in a net increase in daily energy expenditure and physical activity in the elderly. This question has not been vigorously assessed because of methodological weaknesses in the assessment of physical activity in free-living older individuals. From an energetic and cardio-vascular standpoint, one of the goals is to prescribe exercise that significantly enhances physical activity during the nonexercising time of the day. The energy expenditure during the nonexercising time of the day is large (100 to 1000 kcal/d) and can actually be greater than energy expenditure associated with the direct caloric cost of exercise. Exercise that results in maximal increases in total daily energy expenditure would ultimately be most beneficial in preserving fat-free mass and promoting loss of fat mass in older individuals.

In a recent study, elderly volunteers were submitted to a progressive 8-week endurance program in which the energy expenditure was gradually increased to an additional net energy expenditure of 1000 kcal per week (20). We used doubly labeled water to measure total daily energy expenditure before and during the last 10 days of a 2-month exercise program. Resting metabolic rate and the energy cost of the exercise program were determined by indirect calorimetry. Results showed that training increased $\dot{V}O_2$max by 9% and resting metabolic rate by 11%, but no significant change in total daily energy expenditure was noted (figure 13.1). The absence of change in total daily energy expenditure was explained by a 62% reduction in physical activity energy expenditure (571 ± 386 kcal/d to 340 ± 452 kcal/d), that is, older men and women decreased their spontaneous physical activity levels as a result of participation in the exercise program. The important point is that exercise may actually be counterproductive in increasing physical activity during the nonexercising portion of the day. The absence of an increase in physical activity energy expenditure in this study suggests an "energy conserving" reduction in spontaneous physical activity and/or a reduction in voluntary physical activities during the 10-day measurement period.

Decreases in spontaneous physical activity have been shown to occur in response to strenuous physical activity in animal studies (40, 41). It is conceivable that the level of exercise during the last week of training (3 hours per week at 85% of $\dot{V}O_2$ max) was too vigorous and thus fatigued the elderly participants during the

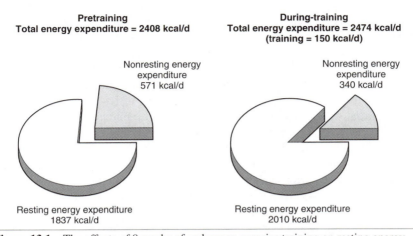

Figure 13.1   The effects of 8 weeks of endurance exercise training on resting energy expenditure and nonresting energy expenditure in older individuals. Adapted from reference 20.

remainder of the day. Another possibility was that the exercise program was too short for older individuals to adapt to the increased energy demands. At this time, our findings should not be interpreted to indicate that all levels of endurance training blunt increases in physical activity energy expenditure. Exercise prescriptions of lower intensity and varying durations should be examined to verify their influence on physical activity energy expenditure. This study, however, raises new questions regarding the lack of an effect of endurance exercise on total daily energy expenditure in older individuals.

Resistance training is becoming increasingly popular as a therapeutic intervention to offset the decline in skeletal muscle mass. Advancing age and low levels of physical activity are implicated in the loss of skeletal muscle and decline in functional independence. We surveyed the literature for studies that examined the effects of resistance training on physical activity in older individuals. Although several studies have investigated the effects of resistance training on resting energy expenditure and 24-hour energy expenditure in a room calorimeter (10, 36, 44, 45), no studies were identified that used doubly labeled water methodology. Fiatarone and colleagues (14) examined the effects of a progressive-resistance exercise program on physical activity in 100 frail nursing home residents. Free-living physical activity was estimated with integrated activity monitors worn around both ankles during a 72-hour period. The exercise program significantly increased overall physical activity level, with improved gait velocity and stair-climbing ability. Campbell and colleagues also reported that resistance training increased daily energy needs in older adults (10).

We identified only one study that examined the effects of resistance training on total daily energy expenditure using doubly labeled water (48). These investigators studied resistance training in 18 healthy sedentary young men (23 to 41 years of age) for 18 weeks. A subsample of 12 individuals had total daily energy expenditure measured before and after the training period using doubly labeled water. Body composition was measured using a three-compartment model using underwater weight and total body water. The volunteers trained 2 times per week; their training consisted of 3 sets of 15 repetitions for the major muscle groups. After 18 weeks, volunteers lost 2.0 kg of fat mass and gained 2.1 kg of fat-free mass. The mean increase in daily energy expenditure after the 18-week training program approached 260 kcal per day, of which 40-50% was due to the direct caloric cost of the resistance-training program. Sleeping metabolic rate and free-living physical activity (independent of the direct energy cost of the training program) did not change in response to resistance training. The absence of an increase in free-living physical activity was confirmed with measurements from the triaxial accelerometer. Thus, this well-controlled study does not support the hypothesis that resistance training, resulting in moderate changes in body composition, increases resting energy expenditure or the energy cost of physical activity in younger individuals. Collectively, the findings of these studies are disappointing from a public health perspective. The finding that endurance and resistance training did not increase free-living physical activity, despite increases in $\dot{V}O_2$max and muscular strength, may partially explain the poor track record of exercise (without caloric restriction) to substantially impact weight loss. Moreover, these findings further emphasize the importance of measuring free-living total daily energy expenditure when examining the effects of exercise on body weight regulation and energy expenditure.

Physical Activity in Minority Populations

Relatively little information is available on physical activity levels and their relationship with cardiovascular and disease risk in older minority populations. Differences in education, income, cultural values, and availability of recreational outlets, may limit physical activity–related opportunities in older minority populations and increase their risk for metabolic and cardiovascular disease. For example, the prevalence of obesity in African-American women is estimated to be twice that of Caucasian women. Moreover, the associated rate of hypertension and mortality due to heart disease, stroke, and diabetes in African-American women is 1.5 to 2.0 times that of Caucasian women (13, 25). We recently examined daily energy expenditure and physical activity in older African-American men and women using doubly labeled water methodology and indirect calorimetry (11). Individuals ranged in age from 55 to 80 years. The most striking finding was the diminished levels of total daily energy expenditure in older African-American women. They showed a lower total daily energy expenditure when compared with Caucasian women (12%), African-American men (18%), and Caucasian men (24%); see figure 13.2. The lower total daily energy expenditure was due to lower rates of resting metabolic rate and physical activity–related energy expenditure. These findings persisted after control for body composition, education, and socioeconomic status. These results suggest that older African-Americans, particularly African-American

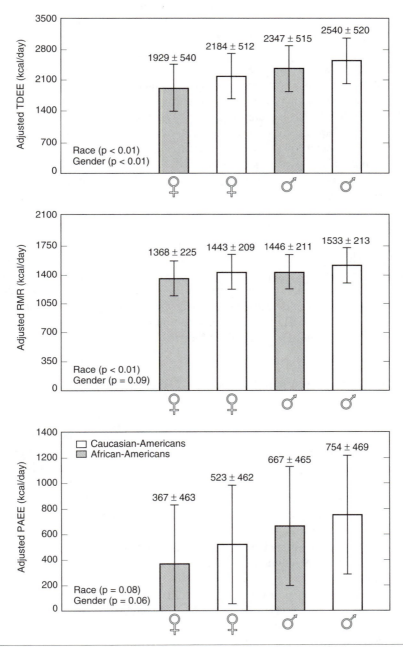

Figure 13.2 Differences in adjusted total daily energy expenditure (TDEE), resting metabolic rate (RMR), and physical activity energy expenditure (PAEE) between older African-American and Caucasian men and women.

Reprinted, by permission, from W.H. Carpenter, T. Fonong, M.J. Toth, P.A. Ades, J. Calles-Escandon, J.D. Walston, and E.T. Poehlman, 1998, "Total daily energy expenditure in free-living African-Americans and Caucasians," *American Journal of Physiology* 273: 96-101.

women, require fewer calories for their body size to maintain body weight. Low rates of total daily energy expenditure, particularly physical activity, may be important metabolic factors predisposing African-American women to obesity and its metabolic consequences. These findings suggest that intervention strategies aimed at increasing physical activity, particularly in older African-American women, should be a major public health priority.

What are the factors that determine low physical activity in African-Americans? In a follow-up study (38), we examined determinants of physical activity energy expenditure in older (64 ± 7 years) African-Americans (28 men and 37 women) using doubly labeled water methodology. We found that physical activity energy expenditure was 246 kcal per day lower in older African-American women than men. These differences persisted after taking into account differences in body size and composition. We noted an inverse relationship ($r = -0.44$; $p < 0.05$) between age and physical activity energy expenditure in older African-American women, although this relationship was less robust in older African-American men ($r = -0.28$; $p > 0.05$). Thus, aging is associated with low physical activity, even within an elderly cohort of African-Americans, possibly due to a greater loss of skeletal muscle mass. Reduced cardiorespiratory fitness was also associated with low physical activity energy expenditure in older African-American men ($r = 0.39$) and women ($r = 0.31$). Taken together, age and low levels of cardiorespiratory fitness appear to contribute to low physical activity levels in older African-Americans.

Physical Activity and Body-Fat Distribution

A review of recent data suggests that the average increase in body-fat content between the ages of 20 and 70 in the U.S. is approximately 15 kilograms (24). Moreover, this increase in body fatness does not appear to be uniform, but preferentially stored in central regions of the body. Abdominal obesity is a strong predictor of several risk factors for heart disease, including insulin resistance, dyslipidemia, and hypertension (6). It is unclear whether the accumulation of central body fat is an immutable effect of aging, a change in lifestyle (i.e., decreased physical activity), or a combination of both. We addressed this question by comparing younger and older individuals who have remained physically active throughout their lives.

We showed that the increase in the waist circumference with aging is greater in women (slope $= 0.28$ cm/yr) than men (slope $= 0.18$ cm/yr; $p < 0.01$) (31). Moreover, the increasing waist circumference was strongly associated with declines in leisure-time physical activity and peak $\dot{V}O_2$. Interestingly, we observed no independent contribution of resting metabolic rate, respiratory quotient, energy, or macronutrient to the age-related increase in waist circumference. Cross-sectional comparisons of younger and older athletes suggest that regular physical activity decreases cardiovascular risk (39), but cannot totally prevent the increase in abdominal fat;

however, the differences between younger and older athletes are much smaller than those between younger and older sedentary individuals (12, 24). Longitudinal 20- and 22-year follow-up studies of male endurance-trained athletes provide evidence of a protective effect of regular physical activity on total and central body-fat accumulation (34, 43). In both studies, men who continued to train for competition marginally increased their body-fat content over 20 to 22 years. This is in contrast to the 16-kg weight gain that occurred in the men who stopped training, in whom body-fat content increased from 9% to 22% of body weight. Waist circumference increased only 4 to 5 cm in men who continued to exercise for fitness, compared with 12 cm in the more sedentary men. Furthermore, results from exercise intervention studies showed that endurance training selectively reduced body fat in central regions in older individuals (37). These results raise interesting questions regarding the utility of endurance and resistance exercise to selectively reduce abdominal obesity (37, 44, 45) in older persons.

The menopause transition is another critical period that may accelerate the decline in physical activity and increase central adiposity in aging women. We recently examined the effects of the normal menopause transition on body composition, fat distribution, and leisure-time physical activity in a cohort of 35 premenopausal women (33). After 6 years of follow-up, 18 women had spontaneously stopped menstruating, whereas 17 women remained premenopausal. No women received hormone replacement therapy. Changes in several metabolic variables were compared. Women who experienced menopause lost more fat-free mass than women who remained premenopausal (-3.0 ± 1.1 kg and -0.5 ± 0.5 kg, respectively) and had greater increases in fat mass (2.5 ± 2 kg and 1.0 ± 1.5 kg) and waist-to-hip ratios (0.04 ± 0.01 and 0.01 ± 0.01; figure 13.3). Moreover, leisure-time physical

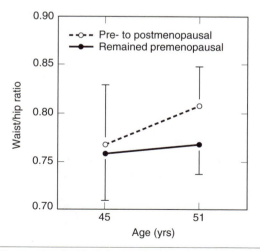

Figure 13.3 Six-year longitudinal changes in the waist-to-hip ratio in women who transitioned from pre- to postmenopausal status and in women who remained premenopausal over this time period. Adapted from reference 33.

activity decreased more in women who became postmenopausal (-127 ± 79 kcal/d) than those who remained premenopausal (64 ± 60 kcal/d). These findings suggest that the decline in physical activity may accelerate the accumulation of central body fat in women during the menopause transition. This finding is particularly noteworthy given that a high waist-to-hip ratio has been found to be positively associated with risk of mortality (17). These findings suggest that menopause may accelerate the decline in physical activity and increase central adiposity more so than the aging process. Dietary, exercise, and hormonal interventions particularly targeted at premenopausal women may help offset the worsening cardiovascular and metabolic risk profile associated with menopause.

Physical Activity and Fat Oxidation

The mechanisms that determine the age-associated changes in body composition are not well-defined, but the evidence so far suggests that the loss of fat-free mass and gain in body fatness is at least partially attributable to physical inactivity (22). Another possible mechanism that may contribute to alterations in body composition and body-fat distribution are related changes in substrate oxidation. Specifically, an age-related alteration in fatty acid metabolism in which there is a shift toward nonoxidative disposal and away from oxidation may be implicated (15, 16). Macronutrient balance is achieved when there is an equilibrium between intake and oxidation for protein, carbohydrate, and lipid. It is generally accepted that protein and carbohydrate balances are tightly regulated by mechanisms that couple oxidation to intake. On the other hand, changes in fat intake do not result in acute compensatory adjustments in oxidation. Thus, an increase in fat intake or a decrease in fat oxidation may promote the maintenance of body-fat stores in older individuals. This area has recently been reviewed and will be summarized in this chapter (8).

How does fat oxidation change with advancing age, and what are the physiological factors that determine fat oxidation? Recent studies found that fat oxidation decreases as a function of age in women (8). This decrease in fat oxidation was principally determined by a decrease in fat-free mass, and not age, per se. This finding has several implications. Interventions that increase the quantity of fat-free mass (i.e., resistance training) may also increase fat oxidation and therefore be helpful in minimizing the increase in adiposity in older individuals. The mechanisms by which maintenance and/or accretion of fat-free mass might accomplish this objective are an increase in resting metabolic rate (32) and an increase in muscle mass (3), which is the main consumer of fatty acids in the human body. Thus, the increase in fat-free mass may exert favorable changes from the energetic as well as the substrate utilization points of view.

Fatty acid mobilization and oxidation from adipose tissue triglyceride stores is generally considered a predominant source of fuel in working muscles, particularly during long-term exercise. However, it is unclear as to which *in vivo* factors are involved in the regulation of fat oxidation in response to endurance training in the elderly. Although sympathetic nervous system activity probably plays a regulatory role in fat mobilization and utilization in adult humans (2), the metabolic link between changes in fat oxidation and sympathetic nervous system activity in response to exercise training in the elderly remains unclear. A reduced capacity to oxidize stored lipids has been found in the elderly in response to short-term fasting and catecholamine stimulation (23, 26). We recently submitted older individuals to a weight-loss program (4). We hypothesized that weight reduction may restore "youthful" levels of fat oxidation. Contrary to our hypothesis, we found that weight loss actually decreased fat oxidation. This decrease in fat oxidation could explain, at least partially, the tendency for body-fat accretion so commonly observed after weight-reducing programs.

Very few studies have examined the effects of chronic exercise training on fat oxidation and sympathetic nervous system in the elderly. In a cross-sectional study, we found that younger aerobically trained individuals have a decrease in the ratio of nonoxidative to oxidative disposal, when compared with younger individuals matched for age and body weight who are not aerobically trained (9). This decrease in the ratio of nonoxidative to oxidative disposal of fatty acids may explain, in part, some of the beneficial effects of exercise in weight-loss programs, by shifting the utilization of fatty acids toward oxidation and away from recycling. We recently conducted a short-term trial of exercise training in elderly individuals (8 weeks) to examine changes in fatty acid metabolism (30). Individuals did not change body weight or composition at the end of the training period. An increase in peak oxygen consumption was observed, documenting a training effect. We found that whole-body fuel utilization patterns shifted toward an increase in the rate of fat oxidation. This increase was independent of the rate of lipolysis, and was related to an increase in the activity of the sympathetic nervous system, as measured by the rate of norepinephrine appearance. These results suggest that the lower rate of fat oxidation in older individuals is at least partially correctable by endurance training.

Summary

In summary, physical activity is a major modifiable component of total daily energy expenditure in older individuals. Preliminary evidence suggests that regular physical activity may enhance fat oxidation and partially prevent the age-related increase in central body fatness. Preliminary evidence using doubly labeled water suggests that endurance and/or resistance training does not increase total daily energy expenditure, although these results await confirmation. Problems of compliance and adherence to physical activity still represent major barriers to adopting a physically active lifestyle in older individuals, particularly in older minority populations.

Acknowledgments

Work in this review was supported by grants from the National Institute of Aging (AG-00564, AG-13978) and Department of Defense (UIS #DE 9502261) to Dr. Poehlman. The Nursing, Physiology and Staff support from the GCRC RR-109 is gratefully acknowledged. Appreciation is expressed to Philip A. Ades, MD for his helpful comments on the manuscript.

References

1. American College of Sports Medicine. 1995. *ACSM's Guidelines for Exercise Testing and Prescription*. American College of Sports Medicine, 5th edition, Williams and Wilkins, Medina, Pennsylvania.
2. Arner, P. 1988. Control of lipolysis and its relevance to development of obesity in man. *Diabetes Metabolism Reviews* 4: 507-515.
3. Ballor, D.L., and E.T. Poehlman. 1994. Exercise-training enhances fat-free mass preservation during diet-induced weight loss: a meta-analytic finding. *International Journal of Obesity* 18: 35-40.
4. Ballor, D.L., J.R. Harvey-Berino, P.A. Ades, J. Cryan, and J. Calles-Escandon. 1996. Contrasting effects of resistance and aerobic training on body composition and metabolism after diet-induced weight loss. *Metabolism* 45: 179-183.
5. Baumgartner, R.N., P.M. Stauber, D. McHugh, K.M. Koehler, and P.J. Garry. 1995. Cross-sectional age differences in body composition in persons 60+ years of age. *Journal of Gerontology* 50A: M307-M326.
6. Bjorntorp, P. 1997. Body fat distribution, insulin resistance, and metabolic diseases. *Nutrition* 13: 795-803.
7. Blair, S.N., H.W. Kohl, C.E. Barlow, R.S. Paffenbarger, L.W. Gibbons, and C.A. Macera. 1995. Changes in physical fitness and all-cause mortality. *Journal of the American Medical Association* 273: 1093-1098.
8. Calles-Escandon, J., and E.T. Poehlman. 1997. Aging, fat oxidation and exercise. *Aging Clinical and Experimental Research* 9: 57-63.
9. Calles-Escandon, J., and P. Driscoll. 1994. Free fatty acid metabolism in aerobically fit individuals. *Journal of Applied Physiology* 77: 2374-2379.
10. Campbell, W.W., M.C. Crim, V.R. Young, and W.J. Evans. 1994. Increased energy requirements and changes in body composition with resistance training in older adults. *American Journal of Clinical Nutrition*. 60: 167-175.
11. Carpenter, W.H., T. Fonong, M.J. Toth, P.A. Ades, J. Calles-Escandon, J.D. Walston, and E.T. Poehlman. 1998. Total daily energy expenditure in free-living older African-Americans and Caucasians. *American Journal of Physiology* 37: 96-101.

12. Davy, K.P., S.L. Evans, E.T. Stevenson, and D.R. Seals.1996. Adiposity and regional body fat distribution in physically active young and middle-aged women. *International Journal of Obesity* 20: 777-783.

13. Dawson, D.A. 1988. Ethnic differences in female overweight: data from the 1985 National Interview Survey. *American Journal of Public Health.* 129: 1376-1379.

14. Fiatarone, M.A., E.F. O'Neill, N.D. Ryan, K.M. Clements, G.R. Solares, M.E. Nelson, S.R. Roberts, J.J. Kehayias, L.A. Lipsitz, and W.J. Evans. 1994. Exercise training and nutritional supplementation for physical frailty in very elderly people. *New England Journal of Medicine* 330: 1769-1775.

15. Flatt, J.P. 1987. Dietary fat, carbohydrate balance, and weight maintenance: effects of exercise. *American Journal of Clinical Nutrition* 45: 296-306.

16. Flatt, J.P. 1995. Use and storage of carbohydrate and fat. *American Journal of Clinical Nutrition* 61(suppl): 952S-959S.

17. Folsom, A.R., S.A. Kaye, T.A. Sellers, C.P. Hong, J.R. Cerhan, J.D. Potter, and R.J. Prineas. 1993. Body fat distribution and 5-year risk of death in older men. *Journal of the American Medical Association* 269: 483-487.

18. Forbes, G.B., and J.C. Reina. 1970. Adult lean body mass declines with age: some longitudinal considerations. *Metabolism*19: 653-656.

19. Ford, E.S., and W.H. Herman. 1995. Leisure-time physical activity patterns in the US diabetic population. *Diabetes Care* 18: 27-33.

20. Goran, M.I., and E.T. Poehlman. 1992. Endurance training does not enhance total energy expenditure in healthy elderly persons. *American Journal of Physiology* 263: E950-E957.

21. Hakim, A.A., H. Petrovitch, C.M Burchfiel, G.W. Ross, B.L. Rodrigues, L.R. White, K. Yano, J.D. Curb, and R.D. Abbott. 1998. Effects of walking on mortality among nonsmoking retired men. *New England Journal of Medicine* 338: 94-99.

22. Holloszy, J.O., and W.M. Kohort. 1995. In *Handbook of Physiology - Aging,* ed. E.J. Masoro, 633-666. Oxford: University Press.

23. Klein, S., V.R. Young, G.L. Blackburn, B.R. Bistrian, and R.R. Wolfe. 1990. Palmitate and glycerol kinetics during brief starvation in normal weight young adult and elderly subjects. *Journal of Clinical Investigation* 78: 928-933.

24. Kohrt, W.M., M.T. Malley, G.P. Dalsky, and J.O. Holloszy. 1991. Body composition of healthy sedentary and trained, young and older women. *Medicine and Science in Sports and Exercise* 24: 822-837.

25. Kumanyika, S. 1987. Obesity in black women. *Epidemiological Reviews* 9: 31-50.

26. Lonnqvist, F., B. Nyberg, H. Wahrenberg, and P. Arner. 1990. Catecholamine-induced lipolysis in adipose tissue of the elderly. *Journal of Clinical Investigation* 85: 1614-1621.

27. Pate, R.R., M. Pratt, S.N. Blaor, W.L. Haskell, C.A. Macera, C. Bouchard, D. Buchner, W. Ettinger, G.W. Heath, A.C. King, A. Kriska, A.S. Leon, B.H.

Marcus, J. Morris, R.S. Paffenbarger, Jr., K. Patrick., M.L. Pollock, J.M. Rippe, J. Sallis, and J.H. Wilmore. 1995. Physical activity and public health. A recommendation from the Centers for Disease Control and Prevention and the American College of Sports Medicine. *Journal of the American Medical Association* 273: 402-407.

28. Poehlman, E.T. Regulation of energy expenditure in aging humans. 1993. *Journal of American Geriatric Society* 41: 552-559.

29. Poehlman, E.T., P.J. Arciero, and M.I. Goran. 1994. Endurance exercise in aging humans: effects on energy metabolism. In *Exercise and Sports Science Reviews,* ed. J.O. Holloszy, 250-284. Baltimore: Williams and Wilkins.

30. Poehlman, E.T., A.W. Gardner, M.I. Goran, and J. Calles-Escandon. 1994. Effects of endurance training on total fat oxidation in elderly persons. *Journal of Applied Physiology* 76: 2281-2287.

31. Poehlman, E.T., M.J. Toth, L.B. Bunyard, A.W. Gardner, K.E. Donaldson, E. Colman, T. Fonong, and P.A. Ades. 1995. Physiological predictors of increasing total and central adiposity in aging men and women. *Archives of Internal Medicine* 155: 2443-2448.

32. Poehlman, E.T., T.L. McAuliffe, D.R. VanHouten, and E. Danforth. 1990. Influence of age and endurance training on metabolic rate and hormones in healthy men. *American Journal of Physiology* 259: E66-E72.

33. Poehlman, E.T., M.J. Toth, and A.W. Gardner. 1995. Changes in energy balance and body composition at menopause: a controlled longitudinal study. *Annals of Internal Medicine* 123: 673-675.

34. Pollock, M.L., L.J. Mengelkoch, J.E. Graves, D.T. Lowenthal, and M.C. Limacher, et al. 1997. Twenty-year follow-up of aerobic power and body composition of older track athletes. *Journal of Applied Physiology* 82: 1508-1516.

35. Powell, K.E., and S.N. Blair. 1994. The public health burdens of sedentary living habits: theoretical but realistic estimates. *Medicine and Science in Sports and Exercise* 26: 851-856.

36. Pratley R., B. Nicklas, M. Rubin, J. Miler, A. Smith, M. Smith, B. Hurley, and A. Goldberg. 1994. Strength increases resting metabolic rate and norepinephrine levels in healthy 50-to 65 yr. old men. *Journal of Applied Physiology* 76: 133-137.

37. Schwartz, R.S., W.P. Schuman, V.L. Bradbury, K.C. Cain, G.W. Gellingham, J.C. Beard, S.E. Kahn, J.R. Stratton, M.D. Cerquiera, and I.B. Abrass. 1991. The effect of intensive endurance exercise training on body fat distribution in young and older men. *Metabolism* 40: 545-551.

38. Starling, R.D., M.J. Toth, D.E. Matthews, and E.T. Poehlman. 1998. Energy requirements and physical activity of older free-living African-Americans: A doubly labeled water study. *Journal of Clinical Endocrinology and Metabolism* 83: 1529-1534.

39. Stevenson, E.T., K.P. Davy, and D.R. Seals.1995. Hemostatic, metabolic and androgenic risk factors for coronary heart disease in physically active and

less active postmenopausal women. *Arteriosclerosis Thrombosis Vascular Biology* 15: 669-677.

40. Stevenson, J.A.F., B.M. Box, B.V. Feleki, and J.R. Beaton. 1966. Bouts of exercise and food intake in the rat. *Journal of Applied Physiology* 21: 118-122.

41. Thompson, B.M., and A.T. Miller Jr. 1958. Adaptation to forced exercise in the rat. *American Journal of Physiology* 193: 350-354.

42. Toth, M.J., and E.T. Poehlman. 1996. Effects of exercise on daily energy expenditure. *Nutrition Reviews* 54: S140-148.

43. Trappe, S.W., D.L. Costill, M.D. Vukovich, J. Jones, and T. Melham. 1996. Aging among elite distance runners. *Journal of Applied Physiology* 80: 285-290.

44. Treuth, M.S., G.R. Hunter, T. Kekes-Szabo, W.L. Weinsier, M.I. Goran, and L. Berland. 1995. Reduction in intra-abdominal adipose tissue after strength training in older women. *Journal of Applied Physiology* 8: 1425-1431.

45. Treuth, M.S., G.R. Hunter, R. Weinsier, and S. Kell. 1995. Energy expenditure and substrate utilization in older women after strength training: 24 hour calorimeter results. *Journal of Applied Physiology* 78: 2140-2146.

46. Tzankoff, S.P., and A.H. Norris. 1977. Effect of muscle mass decrease on age-related BMR changes. *Journal of Applied Physiology* 43: 1001-1006.

47. U.S. Department of Health and Human Services. 1998. *Physical activity and Health: A report of the Surgeon General.* Atlanta, GA: U.S. Department of Health and Human Services, Centers for Disease Control and Prevention, National Center for Chronic Disease Prevention and Health Promotion.

48. Van Etten, L.M.L.A., K.R. Westerterp, F.T.J. Verstappen, B.J.B. Boon, and W.H.M. Saris. 1997. Effect of an 18-wk weight-training program on energy expenditure and physical activity. *Journal of Applied Physiology* 82: 298-304.

CHAPTER 14

Physical Activity Level and Weight Control During Pregnancy

Stephan Rössner

Obesity Unit, MK-division, Huddinge University Hospital, Huddinge, Sweden

Pregnancy and body weight are related in complicated ways. In general, there is little solid evidence that pregnancy is an important risk factor for overweight and obesity. On the other hand, some studies suggest that overweight mothers often describe their pregnancies as an important risk period for future weight increase (4, 28). Of the obese women at our Obesity Unit, 73% reported that they had retained more than 10 kg in connection with a pregnancy.

Age and parity (the history of previous childbirth) are factors statistically associated with weight gain during pregnancy; however, weight development during pregnancy is extremely variable and surprisingly difficult to predict. Some cross-sectional studies (5, 8, 14) have suggested that maternal body weight is higher in women who have had more children. Such studies, however, can be criticized on grounds of numerous confounding factors. Longitudinal studies, comparing the weight development in a mother throughout pregnancy and afterwards, avoid these problems, but can cause other methodological complications. As suggested by Harris (13), there are three main problems that can produce an overestimation of the effect of pregnancy on maternal weight development.

• Most women underreport their body weight (32). This means that studies using retrospective self-reports of prepregnancy weight overestimate the total amount of weight gain throughout a pregnancy.

• It is crucial to decide when after-delivery measurements of body weight should be used for calculations and comparisons. If this weight is taken soon after delivery, weight fluctuations associated with pregnancy processes may not yet have stabilized. Lactation, although with little overall effect on body weight development, will induce a non-steady-state condition as well as behavioral aspects that may bring about a new eating behavior (23). On the other hand, if the time for comparison is chosen too late after delivery, new lifestyle habits may have appeared that complicate the interpretation. One such life event may actually be a new pregnancy.

- Because body weight increases over time in this age group of fertile women, the aging process itself will result in weight increase over time. Such development can and should be appropriately corrected in analysis of weight gain during pregnancy (23).

The importance of correcting for these confounding variables is demonstrated by the study of Harris and Ellison (12), who argued that out of 71 longitudinal studies only 3 satisfied all the criteria, providing an adequate estimate of maternal weight gain. With or without these corrections, the mean weight retention after a pregnancy has been found to be quite variable, generally ranging from 0.6 to 3.0 kg (30).

Physical Activity, Fitness, and Health in Women

There is no general scientific evidence that physical activity induces reproductive morbidity in women (19). However, some data suggest that abruptly imposed prolonged and intense physical activity can induce reproductive disturbances, at least in some women. The relationship between such a stress situation and the neuroendocrinological consequences is not well established. Stress may activate the adrenal axis, and physical activity may well be one of those stress signals. One mechanism by which such stress could cause damage to reproductive function might be by affecting eating behavior. Failure to compensate for the high energy losses associated with physical activity with an increased energy intake may be one such hypothetical mechanism.

Physical activity may influence reproductive function throughout the life of any woman (9). In most cases these changes seem to be the result of a hypothalamic dysfunction and suppression of the gonadotropin-releasing hormone (GnRH) pulse generator. Young women who are engaged in athletic activities commonly have menarche delayed. Abraham (1) suggested that for young ballet dancers menarche is delayed by two years. A similar delay, although less marked, has been described for runners, track and field athletes (21), volleyball players (22), skaters (34), and swimmers (31). In these groups of athletes menstrual irregularities are also common. In the general population the prevalence is generally below 5% (29), whereas in many types of athletes, most marked in dancers, irregularities are often found in 50% or more of these young women.

Numerous maternal cardiovascular adaptations to pregnancy occur, and these changes generally take place in the first trimester. Plasma volume, red cell count volume, and cardiac output and stroke all increase initially (6). Plasma volume increases by almost 50% at the time of delivery (15, 20). Hemoglobin concentration falls during later stages of pregnancy due to the dilution effect (20).

Of particular interest is the effect of pregnancy on exercise efficiency. Treadmill exercise studies during the latter half of pregnancy suggest that the caloric requirements for specific work increases in pregnancy if expressed as total requirements, but remains unaffected if oxygen uptake is expressed per kg body weight (3). However, in other studies this increase in caloric requirements during pregnancy has not been confirmed (for example, 17).

Pregnancy and Lactation

On theoretical grounds it can be assumed that the energy requirement for an entire pregnancy is ~80,000 calories or ~300 calories per day covering the needs for fetal growth and adipose tissue storage. Lactation has been assumed to facilitate weight loss, in particular if the period of lactation exceeds 2 months. Lactation has been calculated to increase the energy requirements by ~500 calories per day, which also takes into account that production of milk is an energy-requiring process.

Studies by Rebuffé-Scrive et al. (26) and others have demonstrated that the characteristics of the adipose tissue change dramatically during pregnancy. Adipose tissue lipolysis in the femoral regions is limited during pregnancy, but can easily be stimulated hormonally during lactation. This seems to be a functional adaptation of the adipose tissue to the needs of the mother and the newborn child. However, the relationships between lactation, energy needs, and energy balance are not fully clarified. In some studies no difference in energy intake was found between lactating and nonlactating women. Other studies suggest that lactation plays a minor role in weight reduction after delivery (16, 23).

Three mechanisms may explain the lack of weight loss after delivery: reduced basal metabolism, impaired thermogenesis, and reduced physical activity (25). These mechanisms could account for a reduction in energy requirements of up to ~240 calories per day. Thus it is possible that the energy need during lactation can be provided from different sources, such as an increased energy intake, utilization of adipose tissue in storage, and a metabolic adaptation.

Smoking and Pregnancy

Smoking increases the basal metabolic rate by ~10% (27). The natural weight development pattern in pregnancy is confounded by the fact that many women smokers fortunately give up this habit when they learn of the pregnancy. In this situation, the weight gain is influenced by an additional factor apart from the growth of the fetus. Smokers gain less weight during pregnancy than nonsmokers, whereas those who give up smoking before delivery increase more in weight. It is reasonable to assume that the smoking habits of a pregnant woman for both physiological and psychological reasons will affect her physical activity habits. However, little scientific information is available to document these possible interrelationships. It is likely that pregnant women, as with others who give up smoking, will find physical activity more rewarding and less strenuous and will therefore be more inclined to be physically active.

Body Image Change During Pregnancy

A woman's image of her body is an important factor in terms of energy intake and expenditure. In general, overweight and obesity is stigmatized in many societies, but as weight increases during pregnancy these attitudes may change. Fox and Yamaguchi (11) have described the feelings of 76 primigravidas of at least 30 weeks' gestation. Women who were of normal body weight at conception were more likely to hold a negative attitude toward their changing body during pregnancy (62% of the women), whereas those who were overweight before pregnancy were more likely to have a positive attitude (again, 62% of the women). Overweight women probably experience a positive attitude because they perceive the benefits of reduced self-consciousness, stigma, and the pressure to adhere to a diet. The pregnancy itself allows them to be overweight without feeling compelled to conform with an ideal slender body shape that is dictated by society (35). Such a finding will of course affect the tendency of women of varying prepregnancy body shapes to incorporate regular physical activity into their everyday lives. A more relaxed attitude toward weight increase makes it more likely that women will refrain from engaging in regular physical activity, unless they perceive a positive value in the exercise for other reasons than body weight control.

There is little information about the role of physical activity in affecting body image during pregnancy. Fox and Yamaguchi (11) found that multigravidas were less likely than primigravidas to experience body-shape concern and more likely to have developed a greater tolerance of their changing bodies, especially if the current pregnancy was planned to be their last one. Extremely obese women may not see the body-shape change as a positive development, because they were frustrated with their weight situation prior to pregnancy. Finally, it is important to note that in different ethnic groups body shape is valued differently. Black American women, compared with white women, have a wider tolerance in their definition of overweight and obesity and are also more inclined to value free time as leisure time and not as time for physical activity (33). These data further underscore the role of physical activity in affecting weight development during pregnancy. Also, physical activity, which is one way to affect body image, is of great importance.

In Harris' study (13) of behavioral factors that predispose parous women to long-term weight gain, the role of physical activity was briefly addressed (13). Mothers who reported in a questionnaire that they exercised less after pregnancy than before had significantly greater long-term weight gain ($+2.1 \pm 0.9$ kg) than those who reported a similar degree of activity before and after pregnancy (weight change -1.0 ± 1.0 kg, $p = 0.03$). However, no significant difference in response was found to three other questions addressing physical activity outdoor time, television watching, and walking time. Behavioral aspects again turned out to be important determinants of weight development; change in body image was most closely associated with change in physical activity (p < 0.001).

The Stockholm Pregnancy and Weight Development Study

As part of the Stockholm Pregnancy and Weight Development Study my colleagues and I studied trends in eating patterns, physical activity, and sociodemographic factors in relation to postpartum body weight development (24). In this study, factors affecting weight development were studied in 1423 pregnant women retrospectively from conception until delivery, and then at postpartum checkups at 2.5 months, 6 months, and 1 year. Weight development data were collected in a standardized fashion (figures 14.1-14.3). Furthermore, a trend method was constructed to identify a number of predefined major patterns of behavior, including physical activity. Table 14.1, table 14.2, and figure 14.4 summarize the results from this study.

Groups with different degrees of physical activity in leisure time before pregnancy or at work before pregnancy did not differ in prepregnancy body mass index

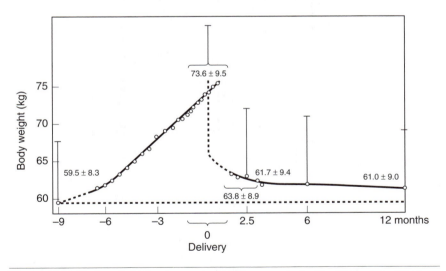

Figure 14.1 Body weight development of 1423 women in the Stockholm Pregnancy and Weight Development Study from onset of pregnancy until 12 months postpartum. Data for mean values ± one standard deviation at conception, delivery, and 2.5, 6, and 12 months postpartum are shown. In addition, circles show mean data from women weighed that particular week.

Reprinted, by permission, from A. Öhlin and S. Rössner, 1990, "Maternal weight development after pregnancy," *International Journal of Obesity* 14: 159-173.

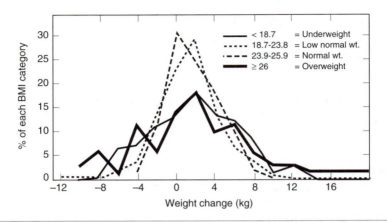

Figure 14.2 Distribution of weight changes in different prepregnancy weight groups. Reprinted, by permission, from A. Öhlin and S. Rössner, 1990, "Maternal weight development after pregnancy," *International Journal of Obesity* 14: 159-173.

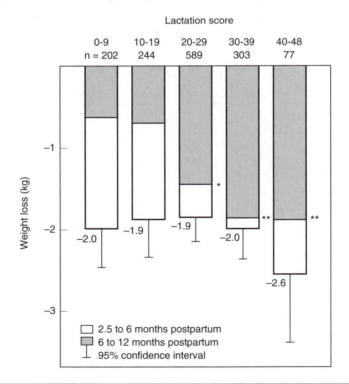

Figure 14.3 Weight loss pattern (kg) from 2.5 to 12 months postpartum in groups with different lactation scores.
* = Differed from (0-9) + (10-19) at $p < 0.05$, ** = Differed from (0-9) + (10-19) at $p < 0.01$ (23).
Reprinted, by permission, from A. Öhlin and S. Rössner, 1990, "Maternal weight development after pregnancy," *International Journal of Obesity* 14: 159-173.

Table 14.1   Results From the Stockholm Pregnancy and Weight Development Study

Factors significantly associated with weight 1 year after delivery (from reference 23)

Weight increase during pregnancy	r = .36***
Lactation score	r = −.09**
Age	r = .06*
Irregular eating habits after delivery	r = −.07*
Leisure time physical activity after delivery	r = −.05*

Main lifestyle changes associated with weight retention 1 year after delivery (kg)

Smoking cessation	+3.4 kg**
Marked weight gain (≥ 6 kg) after previous pregnancy	+2.5 kg*

Factors not significantly associated with body weight retention 1 year after delivery

Initial body weight, parity, previous use of contraceptive pill, social class, occupation, marital status, nationality, dietary advice during pregnancy

* = p < 0.05, ** = p < 0.01, *** = p < 0.001.

Table 14.2   Retrospective Analysis of Women With ± 3 kg Weight Changes 1 Year After Delivery Compared to Prepregnancy Weight vs. All Women Retaining ≥ 5 kg, Compared to Their Prepregnancy Body Weight

Weight development	Net weight changes, one year postpartum		
	± 3 kg	≥ 5 kg	p-value
Prepregnancy BMI (kg/m²)	21.0	22.4	< 0.001
Pregnancy weight gain (kg)	13.5	17.2	< 0.001
Weight change between 2.5 and 12 months postpartum (kg)	−2.3	+0.4	< 0.001
Lactation score (points)	22.7	20.7	< 0.05
Eating scores			
Global eating pattern score (points)	68.2	66.1	< 0.05
Global meal quality score (points)	31.0	31.2	n.s.
Eating habits			
Regular breakfast habit (% of women)	82	74	< 0.05
Regular lunch habit (% of women)	39	9	< 0.01

Data from Öhlen and Rossner 1990(24)

(BMI) or change in weight (Δ weight). Women with different levels of physical activity at work before and during pregnancy did not differ in weight; nor did women in different weight categories differ in degree of physical activity at work.

Fewer of the high weight-retainers (weight = 5 kg) were physically active in their leisure time during all study periods (at least 4-6 hours light activity/week) compared to the women with lower weights (table 14.3). Of the most pronounced weight gainers (weight = 10 kg), 23% were inactive in their leisure time throughout the study, compared with 4% of the women with lower weight (p < 0.001). Groups with different trends of physical activity in leisure time did not differ significantly in weight (figure 14.4). Activity score in leisure time during 7-12 months postpartum correlated with weight (r = -0.05, p < 0.05).

How is physical activity during and after pregnancy associated with weight-change patterns? Physical inactivity before, during, and after pregnancy was reported in 48% of Stockholm women with BMI < 24 kg/m² and in 58% of women with BMI ≥ 24 (p < 0.05). More of the women with BMI < 24 returned to a higher level of physical activity after pregnancy than among the more overweight women (23% vs. 15%, p < 0.005).

Thirteen percent of the women in the Stockholm Study reported that they had intentionally slimmed down during the year after delivery. Between 3 and 12 months postpartum these slimmers lost 3.7 kg, while the other women lost 1.7 kg. The

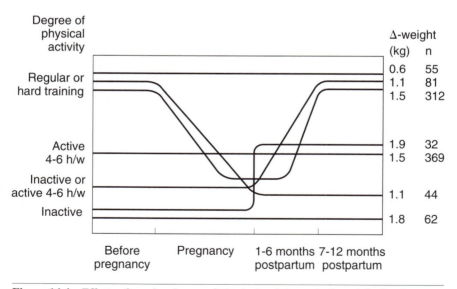

Figure 14.4   Effects of varying degree of physical activity on body weight development before pregnancy, during pregnancy, and during the first and second half year after delivery. Change in weight (Δ weight) and number of women (n) in each group that exhibit a particular pattern of physical activity are shown.

Table 14.3   Factors That Were Represented in Significantly Different Proportions in a Group of Women With Weight Retention (Δ-weight) ≥ 5 kg Compared With Women With Δ-weight < 5 kg During and After Pregnancy

	Women following the described trends (%)		Statistical significance of difference*
Wt retention (kg) . . .	< 5	≥ 5	(p = value)
n . . .	1434	189	
Eating habits			
Perceived increase in meal sizes and/or frequency of snacking during and after pregnancy	26	37	< 0.01
Breakfast daily throughout the study	82	74	< 0.05
Lunch daily throughout the study	39	29	< 0.05
Reduction to 0-4 lunches/week from during or after pregnancy until late follow-up	17	24	< 0.01
Other factors			
Light leisure time physical activity at least 4-6 h/week throughout the study	54	46	< 0.05
Living in suburbs	42	53	< 0.05
Lower clerk employment	22	36	< 0.05

*Chi-square analysis.

Data from the Stockholm Pregnancy and Weight Development study (24).

slimmers lost their weight mainly during the second 6-month period postpartum, while the nonslimmers lost their weight during the first 6 months. The frequencies of reported behavior were very similar between the slimmers and the nonslimmers. The only difference was that more slimmers had increased their physical activity again after the delivery (26%) compared to the other women (13%) (p < 0.0001).

Neither the eating habits or the level of physical activity before or during the pregnancy could be used as predictors for subsequent weight retention. We found only a nonsignificant tendency for women who had a physically demanding occupation before pregnancy to have a higher weight, which might be explained by a lowered energy expenditure during and after the pregnancy than while working.

Does body perception affect weight development after pregnancy? Using the WHO/FAO/UNO BMI cutoff limits for women within the normal weight range, we found that 1% underestimated their true body size and 12% overestimated their body size. Women over the whole weight spectrum who underrated their body size had retained 2.0 kg more after 6 months than those with a more realistic body-size perception (p < 0.01). After 12 months these women still retained 1.2 kg more than the others (p < 0.05). Women underrating their body size were on average 2.0 years older (p < 0.01) and had a significantly higher score for leisure time activity (p < 0.05) throughout the whole 21-month period than the others. Twenty-six percent of those who overrated their body weight had smoked throughout pregnancy compared to 17% in other groups (p < 0.01). Body-size perception was not related to social class.

Our Stockholm data support the notion that body weight changes, as measured 1 year after delivery, are much more determined by lifestyle changes during and after pregnancy than by pre-conception body weight. The return to pre-pregnancy weight seems to be more likely to occur in women who try to adhere to a generally "healthful" lifestyle, i.e., eating regularly and exercising, with one exception—smoking. Cessation of smoking, which certainly is one of the most positive actions a mother can undertake for herself and her child, is one of the behavioral changes associated with the highest increase in weight during and after pregnancy (23, 27).

Overall, there was a reasonable agreement between body-size perception and BMI values. Although overrating of body size was occasionally found at BMI levels as low as 18 to 20 kg/m^2, such ratings were generally observed among women in the upper part of the normal BMI range. It is possible that the underrating found in 4% of all women can be partly explained by the fact that these women were more muscular and thus rated their body size as less overweight than the BMI value suggests.

Living in a suburb and being a clerk were the only sociodemographic factors that were identified as risk factors for increased weight gain. The association between weight development and these factors could be due to either a lower level of education or differences in social structure, or that the staff acted differently at different maternity clinics. However, social group did not correlate with weight, and the social group profile was similar in suburbs near and far from the city.

Physical activity during the follow-up year thus seems to be of importance for facilitating postpartum weight control. The degree of leisure activity might, besides the effect on energy expenditure, also indirectly reflect the health consciousness of the woman.

Weight Control During and After Pregnancy—Realistic Expectations

It is likely that incorporation of regular physical activity will be of value for the woman with an uncomplicated pregnancy. An active lifestyle would increase her control over her body weight development and also have other important metabolic benefits. However, some physicians that care for their pregnant clients have little knowledge about the role of physical activity in a healthy lifestyle. Kristeller and Hoerr (18) have recently summarized data from a mail survey of 1,222 physicians from six specialities including gynecology. Data from this study must however be interpreted with caution. In particular, the response rate of 43% is rather low. Table 14.4 summarizes some of their findings. The overall impression from this study is that speciality groups shared a high concern for health risks of moderate and morbid obesity, but that the gynecologists were not very active in intervening in weight-related issues and were more likely to refer their patients to other types of health support. As a matter of fact, the gynecologists were found to be among the specialists least inclined to be involved in caring for their obese patients, providing a counseling plan, or seeing the need for patient skill training. This study does not address the particular question of the actual role of physical activity in weight control during pregnancy, but serves to illustrate that gynecologists may not advise pregnant women to prevent excessive weight gain through physical activity. As stated by Kristeller and Hoerr, "Women who see gynecologists as their primary care provider may not be receiving comparable benefit of preventive care."

The effects of strenuous exercise or heavy physical work during pregnancy on weight development and fetal outcome has not been studied extensively (2). There are several methodological reasons for the paucity of scientific information. In women undergoing extremely high levels of exercise during pregnancy, a low maternal weight gain and low fetal birth weights have, however, been reported (10). In general, the studies available have not examined weight gain in relationship to physical activity in detail, but rather as a general component in the overall energy balance. In the summary by Abrams (2), the various individual types of outcome during a pregnancy regarding energy balance is thus summarized: "For some women increased amounts of maternal weight gain may improve fetal growth and therefore improve fetal health. For other women high gains . . . may enhance fetal growth to the point where numerous health complications may occur. Balancing the health benefits and the risk of maternal weight gain is a challenge that mothers and care providers will continue to face in the future." This quotation further illustrates that the role of physical activity in determining the final outcome of weight development during pregnancy is a highly variable component in the energy balance equation.

Summary

In spite of statistically significant associations, it remains surprisingly difficult to predict weight changes during and after pregnancy in a clinically meaningful way. Women who will increase more in weight during pregnancy start to do so early, and such women should be identified for possible intervention. Lifestyle changes,

Table 14.4   Physician Attitude Toward Managing Obesity

	Family practice	Internal medicine	Endocrinology	Cardiology	Gynecology	Orthopedics
Respond to questionnaire (%)	41	38	54	33	49	39
Overall involvement (1-7)*	4.4	4.4	4.4	3.8	2.5	2.7
Provide counseling plan (1-5)*	3.7	3.5	3.7	3.1	3.1	1.8
Need for patient skill training (1-5)*	3.8	3.8	3.7	3.3	2.8	2.0

*Range of involvement, 1 = least, 5 (7) = most.

Adapted from Kristeller and Hoerr 1997 (18).

including physical activity, seem to be important determinants of body weight changes after pregnancy; this finding has practical implications for advice in the maternity unit. Changes in eating behavior patterns after delivery seem to counteract the inherent weight controlling potential of breastfeeding. Slightly disappointing was our finding in Stockholm that women receiving dietary advice had the same weight outcome as those who received no such support (23). We do not know if encouragement of physical activity during pregnancy will be of equally low impact. Recommendations for pregnancy weight gain in other societies than those studied so far will have to include population-specific strategies. Methodological problems make it difficult to isolate the true effect of physical activity on weight development during and after pregnancy.

References

1. Abraham, S.F., P.J.V. Beaumont, I.S. Fraser, and D. Llewellyn-Jones. 1982. Body weight, exercise and menstrual status among ballet dancers in training. *British Journal of Obstetrics and Gynaecology* 89:507-510.

2. Abrams, B. 1994. Weight gain and energy intake during pregnancy. *Clinical Obstetrics and Gynecology* 37: 515-527.

3. Blackburn, M.W., and D.H. Calloway. 1976. Basal metabolic rate and work energy expenditure of mature, pregnant women. *Journal of the American Dietetic Association* 69: 24-28.

4. Bradley, P.J. 1985. Conditions recalled to have been associated with weight gain in adulthood. *Appetite* 6: 235-241.

5. Brown, J.E., S.A. Kaye, and A.R. Folsom. 1992. Parity-related weight change in women. *International Journal of Obesity* 16: 627-631.

6. Capeless, E., and J. Clapp. 1989. Cardiovascular changes during pregnancy. *Society of gynecologic investigation scientific program and abstracts 36th annual meeting,* 165.

7. Carpenter, M.W. 1994. Physical activity, fitness, and health of the pregnant mother and fetus. In *Physical activity, fitness, and health,* eds. C. Bouchard, R.J. Shephard, and T. Stephens, 967-979. Champaign, IL: Human Kinetics.

8. Cederlöf, R, and L. Kaij. 1970. The effect of childbearing on body weight. A twin control study. *Acta Psychiatrica Scandinavica* 219 (Suppl): 47-49.

9. Constantini, N.W., and M.P. Warren. 1994. Physical activity, fitness, and reproductive health in women: clinical observations. In *Physical activity, fitness, and health,* eds. C. Bouchard, R.J. Shephard, and T. Stephens, 955-966. Champaign, IL: Human Kinetics.

10. Dewey, K., and M. McCrory. 1994. Effects of dieting and physical activity on pregnancy and lactation. *American Journal of Clinical Nutrition* 59(suppl): 446-453.

11. Fox, P., and C. Yamaguchi. 1997. Body image change in pregnancy: A comparison of normal-weight and overweight primigravidas. *BIRTH* 24(1): 35-40.
12. Harris, H.E., and G.T.H. Ellison. 1997. Do the changes in energy balance that occur during pregnancy predispose parous women to obesity? *Nutrition Research Reviews* 10: 57-81.
13. Harris, H.E., G.T.H. Ellison, M. Holliday, and E. Lucassen. 1997. The impact of pregnancy on the long-term weight gain of primiparous women in England. *International Journal of Obesity* 21: 747-755.
14. Heliövaara, M., and A. Aromaa. 1981. Parity and obesity. *Journal of Epidemiology and Community Health* 35: 197-199.
15. Hytten, F.E., and D.B. Paintin. 1963. Increase in plasma volume during normal pregnancy. *Journal of Obstetrics and Gynaecology of the British Commonwealth* 70: 402.
16. Illingworth, P.J., R.T. Jung, P.W. Howie, and T. E. Isles. 1987. Reduction in postprandial energy expenditure during lactation. *British Medical Journal* 294: 1573-1576.
17. Knuttgen, H.G., and K. Emerson. 1974. Physiological response to pregnancy at rest and during exercise. *Journal of Applied Physiology* 36: 549-553.
18. Kristeller, J.L., and R.A. Hoerr. 1997. Physician attitudes toward managing obesity: Differences among six specialty groups. *Preventive Medicine* 26: 542-549.
19. Loucks, B.B. 1994. Physical activity, fitness, and female reproductive morbidity. In *Physical activity, fitness, and health,* eds. C. Bouchard, R.J. Shephard, and T. Stephens, 943-954. Champaign, IL: Human Kinetics.
20. Lund, C.J., and J.C. Donovan. 1967. Blood volume during pregnancy. *American Journal of Obstetrics and Gynecology* 98: 393-403.
21. Malina, R.M., A.B. Harper, H.H. Avent, and D.E. Campbell. 1973. Age at menarche in athletes and non-athletes. *Medicine and Science in Sports and Exercise* 5: 11-13.
22. Malina, R.M., W.W. Spirduso, C. Tate, and A.M. Baylor. 1978. Age at menarche and selected menstrual characteristics in athletes at different competitive levels and in different sports. *Medicine and Science in Sports and Exercise* 190: 218-222.
23. Öhlin, A., and S. Rössner. 1990. Maternal weight development after pregnancy. *International Journal of Obesity* 14: 159-173.
24. Öhlin, A., and S. Rössner. 1994. Trends in eating patterns, physical activity and socio-demographic factors in relation to postpartum body weight development. *British Journal of Nutrition* 71: 457-470.
25. Prentice, A.M., and A. Prentice. 1988. Energy costs of lactation. *Annual Review of Nutrition* 8: 63-79.
26. Rebuffé-Scrive, M., L. Enk, P. Crona, et al. 1985. Fat cell metabolism in different regions in women. Effects of menstrual cycle, pregnancy and lactation. *Journal of Clinical Investigation* 75: 1973-1976.

27. Rössner, S. 1986. Cessation of cigarette smoking and body weight increase. *Acta Medica Scandinavica* 219: 1-2.
28. Rössner, S. 1992. Pregnancy, weight cycling and weight gain. *International Journal of Obesity* 16: 145-147.
29. Singh, K.B. 1981. Menstrual disorders in college students. *American Journal of Obstetrics and Gynecology* 1210: 299-302.
30. Smith, D.E., C.E. Lewis, J.L. Caveny, L.L. Perkins, G.L. Burke, and D.E. Bild. 1994. Longitudinal changes in adiposity associated with pregnancy. *Journal of the American Medical Association* 271: 1747-1751.
31. Stager, J.M., and L.K. Hatler. 1988. Menarche in athletes: The influence of genetics and prepubertal training. *Medicine and Science in Sports and Exercise* 20: 369-373.
32. Stevens-Simon, C., K.J. Roghmann, and E.R. McAnarney. 1992. Relationship of self-reported prepregnant weight and weight gain during pregnancy to maternal body habits and age. *Journal of the American Dietetic Association* 92: 85-87.
33. Story, M., S. French, M. Resnic, and R. Blum. 1995. Ethnic/racial and socioeconomic differences in dieting behaviors and body image perceptions in adolescents. *International Journal of Eating Disorders* 18: 173-179.
34. Warren, M.P., and B. Brooks-Gunn. 1989. Delayed menarche in athletes: The role of low energy intake and eating disorders and their relation to bone density. In *Hormones and sport,* eds. Z. Laron, and A.D. Rogol. Vol 55: 41-54. New York: Serono Symposia Publications from Raven Press.
35. Wiles, R. 1993. "I'm not fat, I'm pregnant." The impact of pregnancy on fat women's body language. *Health Psychological Update* 12: 16-21.

CHAPTER 15

Physical Activity Level and the Treatment of Severe Obesity

Walker S. Carlos Poston II, MPH, PhD [1];
Richard R. Suminski, PhD, MPH [2]; **and John P. Foreyt, PhD** [3]

[1] Department of Psychology, University of Missouri-Kansas City and the Mid America Heart Institute, St. Luke's Hospital, Kansas City, Missouri, U.S.A.;
[2] Department of Health and Human Performance, University of Houston, Houston, Texas, U.S.A.; and [3] Behavioral Medicine Research Center, Department of Medicine, Baylor College of Medicine, Houston, Texas, U.S.A.

Definition and Prevalence of Severe Obesity

Obesity has become epidemic in the United States and other industrialized nations (24, 44, 53). Currently, overweight and obesity are defined as a body mass index (BMI) ≥ 25 and ≥ 30, respectively (44). Within the category of obesity, three severity levels have been developed: Class I (BMI 30-34.9), Class II (BMI 35-39.9), and Class III (BMI ≥ 40) (24, 44). The age-adjusted prevalence rates of Class I, Class II, and Class III obesity in U.S. adults are estimated to be 14.4%, 5.2%, and 2.9%, respectively (24, 44; see figure 15.1).

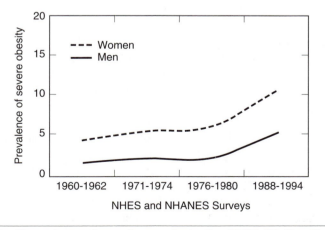

Figure 15.1   Change in prevalence of severe obesity (i.e., BMI ≥ 35; Class II + III obesity). Adapted from Flegal et al., 1998; NIH/NHLBI, 1998 (24, 44).

These estimates represent a 30%, 40%, and 55% increase in the respective obesity classes over just the last decade. While there is no precise definition of "severe" obesity, many studies define severe obesity as either a BMI ≥ 35 or ≥ 40. Thus, the prevalence is approximately 8.0% (37). This definition of severe obesity, which will be used in this paper, includes Class II and Class III obesity: this chapter focuses on physical activity for this population.

Health Concerns Associated With Severe Obesity

Severe obesity is associated with increased risk for numerous medical problems and health hazards, including hypertension, dyslipidemia, coronary heart disease, non-insulin-dependent diabetes (NIDDM), gallbladder disease, sleep apnea, osteoarthritis (OA), and various forms of cancer (6, 13, 18, 33, 44, 55, 57). For example, severely obese men are 42 times more likely to experience NIDDM when compared to men with BMIs less than 23 (17). Severely obese women are significantly more likely to develop and die from cervical, uterine, and endometrial cancers (29). Quesenberry and colleagues (48) found that severely obese individuals (i.e., BMI ≥ 35) are substantially more likely to suffer from a co-morbid medical condition compared to individuals with BMIs in the healthy range (i.e., 20-24.9). Figure 15.2 illustrates the percent increase in prevalence of several co-morbid conditions associated with severe obesity (BMI ≥ 35) compared to BMIs of 20-24.9.

Osteoarthritis can be a significant barrier to physical activity in severely obese individuals because it causes pain, limits range of movement, and reduces the number of viable exercise options. Several cross-sectional and longitudinal cohort studies have found obesity to increase the risk for OA of the knees, hips, and hands (22, 23). For example, severely obese persons have a 20-200% increase in risk for arthritis of the hip and a sevenfold increase in risk for arthritis of the knee (4, 32). Even at slightly lower BMIs (i.e., BMI > 30), there is a substantially greater risk for osteoarthritis of the knee when compared to individuals who are normal weight (i.e., BMI < 25) (22, 23). Severe obesity also significantly increases the risk for arthritis of the hands (16).

The severely obese experience substantially greater risk for mortality (9, 38, 44, 55, 57). Among women who never smoked and recently had stable weight, the relative risks of all-cause mortality were 1.6 [95% CI (confidence interval) = 1.1-2.5] for BMI = 27.0 to 28.9; 2.1 (95% CI = 1.4-3.2) for BMI = 29.0 to 31.9; and 2.2 (95% CI = 1.4-3.4) for BMI ≥ 32 (41). From an examination of data from several 30-year mortality studies of white men, Troiano and colleagues (57) found that the lowest mortality occurred at a BMI of 24 and that a BMI of 32 (the highest level noted) was associated with a 349% increase in mortality risk over the minimal

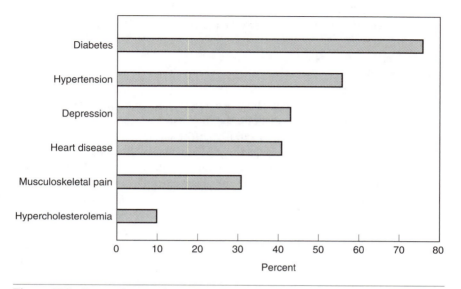

Figure 15.2   Percent increase in prevalence of medical co-morbidities associated with body mass index ≥ 35 compared to individuals with body mass index of 20 to 24.9. Adapted from Quesenberry et al., 1998 (48).

mortality point. More specifically, Ellis and colleagues (20) found that a BMI greater than 35 was associated with a sevenfold increase in risk for death in coronary patients.

Finally, severe obesity is associated with increased health and socioeconomic costs. While several authors have documented the general relationship between obesity, broadly defined, and increasing health care expenditures (11, 61-62), few studies have directly examined the costs associated with severe obesity, i.e., BMI ≥ 35. One study that did examine BMI-specific costs found that severely obese patients had significantly higher visit, pharmacy, laboratory, total outpatient, inpatient, and total health annual service use costs than individuals with BMIs in the 20-24.9 range (48). In general, costs for severely obese patients were 44% higher than for their healthy weight counterparts.

In summary, severe obesity is a substantial burden to the affected individuals and to society, significantly increasing risk for disease, death, and health care costs associated with the increased prevalence of co-morbid conditions. Given these significant costs and the fact that therapeutic weight loss produces short-term improvements in several important risk factors (33, 44), e.g., NIDDM, dyslipidemia, and hypertension, it is no surprise that substantial efforts have been made to develop effective treatments for obesity, including the use of physical activity programs (5, 44, 59).

Benefits of Physical Activity for the Severely Obese

Short-Term Benefits

Physiological responses to a single bout of exercise, although sometimes transient, can be beneficial in terms of metabolic stability, hypertension, and psychological well-being. Acute bouts of aerobic exercise performed at or above a given exercise intensity have been shown to enhance glucose uptake by muscle cells, thus improving the metabolic milieu (43). Reductions in both systolic and diastolic blood pressures have been noted after aerobic and anaerobic (i.e., weight training) bouts of exercise (34, 35). Furthermore, such exercise appears to reduce stress and positively affect mood states (51, 52). Acute responses to exercise provide immediate benefits that may translate into positive health outcomes (i.e., reduced risk of disease).

The acute effects of exercise on physiological and psychological parameters in severely obese subjects have not received the attention noted in nonseverely obese populations. However, it does not appear that severely obese individuals would respond differently to exercise. While there has been a dearth of comprehensive theories concerning metabolic adjustments to exercise, such as improved glucose kinetics or altered lipolytic activity, evidence suggests that certain conditions develop as a consequence of severe obesity that may "interfere," or alter the acute physiological response to exercise.

Pulmonary Function

Pulmonary function and gas exchange are compromised in severe obesity (40, 49). Excessive adipose tissue modifies chest wall mechanics and limits movement of the diaphragm. Hypoventilation of the lower respiratory tract causes a mismatch between ventilation and perfusion leading to subnormal arterial oxygenation (40). This hypoxemic (i.e., oxygen depleted) state is exacerbated by a greater demand for oxygen prompted by altered respiratory mechanisms and the increased energy required for bodily movements (60).

At rest and during exercise at any given workload (i.e., energy output as measured in watts, for example), oxygen consumption ($\dot{V}O_2$) is elevated in severely obese subjects compared with controls (28, 58). During exercise, respiratory frequency is higher and tidal volume (i.e., the volume of air moved during inspiration or expiration) is lower in the severely obese, which results in an increase in the area in the lung where gas exchange does not occur (i.e., dead space) and alveolar hypoventilation (58). Nonetheless, severely obese subjects display an increase in mean arterial oxygen tension (PaO_2) and a decrease in mean alveolar-arterial oxygen difference [$P(A-a)O_2$] when peak cycle ergometer values are compared with those observed at rest in the seated position. However, PaO_2 and $P(A-a)O_2$ values at peak exercise are not significantly different from those obtained at rest in the standing position (3). Despite the alveolar hypoventilation, oxygen removal per liter of air ventilated is similar between controls and severely obese subjects (58). Furthermore, there is no tendency toward delayed CO_2 elimination and CO_2 retention after exercise. In fact, severely obese subjects actually have a greater capacity to recover after a given rise in the O_2 demand (50). This may be attributed to a "training effect" of all daily physical activities associated with increased O_2 demands, especially when moving the body weight is involved (50).

Ventricular Dysfunction

Hypoxia (i.e., low oxygen levels) induced by severe obesity may also produce pulmonary arterial hypertension and clinical manifestations compatible with right ventricular (RV) dysfunction (1). Furthermore, severe obesity may induce structural abnormalities of the left ventricle (LV) (2). Alpert et al. (1) assessed RV systolic function in 22 severely obese patients (112% over ideal body weight) during a symptom-limited cycle ergometry test. Resting RV ejection fraction was below 35% in 5 patients, and RV exercise response failed to increase $\geq 5\%$ during exercise in 22 patients. There was a significant increase in mean RV ejection fraction for the group from 43% at rest to 50% during peak exercise. Nine out of 10 patients with RV internal dimensions ≥ 2 cm had an abnormal RV exercise response. Right ventricular ejection fraction (i.e., the amount of blood pumped by the right ventricle divided by the total amount of blood pumped by the heart) did not increase in the subgroup with RV internal dimension > 2.0 cm (43% at rest and 40% during peak exercise), whereas patients with RV internal dimension < 2.0 cm showed a significant increase in RV ejection fraction (44% at rest vs. 58% during peak exercise). Percent over ideal body weight was negatively correlated with RV exercise response ($r = -0.60$) and positively related to RV internal dimensions ($r = 0.86$).

Thus, exercise may elicit RV systolic function abnormalities not noted at rest; severely obese individuals with an enlarged RV internal dimension are more likely to experience an abnormal RV exercise response. Alpert and colleagues (2) expanded the above protocol to include measurements of LV ejection fractions and dimensions. Again, severely obese patients with increased LV mass did not increase LV ejection fraction during exercise compared with those with a normal LV mass. Thus, hemodynamic stresses (e.g., increased circulating blood volume) on both the right and left ventricles resulting from excessive adipose tissue accumulation may predispose severely obese individuals to abnormal cardiac responses to exercise.

Oxygen Uptake

Severely obese individuals require more oxygen than normal-weight persons to perform a given amount of work. Freyschuss and Melcher (28) studied the relationship between oxygen uptake and workload during seated and supine bicycling and treadmill walking. Severely obese subjects had higher oxygen uptakes than normal controls at any given workload for each of the three exercise modes. In this regard, their mechanical efficiency, as determined by these authors (workload exercise $\dot{V}O_2$), was below that of the controls. However, mechanical efficiency calculated as such does not consider the positive relationship between body weight and exercise $\dot{V}O_2$. An earlier study by Turell and others (58) showed that consideration of body weight when determining mechanical efficiency negates any differences between severely obese and normal subjects.

Weight Loss

The effects of weight loss without exercise on acute physiological responses to exercise also have been investigated. Following weight loss induced by diet or gastric banding, a significant increase in mean oxygen tension at peak exercise was noted while the oxygen cost of ventilation and carbon dioxide recovery time are decreased (50). Oxygen consumption (L/min) and carbon dioxide production ($\dot{V}CO_2$) at peak exercise are unchanged following weight loss; however, $\dot{V}CO_2$ is decreased at submaximal workloads (31). Thus, it appears that weight loss alone promotes positive changes in pulmonary responses to a bout of aerobic exercise. Specifically, ventilation/profusion disturbances are reduced, which improves gas exchange. In addition, dynamic lung function also is improved during exercise.

Long-Term Benefits

Exercise leads to a number of long-term health benefits, including a reduction in body weight and an increase in lean body mass (45). Exercise therapy is used in clinical and nonclinical settings alone or as an adjunct to other interventions to reduce weight in overweight and obese persons. The use of exercise therapy in severely obese populations is not readily undertaken as an exclusive, initial approach to weight loss. Rather, exercise is used in conjunction with or subsequent to other means (e.g., surgery) of promoting weight loss. Therefore, there is a paucity of data concerning exercise intervention outcomes with individuals who are severely obese. Research is needed in severely obese populations with specific reference to the effects of exercise therapy on metabolic control, hypertension, blood lipids, psychological parameters, and functional capacity.

Björvell and Rössner (10) reported very promising results from studies that used behavior modification. They offered severely obese subjects (BMI > 40) either a 6-week behavioral program (Medicine 5) (n = 68), a jaw fixation program (n = 39), or an ordinary weight loss program (control) (n = 16). The Medicine 5 program consisted of lessons on behavior modification, healthy cooking instruction, exercise classes, and other activities (e.g., swimming, "shopping exercise"). The exercise classes were offered three times per week. The authors do not provide the specifics (e.g., intensity) of the exercise program. After the 6-week program, the mean weight loss was 11.1 ± 0.7 kg for the Medicine 5 group and 5.4 ± 0.9 kg for the control group (test of significance not given). At the 1-year follow-up, weight loss was signifi-cantly greater for the Medicine 5 (18.1 kg) and the jaw fixation (20.0 kg) groups, compared with the controls (6.8 kg). The Medicine 5 and jaw fixation groups did not differ with respect to weight change at the 1- or 4- year follow-up (~ -31% decrease in weight for both groups at year 4). Interestingly, in the Medicine 5 group there was a significant positive correlation between the change in weight at the 4-year follow-up and the number of booster sessions provided each year of follow-up. Seventeen percent of the Medicine 5 group and 26% of the jaw fixation group dropped out of the study over the 4-year period. However, the Medicine 5 group did not lose subjects during the initial 6-week intervention. The authors attributed the initial low dropout rate to intensive encouragement. Their data offer encouragement that a multi-component behavioral intervention program with long-term follow-up may result in sustained weight loss in severely obese individuals.

Positive physiological adaptations with exercise training have been reported in severely obese populations. Pronk et al. (47) provided 3-month interventions to severely obese subjects (percent body fat > 45.3). The interventions were a very low-calorie diet (VLCD); VLCD plus endurance exercise; endurance exercise plus resistance exercise; and resistance exercise only. Exercise sessions were conducted 4 days per week at 70% of heart rate reserve and 70-80% of one-repetition maximum for endurance and resistance exercise, respectively. All groups experienced significant and similar decreases in body mass, percent fat, and fat mass. The group that did only resistance exercise was the only group to experience significant gains in strength, while endurance training improved aerobic capacity in all groups receiving this form of training. This study demonstrated that exercise in conjunction with diet does not lead to greater weight loss than with diet alone. Nevertheless, exercise does provide a contribution to fitness that is lacking with diet-only regimens.

Tanco et al. (56) compared the effects of cognitive therapy (CT) (i.e., broad-based treatment aimed at improving emotional well-being and adopting healthy lifestyle behaviors, including exercise) on numerous psychological parameters in severely obese women (BMI ~ 40 kg/m^2). For comparisons, a behavior therapy (BT) group (diet and exercise focus only) and a control (C) group were devised. Both the CT and BT groups increased the percentage of individuals engaged in regular exercise compared with the control group. Both treatment groups demonstrated significant decreases in body weight and BMI at the post-treatment measurement. The CT group, but not the BT and C, experienced significant improvements in depression, state anxiety, trait anxiety, and self-control scores. The CT was the only group to demonstrate improvements across time in several subscales of the Eating Disorder Inventory (EDI), including drive for thinness, bulimia, body dissatisfaction, ineffectiveness, and interoceptive awareness. Several limitations, including the use of multiple therapists, the lack of long-term follow-up data, and the use of self-report questionnaires to assess psychometric parameters were noted. Nevertheless, the results indicate that even in the context of minimal weight loss (1.7 BMI units), a CT intervention can promote increases in physical activity among severely obese women and potentially contribute to improvements in psychological well-being and decreases in eating pathologies.

Developing Physical Activity Programs for the Severely Obese

Barriers to Physical Activity

The severely obese have a higher risk for multiple co-morbid medical conditions (e.g., osteoarthritis) than normal-weight individuals (i.e., those with BMI $\geq$ 19 and < 24.9) that may substantially increase their discomfort during activity and limit their endurance and flexibility (44). For example, Mattsson and colleagues (42) found that severely obese women (i.e., BMI = 44.1 $\pm$ 10.7) who were tested during walking and cycle ergometry walked slower, had lower %$\dot{V}O_2$max/kg and higher $\dot{V}O_2$ and %$\dot{V}O_2$max than normal controls. In addition, the study participants reported high degrees of perceived exertion and pain. Their results suggest that walking, which is generally considered to be a moderate-intensity activity, may actually be perceived as more than moderately intense for severely obese individuals due to the greater relative oxygen cost of walking associated with increased body mass.

Severely obese patients, therefore, may be more likely to perceive even moderate-intensity activity, such as walking, as exerting and painful, and this may negatively impact adherence (42). Thus, the presence of one or more co-morbid medical conditions may be a significant barrier to starting or maintaining a physical activity program. Other barriers include those that affect many individuals, such as lack of time, equipment, facilities, child care, and family support (59).

Another important barrier to physical activity in the severely obese may be program-based shortcomings. For example, the foundation of many exercise programs is focused on cardiopulmonary benefits (14). In addition, weight loss or weight maintenance often are promoted as primary aims of physical activity programs for the obese (56). These training philosophies can have the unfortunate side effect of discouraging severely obese persons from starting activity programs, because that may mistakenly assume that small amounts of low-intensity activity, which is where most severely obese patients will begin, are not of any health value (14). In reality, low- to moderate-intensity physical activity programs for the severely obese may initially be very helpful in improving psychological well-being, functional status, and adherence, paving the way for participation in more intense programs (12, 26, 27, 54).

Implementing a Physical Activity Program for the Severely Obese

Exercise therapy is seldom used as an exclusive, initial procedure for weight loss with severely obese individuals. However, results indicate that for many severely obese patients, exercise training can be a viable component of a comprehensive intervention program (10, 56). As with any population, it is important to develop an individualized exercise prescription for the severely obese individual. Inherent in the exercise prescription are components related to patient safety, physical activity choices, program goals, and strategies to promote adherence and prevent relapse.

Safety Concerns

Obesity, especially severe obesity, often exacerbates the development of other chronic diseases such as diabetes, hypertension, and coronary heart disease (CHD) (44). As such, the adverse circumstances that may arise with exercise in the presence of these co-morbid conditions must be carefully considered. Prior to initiating an exercise program, it is advisable that severely obese patients be examined by a physician (3). Once the patient has been cleared for exercise, the prescription can be developed and the program begun.

Safety concerns may arise during the development of the exercise prescription and/or during the course of the program. Following the physician's examination, a graded exercise test may be conducted. In some circumstances an exercise test may not be warranted due to the very low functional capacities of some severely obese people. In this regard, the prescription may entail only simple and low-intensity exertion such as increasing activities of daily living. For this purpose, an exercise prescription based on an exercise test may not be necessary.

If a more advanced prescription is desired (e.g., exercise sessions) and an exercise test is performed, safety concerns during the test must be considered. Co-morbid conditions should be taken into account. If the patient is deemed "at risk" for CHD, then it is advisable that a 12-lead electrocardiogram be conducted during the exercise test with a cardiologist present. The use of a cycle or arm ergometer may enhance testing capability due to the orthopedic problems associated with weight-bearing exercises in this population. Blood pressure responses, signs of dyspnea, and signs of metabolic disturbances (e.g., hypoglycemia) should be closely monitored.

The safety concerns during exercise training are similar to those noted for testing. Blood pressure, glucose levels, signs of physiological disturbances, and/or orthopedic distress may need to be monitored during the course of the exercise intervention. The initial screening outcomes will dictate the frequency of such monitoring. Orthopedic problems are of specific concern with severely obese individuals. Therefore, joint stress, muscle soreness, and other orthopedic problems should be closely monitored.

Physical Activity Options

Although weight-bearing activities and those involving moderate to vigorous intensities are generally not recommended for severely obese patients, there are many alternative modes of physical activity or exercise they can perform. One very important aspect of an exercise intervention for severe obesity is the expenditure of calories. For this purpose, physical activities, including exercise, can be employed. Low-intensity activities concerned with daily living are extremely useful. They not only provide a means of expending calories, but also emphasize an independent lifestyle that is often devoid in this population. Structured exercise is also a viable option for promoting caloric expenditure and a sense of control over their condition. The exercise prescription may include cycling, stair climbing, or walking, if able.

An alternative form of exercise that has recently been promoted is hydroponics, or water exercises. This mode of exercise is especially appealing for severely obese individuals because it promotes caloric expenditure while being entertaining and non-weight-bearing. These benefits may lead to better adherence and lower drop-out rates; however, scientific evidence in this area is lacking.

Program Goals

An important goal of an exercise program for the severely obese is caloric expenditure, and consequently, weight loss. The American College of Sports Medicine (ACSM) (3) recommends specific guidelines for exercise-induced weight loss. The exercise program should promote an expenditure of 300 to 500 kilocalories (kcal) per session and 1000 to 2000 kcal per week for adults. This goal may not be realistic for the severely obese person; unfortunately, specific caloric expenditure guidelines have not been developed for this population. Therefore, current ACSM guidelines may be useful if the concerns of the severely obese individual are taken into consideration. An important point to remember is that most severely obese individuals are extremely sedentary, and any additional caloric expenditure for them is advantageous.

Social and psychological goals may also be important components of the exercise prescription. Improvements in depression, state anxiety, trait anxiety, and self-control scores may be noted as a result of becoming physically active (56). Furthermore, it may be desirable to set social goals prior to starting the exercise program. Such goals may include leaving the house, going to social functions, or meeting with friends.

Strategies to Promote Adherence

There are several basic strategies that can be utilized to improve adherence to physical activity programs. First, developing good rapport with severely obese patients is very important. This can involve discussing barriers to treatment, providing patient education about the need for intervention, and reviewing the expected course and outcomes. Patient education has been found to be an important part of most compliance-enhancing interventions and addresses many of the patient- and provider-related barriers to adherence (8). Examples of patient education include ensuring that the patient understands the activity prescription by writing it down; providing clear information about obesity and related conditions and how physical activity may be beneficial in improving these conditions; discussing the patient's beliefs about obesity and how physical activity may benefit them; and assessing the patient's treatment expectations. Patient-provider alliance building is a very effective method of improving treatment compliance and addressing important psychosocial barriers to exercise for obese patients (30).

Additionally, treatment regimens can be modified so that they are more easily incorporated into the patient's life. The emphasis on moderate lifestyle activities is one way to make exercise more accessible to obese patients (12). Including patients in the development of treatment goals and encouraging them to increase their activity intensity and duration gradually can be helpful. Giving feedback about treatment progress and providing a way for patients to monitor progress through self-monitoring is one of the most effective methods for enhancing exercise and dietary adherence and promoting weight loss (15, 26). Providing prompts or reminders can also be very effective (36, 59). For example, simple telephone reminders and memory devices to prompt patients to exercise have been found to significantly enhance compliance to a walking program (39).

Finally, the importance of social support should not be overlooked. Involving family members in the treatment process, keeping a list of community resources where patients can be referred for support, or providing group exercise settings or activities can improve long-term adherence. The beneficial effect of social support has been demonstrated for exercise, dietary change, and the treatment of obesity (15, 19, 26, 59). Table 15.1 summarizes important factors that should be addressed when designing activity programs for patients with severe obesity.

Table 15.1   Considerations When Developing Physical Activity Programs for the Severely Obese Patient

Barriers	Physical, psychological, economic, and program based
Safety issues	Co-morbid medical conditions (e.g., hypertension, heart disease, osteoarthritis)
Physical activity options	Low-intensity lifestyle activity (e.g., gardening, light housework); non-weight-bearing activity (e.g., swimming, water aerobics, stretching); walking
	Obese patients should start with moderate levels of physical activity for 30 to 45 min, 3 to 5 d per week. All patients should attempt to eventually accumulate $\geq$ 30 minutes of moderate-intensity physical activity (e.g., brisk walking) on most, and preferably all, days of the week.
Program goals	Caloric expenditure and weight loss, quality of life, improved mood, improved physical functioning
Promoting adherence	Develop good rapport, patient education, address barriers, provide progress feedback, use prompts and rewards, develop social support

Conclusions

Physical activity plays an important role in the treatment of severely obese patients, although traditional treatment philosophies and approaches need to be altered for this population. A stronger emphasis on lower-intensity physical activities is needed, given that severely obese patients may experience greater perceived exertion, pain, and relative oxygen cost (12, 42). In addition, given how difficult it is to achieve long-term weight loss and maintenance (44), we believe that physical activity programs need to shift their emphasis from weight loss to improvements in other health parameters (27, 46, 54, 56). Programs should emphasize short- and moderate-term benefits, such as enhanced quality of life and functional capacity, as well as small improvements in co-morbid medical conditions. Long-term benefits, including reduced risk for mortality, also can be discussed, although the data supporting this benefit are preliminary (7).

Finally, when designing and implementing physical activity programs for the severely obese, greater attention needs to be given to the very real physical, psychological, and social limitations inherent in standard approaches to increasing physical activity and fitness. Thus, clinicians need to be more creative about the types and amounts of activity that they recommend for this population.

Acknowledgments

This work was supported by a Minority Scientist Development Award from the American Heart Association and with funds contributed by the AHA, Puerto Rico Affiliate, and a grant from the National Heart, Lung, and Blood Institute, HL47052.

References

1. Alpert, M.A., A. Singh, B.E. Terry, D.L. Kelly, S. El-Deane, V. Mukerji, D. Villarreal, and A.K. Artis. 1989a. Effect of exercise and cavity size on right ventricular function in morbid obesity. *American Journal of Cardiology* 64: 1361-1365.
2. Alpert, M.A., A. Singh, B.E. Terry, D.L. Kelly, S. El-Deane, D. Villarreal, and V. Mukerji. 1989b. Effect of exercise on left ventricular systolic function and reserve in morbid obesity. *American Journal of Cardiology* 63: 1478-1482.
3. American College of Sports Medicine (ACSM) 1995. *ACSM's guidelines for exercise testing and prescription,* 5th ed. Baltimore, MD: Williams & Wilkins.
4. Anderson, J.J., and D.T. Felson. 1988. Factors associated with osteoarthritis of the knee in the first National Health and Nutrition Examination Survey (HANES I): Evidence for an association with overweight, race, and physical demands of work. *American Journal of Epidemiology* 128: 179-189.
5. Ballor, D.L., E.P. Poehlman, and M.J. Toth. 1998. Exercise as a treatment for obesity. In *Handbook of obesity*, eds. G.A. Bray, C. Bouchard, and W.P.T. James, 891-910. New York: Marcel Dekker, Inc.

6. Barakat, H.A., N. Mooney, K. O'Brien, S. Long, P.G. Khazani, W. Pories, and J.F. Caro. 1993. Coronary heart disease risk factors in morbidly obese women with normal glucose tolerance. *Diabetes Care* 16: 144-149.
7. Barlow, C.E., H.W. Kohl, L.W. Gibbons, and S.N. Blair. 1995. Physical fitness, mortality, and obesity. *International Journal of Obesity and Related Metabolic Disorders* 19 (Suppl. 4): S41-S44.
8. Becker, M.H. 1990. Theoretical models of adherence and strategies for improving adherence. In *The handbook of health behavior change,* eds. R. Schmaker, E.B. Schron, J.K. Ockene, C.T. Parker, J.L. Probsfield, and J.M. Wolle, 5-43. New York: Springer Publishing Company.
9. Bender, R., C. Trautner, M. Spraul, and M. Berger. 1998. Assessment of excess mortality in obesity. *American Journal of Epidemiology* 147: 42-48.
10. Björvell, H., and S. Rössner. 1985. Long term treatment of severe obesity: four year follow up of results of combined behavioural modification programme. *British Medical Journal (Clinical Research Edition)* 291: 379-382.
11. Black, D.R., J.P. Sciacca, and D.C. Coster. 1994. Extremes in body mass index: Probability of healthcare expenditures. *Preventive Medicine* 23: 385-393.
12. Blair, S.N., and J.C. Connelly. 1996. How much physical activity should we do? The case for moderate amounts and intensities of physical activity. *Research Quarterly for Exercise and Sport* 67: 193-205.
13. Bray, G.A. 1992. Pathophysiology of obesity. *American Journal of Clinical Nutrition* 55: (2 Suppl): 488S-494S.
14. Brownell, K.D. 1995. Exercise and obesity treatment: Psychological aspects. *International Journal of Obesity and Related Metabolic Disorders* 19 (Suppl. 4): S122-S125.
15. Burke, L.E., and J. Dunbar-Jacobs. 1995. Adherence to medication, diet, and activity recommendations: From assessment to maintenance. *Journal of Cardiovascular Nursing* 9: 62-79.
16. Carman, W.J., M. Sowers, V.M. Hawthorne, and L.A. Weissfeld. 1994. Obesity as a risk factor for osteoarthritis of the hand and wrist: A prospective study. *American Journal of Epidemiology* 139: 119-129.
17. Chan, J.M., E.B. Rimm, G.A. Colditz, M.J. Stampfer, and W.C. Willett. 1994. Obesity, fat distribution, and weight gain as risk factors for clinical diabetes in men. *Diabetes Care* 17: 961-969.
18. Colditz, G.A., W.C. Willett, M.J. Stampfer, J.E. Manson, C.H. Hennekens, R.A. Arky, and F.E. Speizer. 1990. Weight as a risk factor for clinical diabetes in women. *American Journal of Epidemiology* 132: 501-513.
19. Cousins, J.H., D.S. Rubovits, J.K. Dunn, R.S. Reeves, A.G. Ramirez, and J.P. Foreyt. 1992. Family versus individually oriented intervention for weight loss in Mexican American women. *Public Health Reports* 107: 549-555.
20. Ellis, S.G., J. Elliott, M. Horrigan, R.E. Raymond, and G. Howell. 1996. Low-normal or excessive body mass index: Newly identified and powerful

risk factors for death and other complications with percutaneous coronary intervention. *American Journal of Cardiology* 78: 642-646.

21. Feinleib, M. 1985. Epidemiology of obesity in relation to health hazards. *Annals of Internal Medicine* 103: 1019-1024.

22. Felson, D.T. 1995. Weight and osteoarthritis. *The Journal of Rheumatology* 22: (Suppl 43): 7-9.

23. Felson, D.T. 1996. Weight and osteoarthritis. *American Journal of Clinical Nutrition* 63 (Suppl): 430S-432S.

24. Flegal, K.M., M.D. Carroll, R.J. Kuczmarski, and C.L. Johnson. 1998. Overweight and obesity in the United States: Prevalence and trends, 1960-1994. *International Journal of Obesity and Related Metabolic Disorders* 22: 39-47.

25. Foreyt, J.P., R.L. Brunner, G.K. Goodrick, S.T. St. Jeor, and G.D. Miller. 1995. Psychological correlates of reported physical activity in normal-weight and obese adults: The RENO Diet-Heart Study. *International Journal of Obesity and Related Metabolic Disorders* 19 (Suppl. 4): S69-S72.

26. Foreyt, J.P., and G.K. Goodrick. 1994. Attributes of successful approaches to weight loss and control. *Applied and Preventive Psychology* 3: 209-215.

27. Foreyt, J.P., W.S.C. Poston, and G.K. Goodrick. 1996. Future directions in obesity and eating disorders. *Addictive Behaviors* 21: 767-778.

28. Freyschuss, U., and A. Melcher. 1978. Exercise energy expenditure in extreme obesity: influence of ergometry type and weight loss. *Scandinavian Journal of Clinical and Laboratory Investigation* 38: 753-759.

29. Garfinkel, L. 1995. Overweight and cancer. *Annals of Internal Medicine* 103: 1034-1036.

30. Grilo, C.M. 1995. The role of physical activity in weight loss and weight loss management. *Medicine, Exercise, Nutrition, and Health* 4: 60-76.

31. Hakala, K., P. Mustajoki, J. Aittomaki, and A. Sovijarvi. 1996. Improved gas exchange during exercise after weight loss in morbid obesity. *Clinical Physiology* 16: 229-238.

32. Heliövaara, M., M. Mäkelä, O. Impivaara, P. Knekt, A. Aromaa, and K. Sievers. 1993. Association of overweight, trauma, and workload with coxarthrosis. *Acta Orthopaedica Scandinavica* 64: 513-518.

33. Institute of Medicine (IOM). 1995. *Weighing the options: Criteria for evaluating weight-management programs.* Washington D.C.: National Academy Press.

34. Kelley, G. 1997. Dynamic resistance exercise and resting blood pressure in adults. *Journal of Applied Physiology* 82: 1559-1565.

35. Kelley, G., and P. McClellan. 1994. Antihypertensive effects of aerobic exercise. A brief meta-analytic review of randomized controlled trials. *American Journal of Hypertension* 7: 115-119.

36. King, A.C. 1994. Community and public health approaches to the promotion of physical activity. *Medicine, Science, Sports, and Exercise* 26: 1405-1412.

37. Kuczmarski, R.J., M.D. Carroll, K.M. Flegal, and R.P. Troiano. 1997. Varying body mass index cutoff points to describe overweight prevalence among U.S. adults: NHANES III (1988-1994). *Obesity Research* 5: 542-548.
38. Lee, I., J.E. Manson, C.H. Hennekens, and R.S. Paffenbarger. 1993. Body weight and mortality: A 27-year follow-up of middle-aged men. *JAMA* 270: 2823-2828.
39. Lombard, D.N., T.N. Lombard, and R.A. Winett. 1995. Walking to meet health guidelines: The effect of prompting frequency and prompt structure. *Health Psychology* 14: 164-170.
40. Luce, J.M. 1980. Respiratory complications of obesity. *Chest* 78: 626-631.
41. Manson, J.E., W.C. Willett, M.J. Stampfer, G.A. Colditz, D.J. Hunter, S.E. Hankinson, C.H. Hennekens, and F.E. Speizer. 1995. Body weight and mortality among women. *New England Journal of Medicine* 333: 677-685.
42. Mattsson, E., U. Evers Larsson, and S. Rössner. 1997. Is walking for exercise too exhausting for obese women? *International Journal of Obesity and Related Metabolic Disorders* 21: 380-386.
43. Mikines, K.J., B. Sonne, P.A. Farrell, B. Tronier, and H. Galbo. 1988. Effect of physical exercise on sensitivity and responsiveness to insulin in man. *American Journal of Physiology* 254: E248-259.
44. National Institutes of Health (NIH), National Heart, Lung, and Blood Institute (NHLBI). 1998. *Clinical guidelines on the identification, evaluation, and treatment of overweight and obesity: The evidence report.* Washington D.C.: U.S. Government Press.
45. Pavlou, K.N., W.P. Steffee, R.H. Lerman, and B. Burrows. 1985. Effects of dieting and exercise on lean body mass, oxygen uptake and strength. *Medicine, Science, Sports, and Exercise* 17: 466-471.
46. Poston, W.S.C., J.P. Foreyt, L. Borrell, and C.K. Haddock. 1998. Challenges in obesity management. *Southern Medical Journal* 91: 710-722.
47. Pronk, N.P., J.E. Donnelly, and S.J. Pronk. 1992. Strength changes induced by extreme dieting and exercise in severely obese females. *Journal of the American College of Nutrition* 11: 152-158.
48. Quesenberry, C.P., B. Caan, and A. Jacobson. Obesity, health service use, and health care costs among members of a health maintenance organization. 1998. *Archives of Internal Medicine* 158: 466-472.
49. Ray, C., D.Y. Sue, G. Bray, J.E. Hansen, and K. Wasserman. 1983. Effects of obesity on respiratory function. *American Review of Respiratory Disease* 128: 501-506.
50. Refsum, H.E., P.H. Holter, T. Lovig, J.F.W. Haffner, and J.O. Stadaas. 1990. Pulmonary function and energy expenditure after marked weight loss in obese women: observations before and one year after gastric banding. *International Journal of Obesity and Related Metabolic Disorders* 14: 175-183.
51. Rejeski, W.J., A. Thompson, P.H. Brubaker, and H.S. Miller. 1992. Acute exercise: buffering psychological stress responses in women. *Health Psychology* 11: 355-362.

52. Scully, D., J. Kremer, M.M. Meade, R. Graham, and K. Dudgeon. 1998. Physical exercise and psychological well being: a critical review. *British Journal of Sports Medicine* 32: 111-120.

53. Seidell, J.C. 1995. Obesity in Europe: Scaling an epidemic. *International Journal of Obesity and Related Metabolic Disorders* 19 (Suppl 3): S1-S4.

54. Skender, M.L., G.K. Goodrick, D.J. Del Junco, R.S. Reevers, L. Darnell, A.M. Gotto, and J.P. Foreyt. 1996. Comparison of 2-year weight loss trends in behavioral treatments of obesity: Diet, exercise, and combination interventions. *Journal of the American Dietetic Association* 96: 342-346.

55. Solomon, C.G., and J.E. Manson. 1997. Obesity and mortality: A review of the epidemiologic data. *American Journal of Clinical Nutrition* 66 (Suppl): 1044S-1050S.

56. Tanco, S., W. Linden, and T. Earle. 1998. Well-being and morbid obesity in women: A controlled therapy evaluation. *International Journal of Eating Disorders* 23: 325-339.

57. Troiano, R.P., E.A. Frongillo, J. Sobal, and D.A. Levitsky. 1996. The relationship between body weight and mortality: A quantitative analysis of combined information from existing studies. *International Journal of Obesity and Related Metabolic Disorders* 20: 63-75.

58. Turell, D.J., R.C. Austin, and J.K. Alexander. 1964. Cardiorespiratory response of very obese subjects to treadmill exercise. *Journal of Laboratory and Clinical Medicine* 64: 107-116.

59. U.S. Department of Health and Human Services (USDHHS). 1996. *Physical activity and health: a report of the Surgeon General.* Atlanta, GA: U.S. Department of Health and Human Services, Centers for Disease Control and Prevention, National Center for Chronic Disease Prevention and Health Promotion.

60. Wasserman, K., J.E. Hansen, D.Y. Sue, and O.J. Whipp. 1994. Pathophysiology of disorders limiting exercise. In *Principles of exercise testing and interpretation,* 2nd edition, ed. K. Wasserman, 47-49. Philadelphia, PA: Lea & Febiger.

61. Wolf, A.M., and G.A. Colditz. 1994. The cost of obesity: The U.S. perspective. *PharmacoEconomics* 5 (Suppl. 1): 34-37.

62. Wolf, A.M., and G.A. Colditz. 1996. Social and economic effects of body weight in the United States. *American Journal of Clinical Nutrition* 63 (Suppl): 466S-469S.

CHAPTER 16

Physical Activity and Maintenance of Weight Loss: Physiological and Psychological Mechanisms

Christina Wood Baker, MS, and Kelly D. Brownell, PhD
Department of Psychology, Yale University, New Haven, Connecticut, U.S.A.

Dietary modification combined with increased physical activity is the universal prescription for weight control. Physical activity has been added to diet because there is substantial evidence that it is a critical ingredient for long-term weight maintenance (28, 35). While nearly every weight modification program now highlights the importance of exercise, there is still much to be learned about the mechanisms linking exercise to long-term weight control. For instance, little is known about whether changing physical activity interferes with, interacts with, or behaves in a synergistic way with dietary change. Once the mechanisms are defined, many questions about how best to utilize and apply this knowledge become paramount. It is important to understand the psychological effects of exercise in overweight persons, the means for developing optimal exercise prescriptions based on knowledge about mechanisms, and methods for improving adherence.

These issues are important for more than scientific reasons. With the prevalence of obesity at high levels and rising, and with the difficult nature of treatment, something as widely prescribed as exercise should be understood thoroughly and tested extensively. A great deal more work is needed before this is achieved. The purpose of this chapter is to note what is known about exercise and weight maintenance, to identify gaps in knowledge, and to point the way to research and clinical questions of the future.

The Importance of Physical Activity to Weight Maintenance

Correlational and experimental studies have examined the role of physical activity in weight loss and maintenance. In a review of the literature, Pronk and Wing (35) conclude that both types of studies consistently show that exercise is a significant determinant of long-term weight maintenance.

Correlational Studies

Correlational studies have focused primarily on identifying predictors of long-term weight loss (35). In a typical study, individuals who have participated in a weight-loss program are asked at a follow-up time to retrospectively report their level of physical activity. Investigators then examine the relationship between activity level and changes in weight. These studies do not allow for causal inferences and are limited by the retrospective and primarily self-report nature of the data. Nonetheless, such studies are a first step in exploring the association between activity and weight maintenance, and they find that reported activity is the strongest correlate of weight-loss maintenance (35). The associations between activity and maintenance are consistent across age and gender and occur independently of type of dietary intervention (35).

One frequently cited correlational study (20) attempted to discriminate between three groups of women: "relapsers," "maintainers," and controls. A maintainer was a woman who had maintained her weight loss for at least two years, while a relapser had lost 20% of her weight and subsequently gained it back. Retrospective accounts of weight-loss strategies differed between the groups: 76% of maintainers compared to 36% of relapsers described exercise as a weight-loss method. In the period following weight loss, 90% of weight-loss maintainers and 82% of control subjects exercised regularly (at least 3 times a week for ≥ 30 minutes), in contrast to 34% of relapsers. The relapsers who did exercise reported their activity to be of less intensity and shorter duration than that of maintainers.

A more recent cross-sectional study examined the relationships between exercise calories, frequency of exercise, and maintenance of weight loss in a sample of 45 previously obese participants 2 years after completion of a very low-calorie diet (VLCD; 15). Participants were categorized at follow-up, according to self-reported activity level, as low (< 850 kcals per week), moderate (850-1575 kcals per week), or high (> 1575 kcals per week) in activity. At the 2-year follow-up, participants in the high-active group had maintained greater weight losses and exhibited a lower percent regain. Total exercise calories was a predictor of both weight loss and percentage of weight regained. Another study examined factors that facilitated maintenance following a VLCD (14). Maintainers were more likely to report regular vigorous exercise.

As a final example, Grodstein and colleagues (17) investigated lifestyle factors associated with maintenance of weight loss in a sample of 192 men and women surveyed 3 years after participation in a commercial weight-loss program. Exercise frequency (number of hours of exercise per week) since completing the diet program was the strongest predictor of maintenance. Frequency, as defined above, appeared to be more important than duration of exercise (number of months of regular exercise).

Randomized Controlled Trials

Prospective randomized, controlled trials provide more powerful support for the contribution of activity to weight control. These studies typically compare diet-only to diet-plus-exercise and/or exercise-only groups. Participants are assigned randomly to one of the conditions, followed through the intervention phase, and then contacted for follow-up information at subsequent times. These studies provide information about both weight loss and weight maintenance and afford a stronger basis for making causal inferences. Some (18, 31, 41), but not all (3, 28, 30), studies find that dietary intervention combined with physical activity leads to greater initial weight loss than dietary change alone. Some of the discrepancy in these findings may be due to differences in samples, e.g., gender (48), or types of dietary intervention (29). In contrast, there is little debate regarding the importance of activity after the weight-loss phase. The experimental research is remarkably consistent in illustrating the valuable role of exercise in maintaining weight loss.

Miller and colleagues (28) conducted a meta-analysis of studies from the past 25 years involving the three types of interventions described above: diet, exercise, and diet-plus-exercise. Only studies involving aerobic exercise were included, and mean program duration was 15.6 weeks. Most study samples consisted of middle-aged, moderately obese participants; however, those in the exercise-only studies tended to be younger and to have lower baseline body mass index (BMI) values. The results of the meta-analysis indicated that exercise-only programs were less effective than diet and diet-plus-exercise programs in producing changes in body weight and composition (e.g., weight loss, BMI decrease, reduction in body fat). This result needs to be viewed with caution in light of the population differences. More research is needed to determine whether exercise-alone interventions are useful for long-term weight loss and maintenance (35). Diet-plus-exercise programs did not produce better results than diet-only interventions during the weight loss period. However, at one-year follow-ups, diet-plus-exercise appeared to be the superior intervention. Diet-plus-exercise groups maintained 77% of initial weight loss, whereas diet- and exercise-only groups maintained 56% and 53%, respectively.

The long-term benefits of diet-plus-exercise plans, consistent with findings from correlational studies, have been found across types of dietary interventions (29). Pavlou and colleagues investigated weight loss and maintenance among moderately obese male subjects in a randomized clinical trial involving four different dietary interventions and both exercise/nonexercise conditions. The dietary interventions included a balanced caloric-deficit diet (BCDD), a protein-sparing modified fast (PSMF), a 420-kcal/day liquid VLCD, and an 800-kcal/day liquid VLCD. The exercise component consisted of a 90-minute supervised exercise program that was carried out three times per week and included aerobic activity, calisthenics, and relaxation techniques. Although the added benefit of exercise was not striking during the weight-loss portion of the study, exercisers and nonexercisers differed dramatically at 18-month follow-up, irrespective of type of dietary intervention. All exercise groups maintained weight loss, whereas nonexercise groups regained almost as much weight as had been lost during the program.

One question in need of further research is whether the relationship between exercise and weight control is the same across populations. Wood and colleagues (48) found gender differences in a study comparing diet-alone to diet-plus-exercise plans in overweight men and women. Men, but not women, lost significantly more weight in the diet-plus-exercise group. These results need to be replicated and mechanisms examined. It is possible that the men exercised more vigorously or that there are gender differences in physiological response to combined diet and activity interventions (35). Either way, it is precisely this type of finding that speaks to the need for research in various populations. The presence of a gender discrepancy in response to different interventions could have important implications for exercise prescriptions. Along a similar line, most of the studies have involved somewhat homogeneous populations of middle-aged, moderately obese participants (28). Given the heterogeneity of the obese population (10), it is imperative that we consider response to interventions across broader samples. The effectiveness of exercise programs may differ depending on such factors as weight or demographics, e.g., initial degree of obesity or age (28).

Another factor that warrants attention is exercise prescription. Many of the early studies in this field did not have formal exercise prescriptions (35). Recent studies are more specific in defining exercise interventions. As reviewed by Pronk and Wing (35), most studies involve low-intensity aerobic exercise, usually walking. The majority of studies specify duration and frequency of exercise, with less attention paid to intensity (although some studies prescribe a target heart rate zone). Research is needed on the effects of different types of activity (e.g., planned vs. lifestyle or strength training vs. aerobic) on weight maintenance. We would also benefit from learning about interactions between specific diets and different exercise prescriptions.

We can safely conclude that while there are mixed results concerning the added benefit of physical activity during a weight-loss phase, there is compelling evidence that being physically active is a critical ingredient for long-term weight maintenance. In light of these findings, exercise initiation and adherence become central concerns of the obesity field (see chapter 19). Prescriptions for exercise often lay the groundwork for future adherence. It is imperative that we understand the mechanisms that link exercise to weight control in order to make effective and sustainable exercise prescriptions (7, 8).

Mechanisms Linking Exercise to Weight Control

There is considerable speculation about the mechanisms linking exercise to weight control. Understanding the mechanisms is important on several grounds. To the degree that physical inactivity is implicated in the development of obesity, work on mechanisms might be instructive in understanding etiology. If exercise promotes weight loss through a specific mechanism (e.g., increased metabolic rate), there would be a strong impetus for finding other means (e.g., pharmacotherapy) for producing the same action. Finally, as mentioned above, exercise prescriptions will vary dramatically based on the putative mechanism linking activity to weight control. If preserving or increasing lean body mass is the primary mechanism, strength training would be logical. If increased energy expenditure through the activity itself is the mechanism, aerobic activities might be most suitable. Thus, two entirely different prescriptions would be in order based on these two mechanisms alone. Below, we describe mechanisms that have been proposed and the supporting evidence (summarized in table 16.1). We suggest increased focus on a potentially important but seldom discussed mechanism, the psychological effects of exercise (7, 8). For an overview of mechanisms and proposed pathways, see figure 16.1.

Table 16.1   Summary of Support for Proposed Mechanisms

Mechanism	Support
Energy expenditure	+
Decreased appetite	−
Decreased energy intake/increased dietary adherence*	+
Changes in macronutrient intake**	− +
Preservation of lean body mass	+ +
Decline in RMR	− +

− = no clear support for mechanism; + = some support; + + = strong support;
− + = mixed support; * in overweight samples; **− short term; **+ long term.

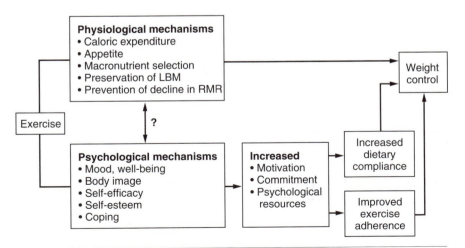

Figure 16.1   Proposed mechanisms and potential pathways linking exercise and weight control.

Energy Expenditure

A primary goal of obesity treatment is to create a negative energy balance. While physical activity can influence energy balance through increased caloric expenditure, it is unlikely that this is a primary mechanism in weight loss or maintenance. Obese individuals are often not able to expend a significant number of calories in a given exercise session, particularly at the beginning of a program, because their level of fitness and excess weight limit both duration and intensity of activity (19, 39). Weight losses in individuals who exercise often exceed what would be predicted from pure energy expenditure (6), and even very vigorous exercise results in relatively small energy expenditure (5). As an example, walking or running a mile expends approximately 100 calories. This level of exercise is beyond most overweight persons, yet consuming 100 calories can occur quite readily. Expending the calories of a single meal from a fast food restaurant consisting of a large hamburger, fries, and a milkshake would require a half-marathon. Thus, while activity may produce small changes in energy balance that could potentially add up over time, energy expenditure is probably one small component of the overall link between exercise and weight maintenance.

Appetite, Energy Intake, and Macronutrient Content of Diet

Exercise, Appetite, and Energy Intake

A second proposed mechanism is that exercise curbs appetite and hence reduces energy intake. Exercise may suppress appetite, but it is also possible that exercise is followed by energy compensation in the form of increased caloric intake.

Investigations have examined both acute and long-term effects of exercise on appetite and diet. In a review of the literature on exercise and appetite control, King, Tremblay, and Blundell (24) concluded that there is no short-term effect of exercise, i.e., on within-day intake. Studies investigating acute exercise and appetite have found evidence for suppression of perceptions of hunger following intense bouts of exercise (22, 23), but the effects appear to be short lived and, more importantly, do not seem to affect actual energy intake (24).

On a long-term basis, evidence suggests a weak association between energy intake and output; that is, increased activity is associated with increased energy intake (24). It seems, however, that this energy compensation occurs among lean, but not obese individuals (24, 32). Woo and colleagues examined exercise and energy intake in a group of severely obese women over a period of 19 days in one study (45) and 57 days in another (46) and compared the findings to those in lean women (47). The results suggested that there was a coupling of energy intake and expenditure among lean women (i.e., food intake was increased to balance energy expended during exercise), but not among obese women. Andersson and colleagues (1) also found no compensatory increase in energy intake in response to physical training in a study involving slightly overweight men and women.

Appetite and energy intake are distinct variables that are not necessarily highly correlated. As described above, people may experience changes in appetite (i.e., perceptions of hunger) without exhibiting changes in caloric intake. Exercise does not appear to systematically influence intake among obese individuals, but does it affect appetite? A recent study involving obese women did not find that the addition of exercise to a weight-loss program affected appetite-related measures. Wadden and colleagues (44) examined the effects of diet alone, diet plus aerobic training, diet plus strength training, and diet plus aerobic and strength training on hunger, satiety, food preoccupation, and intensity of food cravings in a sample of 128 obese women. There were no differences between the diet-only and exercise groups on any of the appetite-related measures. Thus, it appears that exercise does not lead to increased consumption or decreased appetite in overweight individuals.

The absence of a compensatory response to increasing activity could be interpreted as negative, in that it may reflect an underlying asynchrony of energy intake and output. On the other hand, it could be viewed as adaptive in allowing obese individuals to create an energy imbalance in the interest of weight loss. One would expect appetite to increase in response to increases in physical activity in order to maintain a stable energy balance. Thus, by increasing energy expenditure and not increasing intake, obese individuals are creating a negative energy balance (23) that should enhance weight loss and long-term maintenance.

Most of the studies described above have focused on determining whether exercise leads to energy compensation. Few have examined whether exercise could contribute to better dietary compliance, including reduced caloric intake. There have been some interesting findings in this area. For instance, one study (36) assessed the effect of aerobic exercise on dietary compliance among obese women randomly assigned to exercise or nonexercise conditions during a 12-week weight-loss program involving "moderate" energy restriction (prescription of 75% of measured basal metabolic rate). Exercisers completed three 45-minute exercise sessions per week. Daily caloric intake was measured by self-report and by the doubly labeled water method (DLW). Although both groups reported good adherence to prescribed intake levels, the DLW measures indicated that the exercise group exhibited significantly better dietary compliance (i.e., a smaller difference between actual and prescribed intake). Andersson and colleagues (1), in their study of energy intake in response to physical training, found that the most overweight women in the sample exhibited a notable decrease in caloric intake (-329 kcal/24 h). While there may be physiological explanations for these observed energy imbalances, it is also possible that psychological variables mediate the relationship between exercise and dietary intake. These potential relationships will be discussed in the section of the chapter devoted to psychological mechanisms. Either way, the relationship between exercise and dietary intake definitely merits future attention.

Exercise and Macronutrient Intake

Exercise has also been postulated to alter the composition of the diet by influencing food choice and thus, macronutrient selection. The literature on the topic is inconsistent (39). The question is whether physical activity leads to changes in the ratio of carbohydrate (CHO) to fat in the diet. Although cross-sectional research has not found consistent changes in the ratio, a small number of longitudinal studies have shown an association between increased activity intensity and higher intake of CHO-rich foods (39). In their review of exercise and appetite control, King and colleagues (24) report that exercise does not appear to lead to changes in food or nutrient preference in the short term. There is some evidence to suggest that, over time, exercise is associated with an increase in carbohydrate intake. In and of itself this change is probably only important for weight maintenance if the shift in macronutrient intake reduces overall fat or calorie intake. King et al. point out that, similar to overall dietary compliance, it is not clear whether physiological or psychological drives observed macronutrient changes.

In summary, exercise does not appear to result in increased caloric intake, decreased appetite, or significant changes in macronutrient selection. Individuals who are active and do not compensate for the increased energy expenditure from exercise create a negative energy balance. In addition, there are provocative findings indicating that exercise may in fact be associated with decreased intake or at least better adherence to a prescribed caloric level. The relationship between exercise and dietary intake holds promise as an important mechanism, but many questions remain. Is exercise associated with decreased intake or better compliance in the maintenance period following weight loss? Does this relationship occur across populations, e.g., gender? What are the mediating variables, and are they physical or psychological?

Metabolic and Body Composition Variables

Mechanisms related to metabolism and body composition have received extensive attention. The two variables of particular interest are lean body mass (LBM) and resting metabolic rate (RMR). Energy restriction causes a substantive decrease in RMR, due primarily to loss of LBM (21). It has been proposed that exercise combined with dietary restriction leads to greater fat loss and preservation of LBM, which in turn may prevent or reduce decline in RMR (38).

Saris (38) reviewed studies that investigated the added benefits of exercise for protection of fat-free mass, increase in fat loss, and increase in RMR during dietary restriction. He found that the evidence in favor of exercise was clear for protection of fat-free mass and increase in fat loss. In contrast, he concluded that exercise does not prevent decline in RMR. Results of a recent study (37) mirror the conclusions drawn by Saris (38). Racette and colleagues investigated changes in body composition in a sample of 23 obese women randomly assigned to either aerobic exercise or no exercise, and low-fat or low-carbohydrate diet. Among exercisers there was enhanced fat loss relative to loss of lean body mass. In the exercise group, women lost 89.4% of their weight as fat mass compared to 71.3% in the nonexercise group.

Both groups, however, exhibited a decline in RMR, and there was no difference between exercisers and nonexercisers in the magnitude of the decline.

Recent meta-analyses help integrate findings concerning LBM and RMR as mechanisms linking exercise and weight control. A meta-analysis conducted by Ballor and Poehlman (2) supports the contention that exercise provides protection against loss of lean body mass during dietary restriction. The analysis compared diet-only to diet-plus-exercise groups. Diet-plus-exercise groups lose less of their weight as fat-free mass compared to diet-only groups, and the findings hold across gender. Among men, those in the diet-only groups lost 28% of their weight as fat-free mass compared to 13% of weight lost among exercisers. The percentages of weight loss as fat-free mass for women in the diet-only and diet-plus-exercise groups were 24% and 11%, respectively. Of note, the amount and intensity of exercise needed to observe this conservation of fat-free mass was quite moderate (e.g., 3 days/week at low or moderate intensity).

The impact of diet versus diet-plus-exercise on RMR is an extremely important issue with respect to weight maintenance. Resting metabolic rate is the primary determinant of an individual's daily energy expenditure (34). Thus, a reduced RMR from dieting may facilitate weight regain during maintenance. If exercise were to prevent a decline in RMR, this would constitute a viable mechanism linking exercise to long-term weight maintenance.

The findings concerning the role of exercise in preventing diet-induced reductions in RMR remain equivocal even with the publication of two relevant meta-analyses. One meta-analysis (3) did not find significant effects of exercise training (aerobic only) on RMR during weight loss. These findings held irrespective of gender, although statistical power was low for this analysis. One explanation proposed by the authors is that RMR responses to exercise and/or diet interventions may depend on subtype or genotype of obesity (e.g., abdominal or gluteal-femoral), a variable that was not necessarily consistent across all studies.

A second meta-analysis (43) provided slightly different conclusions. The typical sample in this analysis consisted of middle-aged women who were prescribed diets that were generally low in fat and high in carbohydrate. Exercise programs were primarily aerobic and consisted of 31 to 60 minutes of moderate-intensity exercise on 4 to 5 days a week. There were significant decreases in RMR in both diet and diet-plus-exercise groups; however, the decrease in the diet-only groups was significantly greater than in the combined diet and exercise conditions. These results suggest that adding exercise to a dietary intervention may not prevent a decline in RMR, but may minimize the decrease.

Lean body mass and RMR most likely have an important function in the relationship between exercise and weight control. While much of the research has focused on the weight-loss phase, the implications for maintenance are similar. An individual who continues to be active during maintenance is likely to benefit from the preservation of LBM and minimized reductions in RMR. Ballor and Poehlman (3) point out that, even if exercise training does not have a unique impact on RMR, exercise can still have an important effect on daily energy expenditure by increasing energy expended during exercise and through increased physical activity beyond planned exercise. Clearly, more research is needed to explain discrepant findings in this area and to focus more specifically on the maintenance period. Inconsistencies may be due in part to differences across studies in variables such as degree of caloric restriction, type or amount of exercise prescription, or macronutrient content of the diet. There may also be differences based on population variables such as initial level of fitness, degree overweight, obesity genotype, or age.

Psychological Factors

A potentially important mechanism linking exercise to long-term weight loss is a collection of psychological factors, e.g., self-esteem and mood (7, 8). These factors are important to explore for several reasons. Exercise prescription might be affected (8), and there is a possibility, judging from our clinical experience and recent research, that improved well-being and enhanced self-esteem produced by physical activity generalize to other areas of life and lead to improved dietary adherence.

A review of the literature on exercise and psychology suggests several possible effects of physical activity on psychological factors that may be related to weight control. These factors are mood, body image, self-esteem, self-efficacy, and coping (7, 8).

Mood and General Well-Being

There is consistent evidence that exercise is associated with increased well-being and mood. Positive changes in overall well-being, and in specific variables such as depression and anxiety, could lead to a healthier psychological climate in which individuals have more cognitive and emotional resources, as well as motivation and energy, to sustain the long-term commitment to a weight-loss program. In addition, our clinical impression is that negative affective cues are often triggers for eating or bingeing. Improvements in general well-being may reduce the frequency of such cues. This could have important implications for maintenance of weight loss.

Research in this area has considered both acute and long-term effects of exercise in a variety of populations (e.g., clinical, nonclinical). In a review of the post-1975 literature on exercise and affective states, Byrne and Byrne (11) reported that exercise programs have antidepressant, antianxiety, and mood enhancing effects across both clinical and nonclinical populations. The authors described the literature as encouraging but, along with others (33, 49), suggest caution in interpreting data due to methodological problems in many of the studies.

Plante and Rodin (33) reviewed the effects of exercise on four areas of psychological functioning using studies of nonclinical populations. The authors concluded that exercise leads to improved mood and well-being, as well as reduced stress, anxiety, and depression. They qualified their conclusions by noting that the evidence supporting the acute impact of exercise on mood is more compelling than the research on long-term effects.

Lastly, Yeung (49) completed a review focused solely on the acute effects of exercise on mood state, i.e., the influence on mood of participation in a single exercise session. Despite variability in type of exercise, duration, and intensity, over 85% of the studies found some degree of improved mood following exercise. This effect appeared to be consistent across gender and age. Yeung suggests that exercise may be useful as a short-term method for self-regulation of mood.

One could still ask whether these mood-related findings apply specifically to overweight individuals. A recent study addressed this question (16). Foreyt and colleagues assessed psychological variables associated with self-reported physical activity among overweight adults over the course of four years. At years one and five, physical activity was associated with a "more positive psychological profile" in obese participants. Higher levels of activity were related to lower levels of depression. In addition, an increase in activity between year one and year five was associated with enhanced well-being.

The mood-enhancing effects of physical activity merit more attention with respect to the exercise/weight control relationship. Studies are needed that examine affective symptoms and/or reductions in emotionally triggered eating as mediators between activity and weight maintenance. Studies should also look at the role of goal setting as a potential mediating variable. Goals that people set can have a major impact on motivation and persistence. Depressed individuals may set more difficult goals (13) that, particularly with respect to weight loss, can lead to failure and discouragement.

The relationship between exercise and mood may have important implications for relapse, a major concern during the maintenance phase of health behavior change. Emotional states, including depression and anxiety, are associated with relapse (9). Exercise may operate as a relapse prevention strategy through its positive effect on psychological states.

Body Image
Body image is another psychological variable likely to be enhanced by physical activity that may subsequently influence weight control attitudes and behaviors. An individual with poor body image feels extremely dissatisfied with his or her shape and weight. This can lead to frustration with the modest weight losses that most programs produce. This in turn could lead to negative mood, resignation, and, ultimately, relapse.

Surprisingly, there is very little research on the effect of exercise on body image, particularly in the context of an obesity treatment program. However, there is some evidence that exercise influences people's feelings about their bodies. In a prospective study of activity and self-perception, college students who participated in a 10-week exercise/activity program showed improved physical self-perceptions (12). While this study did not specifically measure body image, it suggests that exercise has positive effects on perceptions of one's body. Another study, involving 208 healthy, middle-aged adults, examined the effects of diet-only, exercise-only, diet-plus-exercise, and no intervention on perceptions of the body (42). Over the course of a one-year program, exercise participation led to enhanced body perception. McAuley, Mihalko, and Bane (27) found that a 20-week exercise program for middle-aged adults resulted in improved self-perceptions of physical attractiveness.

Exercise in obese persons is associated with changes that might lead to better body image. Wood and colleagues (48), in their study comparing diet-alone to diet-plus-exercise in overweight men and women, found that the two interventions did not differ in total weight loss among women. However, women in the diet-plus-exercise group significantly reduced their ratio of abdominal to hip girth compared to the control group, whereas those in the diet-only group did not. Thus, exercise can result in physical changes in body shape, which can lead to improvements in an individual's self-image and confidence. These positive changes in body image, irrespective of change in the numbers on the scale, may help prevent the discouragement and resignation described above as a potential outcome for individuals with poor body image.

Negative body image can be a barrier to exercise for overweight individuals. People often describe feeling self-conscious in exercise clothing and worry that, when exercising in public places, others are making negative judgments about them. Improvements in body image that are attributed to exercise may be reinforcing and lead to greater long-term exercise adherence, as well as increased confidence in one's ability to make positive changes related to body shape or weight. More research is needed on the impact of physical activity on body image among overweight individuals, and whether or not this translates into increased adherence or commitment to weight control.

Self-Esteem and Self-Efficacy

On a more global level, increased self-esteem and self-efficacy could also be implicated in the relationship between exercise and long-term weight control. Plante and Rodin (33) included self-esteem as an outcome variable in their review of exercise and psychological functioning. In all relevant studies, there was a positive relationship between exercise and self-esteem. If the term "self-esteem" is used to refer to a person's self-evaluation, then it appears that exercise may lead to a more positive global self-image. As with enhanced mood and well-being, improvements in self-esteem may lead to a more positive psychological climate in which to make lifestyle changes. Increased self-esteem may also allow an individual to view him- or herself as a person able to make and commit to positive changes.

Enhanced self-efficacy could lead to increased confidence in one's ability to make behavioral changes and to lose weight. Perceived self-efficacy refers to "beliefs in one's capabilities to organize and execute the courses of action required to manage prospective situations…. Efficacy beliefs influence how people think, feel, motivate themselves, and act" (4; p. 2). Self-efficacy is an obvious variable to consider in understanding the connection between exercise and weight maintenance because of its widespread application in the field of health behavior. There is a substantial body of research indicating that perceived self-efficacy is associated with behavioral intentions and health behavior changes (40). Self-efficacy beliefs influence cognitive construction of action plans, i.e., helping develop visualizations of scenarios that can guide goal attainment (40). People with a high sense of efficacy may visualize success, while individuals who doubt their self-efficacy may visualize or anticipate failure and may dwell on things that could go wrong (4). In the context of exercise and weight control, being physically active may lead to increased self-efficacy about exercise itself and perhaps about other health behaviors, such as eating, as well.

Few studies have directly examined the effects of exercise participation on exercise self-efficacy. Correlational studies have found that those who persist at an exercise regimen have higher exercise self-efficacy. However, it is very likely that those individuals also had higher baseline levels of confidence about being active. Prospective studies that assess initial level of self-efficacy are needed. One such study (26) examined physical efficacy expectations as outcomes of exercise participation in middle-aged males and females and found increased self-efficacy in response to both acute bouts and long-term participation in exercise. Sorenson and colleagues (42) examined the effects of diet-only, exercise-only, diet-plus-exercise, and no intervention on perceptions of the body and physical mastery and competence. The intervention lasted for one year. Results indicated that exercise led to more positive self-perceptions of physical mastery and ability.

It is plausible that efficacy expectations of diet and exercise are related. Increases in exercise self-efficacy could influence eating self-efficacy and dietary compliance through a more general sense of weight-loss self-efficacy. Foreyt and colleagues (16) assessed psychological correlates of reported physical activity in a longitudinal (four-year) observational study involving close to 400 adults, both normal weight and obese. One of the outcome variables was an eating self-efficacy measure. Among obese participants, higher levels of physical activity were associated with improved eating self-efficacy at years one and five. Moreover, change in physical activity (an increase) was significantly associated with change in eating self-efficacy (improvement).

Coping
Exercise may serve as a helpful coping strategy in the process of behavior change and maintenance. Having coping responses to rely on is particularly important in preventing lapses during the maintenance phase of behavior change (9).

Coping can be conceptualized as a process of responding to a stressful situation or event and can operate at different levels, e.g., in preparation for or in response to stress (25). According to Long, exercise may operate as an emotion-focused strategy, a means of coping with one's reactions to stressors, perhaps through distraction or relaxation. Thus, for individuals who overeat or binge eat in response to negative events, exercise may serve as an important alternate activity.

Speculative Nature of Psychological Mechanisms
While there are many reasons to believe that exercise and weight control may be linked in part by psychological mechanisms, it is important to note that these relationships are largely speculative at this time. Research on exercise and weight

maintenance has generally overlooked the potential role of psychological variables (8). One major impediment to establishing a causal role for psychological variables is the difficulty of separating them from the physiological effects of exercise. The psychological benefits of exercise could be a result of either biological or cognitive factors (8). There may, in fact, be an intricate interplay between physiology and psychology, but we believe that cognitive factors such as self-esteem and self-efficacy are particularly important. We also believe that there is a need for open-mindedness regarding the potential role of psychological variables in the association between activity and weight maintenance. Clearly, there are many research questions that remain to be addressed.

Summary

Long-term maintenance of weight loss is a primary goal in the treatment of obesity. It is crucial that the field identify factors that contribute to sustained weight control. At this time, exercise holds promise as a predictor. Research evidence suggests that individuals who exercise are more likely to maintain weight losses. What is less clear is the reason why this relationship exists. Many mechanisms have been proposed and examined, including caloric expenditure; appetite and macronutrient selection; and preservation of LBM and RMR. While these variables most likely play a role in the exercise/weight control link, psychological factors such as mood, body image, self-esteem, self-efficacy, and coping skills may be equally important to consider. This chapter reviewed available research on these mechanisms and presented a rationale for the inclusion of psychological variables in the search for an understanding of how exercise contributes to long-term weight maintenance.

There are important practical reasons why we need to understand the mechanisms that link exercise to weight maintenance. Knowledge about mechanisms has implications for making effective exercise prescriptions. For example, two different mechanisms call for different prescriptions. If the primary mechanism is caloric expenditure, then the exercise prescription should be aimed at maximizing energy output. On the other hand, if exercise influences weight control through psychological mechanisms, the prescription should be designed to lead to the desired psychological outcomes (8). The practicalities of these two prescriptions could look extremely different.

In making exercise prescriptions, adherence also becomes a major concern (see chapter 19). It is logical that individuals will have greater adherence to an exercise program if they perceive the exercise to be leading toward their goals. They are most likely to feel that the exercise is effective if they are following an exercise prescription that is based on knowledge about mechanisms as related to specific goals.

There are many unanswered research questions on this topic of exercise and weight maintenance. Table 16.2 presents a summary of some of the pressing research questions for the field.

Table 16.2   Recommendations for Future Research

1. Is exercise equally valuable for all obese individuals? Research is needed involving broader populations, e.g., those including more diversity in variables such as age and degree of obesity.

2. What components of exercise (e.g., frequency, intensity, duration, total energy expended, perceived exertion, etc.) form the strongest links with weight control?

3. What are the effects of different types of activity (e.g., planned versus lifestyle, or strength training versus aerobic) on weight loss and maintenance?

4. Are there interactions between specific types of diet and different exercise prescriptions?

5. Are there interactions between population and type of exercise? Is strength training more important for women? To whom should lifestyle activity be prescribed?

6. Do baseline psychological variables (e.g., mood, self-efficacy, or body image) have implications for adherence to different types of programs?

7. Do weight control interventions that include exercise lead to greater long-term improvements in mood, anxiety, body image, self-esteem, self-efficacy, and perceived coping ability?

8. What psychological or physical variables predict adherence to an exercise program?

9. Do psychological variables (e.g., mood, self-efficacy, body image, and coping strategies) mediate the relationship between physical activity and commitment to weight control, specifically dietary adherence?

In light of increasing rates of obesity, a dearth of treatments that produce long-term weight changes, and the positive findings about exercise, there is no doubt as to the need for future investigations focused on increasing our knowledge about exercise and weight maintenance.

References

1. Andersson, B., X. Xu, M. Rebuffe-Scrive, K. Terning, M. Krotkiewski, and P. Bjorntorp. 1991. The effects of exercise training on body composition and metabolism in men and women. *International Journal of Obesity* 15: 75-81.

2. Ballor, D.L., and E.T. Poehlman. 1994. Exercise training enhances fat-free mass preservation during diet-induced weight loss: a meta-analytical finding. *International Journal of Obesity* 18: 35-40.

3. Ballor, D.L., and E.T. Poehlman. 1995. A meta-analysis of the effects of exercise and/or dietary restriction on resting metabolic rate. *European Journal of Applied Physiology and Occupational Physiology* 71: 535-542.

4. Bandura, A. 1995. Exercise of personal and collective efficacy in changing societies. In *Self-efficacy in changing societies,* ed. A. Bandura, 1-45. New York: Cambridge University Press.

5. Bjorntorp, P. 1978. Exercise and obesity. *Psychiatric Clinics of North America* 1: 691-696.

6. Bray, G. A. 1976. *The obese patient.* Philadelphia: Saunders.

7. Brownell, K.D. 1995a. Exercise in the treatment of obesity. In *Eating disorders and obesity: a comprehensive handbook,* eds. K.D. Brownell and C.G. Fairburn, 473-478. New York: Guilford Press.

8. Brownell, K.D. 1995b. Exercise and obesity treatment: psychological aspects. *International Journal of Obesity and Related Metabolic Disorders* 10 (Suppl. 4): 122-125.

9. Brownell, K.D., G.A. Marlatt, E. Lichtenstein, and G.T. Wilson. 1986. Understanding and preventing relapse. *American Psychologist* 41: 765-782.

10. Brownell, K.D., and T.A. Wadden. 1992. Etiology and treatment of obesity: understanding a serious, prevalent, and refractory disorder. *Journal of Consulting and Clinical Psychology* 60: 505-517.

11. Byrne, A., and D.G. Byrne. 1993. The effect of exercise on depression, anxiety and other mood states: a review. *Journal of Psychosomatic Research* 37: 565-574.

12. Caruso, C.M., and D.L. Gill. 1992. Strengthening physical self-perceptions through exercise. *Journal of Sports Medicine and Physical Fitness* 32: 416-427.

13. Cervone, D., D.A. Kopp, L. Schaumann, and W.D. Scott. 1994. Mood, self-efficacy, and performance standards: lower moods induce higher standards for performance. *Journal of Personality and Social Psychology* 67: 499-512.

14. DePue, J.D., M.M. Clark, L. Ruggiero, M.L. Medeiros, and V. Pera, Jr. 1995. Maintenance of weight loss: a needs assessment. *Obesity Research* 3: 241-248.

15. Ewbank, P. P., L.L. Darga, and C. P. Lucas. 1995. Physical activity as a predictor of weight maintenance in previously obese subjects. *Obesity Research* 3: 257-263.

16. Foreyt, J.P., R.L. Brunner, G.K. Goodrick, S.T. St. Jeor, and G.D. Miller. 1995. Psychological correlates of reported physical activity in normal-weight and obese adults: the Reno diet-heart study. *International Journal of Obesity and Related Metabolic Disorders* 19 (Suppl. 4): 69-72.

17. Grodstein, F., R. Levine, L. Troy, T. Spencer, G.A. Colditz, and M.J. Stampfer. 1996. Three-year follow-up of participants in a commercial weight loss program. Can you keep it off? *Archives of Internal Medicine* 156: 1302-1306.

18. Hill, J.O., D.G. Schundlt, T. Sbrocco, J. Pope-Cordle, B. Stetson, M. Kaler, and C. Heim. 1989. Evaluation of an alternating-calorie diet with and without exercise in the treatment of obesity. *American Journal of Clinical Nutrition* 50: 248-254.

19. Institute of Medicine. 1995. *Weighing the options: criteria for evaluating weight management programs.* Washington D.C.: U.S. Government Printing Office.

20. Kayman, S., W. Bruvold, and J.S. Stern. 1990. Maintenance and relapse after weight loss in women: behavioral aspects. *American Journal of Clinical Nutrition* 52: 800-807.

21. Keys, A., J. Brozek, A. Henschel, O. Mickelsen, and H.L. Taylor. 1950. *The biology of human starvation*. Minneapolis: University of Minnesota Press.

22. King, N.A., and J.E. Blundell. 1995. High-fat foods overcome the energy expenditure induced by high-intensity cycling or running. *European Journal of Clinical Nutrition* 49: 114-123.

23. King, N.A., V.J. Burley, and J.E. Blundell. 1994. Exercise-induced suppression of appetite: effects on food intake and implications for energy balance. *European Journal of Clinical Nutrition* 48: 715-724.

24. King, N.A., A. Tremblay, and J.E. Blundell. 1997. Effects of exercise on appetite control: implications for energy balance. *Medicine and Science in Sports and Exercise* 29: 1076-1089.

25. Long, B.C. 1993. A cognitive perspective on the stress-reducing effects of physical exercise. In *Exercise psychology: the influence of physical exercise on psychological processes,* ed. P. Seraganian, 339-357. New York: John Wiley & Sons.

26. McAuley, E., S.M. Bane, and S.L. Mihalko. 1995. Exercise in middle-aged adults: self-efficacy and self-presentational outcomes. *Preventive Medicine* 24: 319-328.

27. McAuley, E., S.L. Mihalko, and S.M. Bane. 1997. Exercise and self-esteem in middle-aged adults: multidimensional relationships and physical fitness and self-efficacy influences. *Journal of Behavioral Medicine* 20: 67-83.

28. Miller, W.C., D.M. Koceja, and E.J. Hamilton. 1997. A meta-analysis of the past 25 years of weight loss research using diet, exercise or diet plus exercise intervention. *International Journal of Obesity and Related Metabolic Disorders* 21: 941-947.

29. Pavlou, K.N., S. Krey, and W.P. Steffee. 1989. Exercise as an adjunct to weight loss and maintenance in moderately obese subjects. *American Journal of Clinical Nutrition* 49: 1115-1123.

30. Pavlou, K.N., W.P. Steffee, R.H. Lerman, and B.A. Burrows. 1985. Effects of dieting and exercise on lean body mass, oxygen uptake, and strength. *Medicine and Science in Sports and Exercise* 17: 466-471.

31. Perri, M.G., W.G. McAdoo, D.A. McAllister, J.B. Lauer, and D.Z. Yancey. 1986. Enhancing the efficacy of behavior therapy for obesity: effects of aerobic exercise and a multicomponent maintenance program. *Journal of Consulting and Clinical Psychology* 54: 670-675.

32. Pi-Sunyer, F.X. 1992. The effects of increased physical activity on food intake, metabolic rate, and health risks in obese individuals. In *Treatment of the seriously obese patient,* eds. T.A. Wadden and T.B. VanItallie, 190-210. New York: Guilford Press.

33. Plante, T. G., and J. Rodin. 1990. Physical fitness and enhanced psychological health. *Current Psychology: Research and Reviews* 9: 3-24.

34. Poehlman, E.T. 1989. A review: exercise and its influence on resting energy metabolism in man. *Medicine and Science in Sports and Exercise* 21: 515-525.

35. Pronk, N.P., and R.R. Wing. 1994. Physical activity and long-term maintenance of weight loss. *Obesity Research* 2: 587-599.

36. Racette, S.B., D.A. Schoeller, R.F. Kushner, and K.M. Neil. 1995. Exercise enhances dietary compliance during moderate energy restriction in obese women. *American Journal of Clinical Nutrition* 62: 345-349.

37. Racette, S.B., D.A. Schoeller, R.F. Kushner, K.M. Neil, and K. Herling-Iaffaldano. 1995. Effects of aerobic exercise and dietary carbohydrate on energy expenditure and body composition during weight reduction in obese women. *American Journal of Clinical Nutrition* 61: 486-494.

38. Saris, W.H.M. 1993. The role of exercise in the dietary treatment of obesity. *International Journal of Obesity* 17 (Suppl. 1): 17-21.

39. Saris, W.H.M. 1996. Physical activity and body weight regulation. In *Regulation of body weight: biological and behavioral mechanisms,* eds. C. Bouchard and G.A. Bray, 135-148. New York: John Wiley & Sons.

40. Schwarzer, R., and R. Fuchs. 1995. Changing risk behaviors and adopting health behaviors: the role of self-efficacy beliefs. In *Self-efficacy in changing societies,* ed. A. Bandura, 259-288. New York: Cambridge University Press.

41. Sikand, G., A. Kondo, J.P. Foreyt, P.H. Jones, and A.M. Gotto. 1988. Two-year follow-up of patients treated with a very-low-calorie-diet and exercise training. *Journal of the American Dietetic Association* 88: 487-488.

42. Sorenson, M., S. Anderssen, I. Hjerman, I. Holme, and H. Ursin. 1997. Exercise and diet interventions improve perceptions of self in middle-aged adults. *Scandinavian Journal of Medicine and Science in Sports* 7: 312-320.

43. Thompson, J.L., M.M. Manore, and J.R. Thomas. 1996. Effects of diet and diet-plus-exercise programs on resting metabolic rate: a meta-analysis. *International Journal of Sport Nutrition* 6: 41-61.

44. Wadden, T.A., R.A. Vogt, R.E. Andersen, S.J. Bartlett, G.D. Foster, R.H. Kuehnel, J. Wilk, R. Weinstock, P. Buckenmeyer, R.I. Berkowitz, and S.N. Steen. 1997. Exercise in the treatment of obesity: effects of four interventions on body composition, resting energy expenditure, appetite, and mood. *Journal of Consulting and Clinical Psychology* 65: 269-277.

45. Woo, R., J.S. Garrow, and F.X. Pi-Sunyer. 1982a. Effect of exercise on spontaneous calorie intake in obesity. *American Journal of Clinical Nutrition* 36: 470-477.

46. Woo, R., J.S. Garrow, and F.X. Pi-Sunyer. 1982b. Voluntary food intake during prolonged exercise in obese women. *American Journal of Clinical Nutrition* 36: 478-484.

47. Woo, R., and F.X. Pi-Sunyer. 1985. Effect of increased physical activity on voluntary intake in lean women. *Metabolism* 34: 836-841.

48. Wood, P.D., M.L. Stefanick, P.T. Williams, and W.L. Haskell. 1991. The effects on plasma lipoproteins of a prudent weight-reducing diet, with or without exercise, in overweight men and women. *The New England Journal of Medicine* 325: 461-466.

49. Yeung, R.R. 1996. The acute effects of exercise on mood state. *Journal of Psychosomatic Research* 40: 123-141.

PART IV

Physical Activity, Fitness, and Health in the Obese State

CHAPTER 17

Physical Activity and the Metabolic Complications of Obesity

Jean-Pierre Després, PhD, and Benoît Lamarche, PhD
Department of Food Sciences and Nutrition, Laval University, and the Lipid Research Center, Laval University Research Center, Ste-Foy, Québec, Canada

The prevalence of obesity is unfortunately increasing in developed and also in developing countries (102, 101). Indeed, the recent report of the World Health Organization emphasized the worldwide trend of an increased prevalence for overweight and obesity in numerous countries (120). Obesity is a complex and multifactorial condition, and several factors contribute to its development. Although there are controversies regarding the critical factors involved, it is generally believed that a marked reduction in daily physical activity combined with our affluent diets, rich in fat and simple sugars, are probably two important contributing factors in the progressive increase in the prevalence of obesity (22).

The objective of the present chapter is to describe the health hazards of obesity, with a particular focus on metabolic risk factors for cardiovascular disease. Pathophysiological aspects of insulin resistance and dyslipidemia commonly found among obese patients will be discussed with a section on the potential role of regular physical activity in improving metabolic parameters and reducing the risk of type 2 diabetes and coronary heart disease.

Obesity As a Health Burden

Although it is commonly accepted that obesity is a health burden and that overweight patients are frequently characterized by hypertension, diabetes, and dyslipidemia, as well as cerebral and vascular diseases (6, 21, 32, 62, 60), not every patient is characterized by these complications (8). Thus, some very obese patients have a fairly normal metabolic risk profile, whereas some moderately overweight individuals are characterized by severe insulin resistance, marked dyslipidemia, hypertension, and by an elevated risk of developing type 2 diabetes or premature coronary heart disease. In this regard, it is now becoming more and more accepted that obesity is not only heterogeneous in terms of its etiology, but also with respect to its related metabolic complications (8, 9, 30, 32, 40, 59, 60, 63, 62). To put this issue in proper historical perspective, it is remarkable to note that in 1947 a French physician, Jean Vague, was the first to suggest that body-fat topography was a more important factor to take into account in the evaluation of the obese patient than excess fatness per se (112). Indeed, Vague suggested that upper-body obesity, a condition that he described as android or male type obesity, was frequently found among his overweight patients with hypertension, diabetes, or clinical signs of coronary heart disease. He also proposed that the typical female pattern of body-fat deposition, which he referred to as gynoid obesity, was not commonly associated with complications. His theory was published in the French medical literature; however, many years went by before these pioneering clinical observations received support from large epidemiological studies.

Two groups of investigators, one located in Milwaukee and led by Ahmed Kissebah and the other located in Gothenburg, Sweden under the direction of Per Björntorp, almost simultaneously reported that a high proportion of abdominal fat was associated with alterations in glucose tolerance as well as with increased fasting insulin and triglyceride levels (65, 63). One year later, results from the prospective study of men and women of Gothenburg confirmed that high proportion of abdominal fat, crudely assessed by an increased waist-to-hip ratio, was not only associated with metabolic complications, but was also predictive of an increased risk of ischemic heart disease (72, 73); this association was found to be completely independent from the concomitant variation in the level of total body fat. In 1985, the Gothenburg Group also reported that a high waist-to-hip ratio combined with an elevated body mass index was associated with more than a 30-fold increased risk of developing diabetes (91). Therefore, these studies published in the mid '80s confirmed the early suggestion of Vague: Among equally obese individuals, excess abdominal fat accumulation, at least crudely estimated by an increased ratio of waist-to-hip circumferences, is predictive of an increased risk of developing type 2 diabetes and ischemic heart disease. These results generated considerable interest from the medical community. Numerous research groups around the world are currently studying the etiology of abdominal fat accumulation as well as pathophysiological aspects of the complications resulting from the presence of abdominal obesity.

Upper Body Obesity and Metabolic Risk Factors

In epidemiological studies, the proportion of upper-body fat was estimated by a very simple anthropometric index: the ratio of waist girth divided by the hip circumference. The rationale for using this ratio is very simple; the more abdominal fat a subject has, the greater is the waistline. However, with the development of imaging techniques such as magnetic resonance imaging or computed tomography, it has been possible to assess with greater accuracy abdominal fat accumulation and to measure selectively the fat located in the abdominal cavity as well as the subcutaneous adipose tissue (18, 49, 51, 66, 97, 104, 111). By using computed tomography, we found that among equally obese men and women matched for the level of total body fat, the subgroup of obese patients with a high accumulation of visceral adipose tissue were character-ized by the most severe metabolic disturbances, whereas obesity per se (in the absence of excess visceral adipose tissue accumulation) was associated with much less severe insulin resistance and dyslipidemic states compared to lean controls (27, 40, 30, 97).

However, our observations did not suggest that subcutaneous obesity is not a health hazard. Indeed, we found that subcutaneous fat accumulation is a good correlate of total body fat (78). It is also very well known that excess fatness is associated with insulin resistance and with some significant alterations in the plasma lipid lipoprotein profile (60, 61). Therefore, when comparing obese indi-viduals with a substantial accumulation of subcutaneous fat to lean controls, one should expect a moderate state of insulin resistance as well as differences in the plasma lipoprotein lipid profile. However, among equally obese individuals, those with the highest accumulation of visceral adipose tissue are those characterized by the most severe deterioration in metabolic risk variables (27, 40, 30, 97). Several lines of evidence support the notion that excess visceral adipose tissue accumulation exacerbates the metabolic risk profile of obese patients. First, substantial gender differences in visceral adipose tissue accumulation are documented, with men being on average more prone to visceral fat deposition than premenopausal women (78). Accordingly, men are on average characterized by a more disturbed metabolic risk profile than premenopausal women (77). However, after matching men and women for the same amount of visceral adipose tissue, we reported that gender differences in the metabolic risk profile were largely abolished (77), suggesting that the sex difference in visceral adipose tissue accumulation is a very important factor involved in the gender difference in the metabolic risk profile. Furthermore, we have also reported that there is, with age, a selective deposition of visceral adipose tissue in both men and women (79, 80). Indeed, middle-aged men and women are characterized by a higher accumulation of visceral adipose tissue than young adults (79, 80) for any level of total body fat. In men it appears that there is a progressive increase in visceral adipose tissue accumulation with age. In women, however, although there is also an age-specific effect on visceral adipose tissue deposition, the estrogen deficiency that occurs at menopause has been suggested to be associated with an acceleration in visceral adipose tissue deposition that would lead to the progressive development of insulin resistance and to a dyslipidemic profile (110). Finally, women on hormone replacement therapy are characterized by a lower relative accumulation of abdominal adipose tissue than are postmenopausal women not on hormone replacement therapy (108). Accordingly, postmenopausal women on hormone replacement therapy are not only characterized by less abdominal fat, but also by a more favorable metabolic risk profile than are postmenopausal women not on hormone replacement therapy (108).

Taken together, these results support the notion that visceral adipose tissue is an important fat depot to consider in the evaluation of the risk associated with obesity. Thus, excess fatness has, to a certain extent, a deleterious impact on *in vivo* insulin action and on plasma lipoprotein levels. However, it appears from our studies that obese subjects with the highest accumulation of visceral adipose tissue are characterized by the most deteriorated metabolic risk profile. We believe that this interpretation reconciles our results with those of studies (1) that have suggested that both visceral adipose tissue and excess subcutaneous fat (a good marker of total body fat) are significant correlates of an impaired *in vivo* insulin action.

The Athero-Thrombotic Metabolic Risk Profile of Visceral Obesity

Excess visceral adipose tissue accumulation in obese patients is associated with *in vivo* insulin resistance, which results from an impaired insulin action in the skeletal muscle, in adipose tissue, and in the liver (9, 108, 62). Reduced glucose transport and storage has been found in insulin-resistant abdominally obese individuals (9, 108, 61) and it appears that the ability of insulin to inhibit very low-density lipoprotein (VLDL) secretion may be impaired in visceral obesity (82, 83). This impaired hepatic metabolism may be the consequence of the hyperlipolytic state resulting from the expanded visceral adipose tissue mass (10, 9, 61, 62, 32). Indeed, studies have shown that intra-abdominal adipocytes are resistant to the antilipolytic effect of insulin as opposed to subcutaneous adipocytes (17, 86). Therefore, in the presence of visceral obesity, the expanded visceral adipose tissue mass is associated with a hyperlipolytic state despite the presence of hyperinsulinemia. Through the portal circulation, this high lipolytic flux exposes the liver to high concentrations of free fatty acids that have been suggested to protect apolipoprotein B against its degradation, leading to an overproduction of triglyceride-rich lipoproteins (32, 10). Furthermore, this increased portal flux of free fatty acids contributes to the reduced hepatic extraction of insulin (55, 107), thereby exacerbating the systemic hyperinsulinemia in visceral obesity. Finally, hepatic glucose production is stimulated by the increased lipid oxidation, which provides precursors for the gluconeogenesis (56) that contributes to the impairment of glucose tolerance and to hyperglycemia. Therefore, one can relate the expanded visceral adipose tissue mass and the hyperlipolytic state of visceral adipocytes to the dyslipidemic, hyperinsulinemic, and hyperglycemic states commonly found in visceral obese patients. Furthermore, there is evidence suggesting that visceral obese, insulin-resistant patients are characterized by a reduced post-heparin plasma lipoprotein lipase activity and by a marked increase in hepatic triglyceride lipase activity (33).

The overproduction of VLDL in the fasting state should also compete with dietary fat (chylomicrons) for the catalytic activity of lipoprotein lipase. This phenomenon may contribute to the impaired tolerance to dietary fat, which has been reported in visceral obesity (25). Thus, the ability of visceral obese patients to clear triglyceride-rich lipoproteins appears to be impaired, and it has been proposed that it is largely the consequence of an overproduction of triglyceride-rich lipoproteins by the liver (82). The hypertriglyceridemia found in fasting as well as in postprandial states among visceral obese patients also favors the exchange of lipids from triglyceride-rich lipoproteins to LDL and HDL particles, a process which leads to triglyceride enrichment of HDL and LDL fractions to the expense of cholesterol (42, 33). The increased activity of hepatic lipase and the high affinity of this enzyme for triglyceride-rich lipoproteins of smaller size lead to the formation of small, dense HDL and LDL particles, these alterations having been reported among visceral obese insulin-resistant and dyslipidemic patients (109). Therefore, these complex metabolic disturbances lead to insulin resistance; hyperinsulinemia; impaired glucose tolerance; hypertriglyceridemia; elevated apo B concentration; increased proportion of small, dense LDL particles; and reduced HDL particle size and cholesterol content that have been documented in visceral obese patients (27, 28, 29, 31, 32, 33, 37, 38, 39, 41, 40, 43, 44, 45) (figure 17.1).

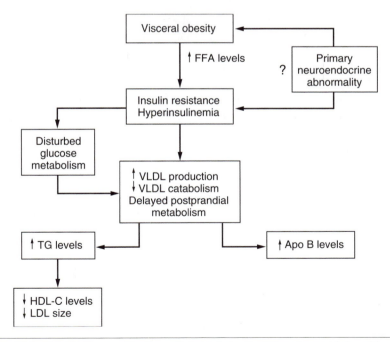

Figure 17.1   Metabolic complications of visceral obesity. TG = plasma triglyceride levels; HDL-C = high density lipoprotein cholesterol; LDL = low density lipoprotein; VLDL = very low-density lipoproteins; apo = apolipoprotein; FFA = free fatty acid.

The insulin resistance dyslipidemic state found among obese subjects with excess visceral adipose tissue accumulation is not commonly associated with a marked increase in plasma cholesterol and LDL-cholesterol levels (30, 31, 29, 39, 38). Thus, one should not focus on plasma cholesterol or LDL-cholesterol levels to adequately assess coronary heart disease risk in these patients. As there is one apo B molecule per LDL particle, it has been suggested that the measurement of apo B in the LDL fraction, or at least in the total plasma, could provide a better estimate of the concentration of atherogenic particles than plasma cholesterol or LDL-cholesterol levels especially among insulin-resistant visceral obese patients. With the use of rocket immunoelectrophoresis, we have reported an approximately 20% increase in apo B or LDL apo B concentration in visceral obese patients in the absence of any difference in LDL-cholesterol concentration (37), suggesting the presence of an increased proportion of small dense, cholesteryl ester-depleted LDL particles. Indeed, when we measured LDL particle size in visceral obese patients with the use of 2-16% polyacrylamide gel electrophoresis, we reported that excess visceral adipose tissue accumulation associated with insulin resistance, hypertriglyceridemia, and reduced HDL-cholesterol levels was also related to the presence of small LDL particles on the gel (109).

It is important to point out that the LDL peak particle diameter assessed with this technique was not correlated with plasma cholesterol nor with LDL-cholesterol levels (109). Additional multivariate analyses revealed that fasting triglyceride concentration was the best correlate of LDL particle size (109). Thus, in visceral obese insulin-resistant patients, it should be emphasized that plasma cholesterol and LDL-cholesterol concentrations cannot adequately reflect LDL particle concentration or size. Any intervention study leading to changes in body-fat content, insulin action, and fasting triglyceride concentrations may produce changes in LDL particle concentration and size that may not be adequately described by changes in plasma LDL-cholesterol levels. As we had estimated that about 25% of sedentary adult males may be characterized by the insulin-resistance dyslipidemic syndrome (37), we have suggested that the dyslipidemia of visceral obesity may represent, among sedentary male subjects, the most prevalent cause of coronary heart disease. In this regard, although considerable literature from epidemiological and metabolic studies has emphasized the importance of measuring cholesterol and LDL-cholesterol levels as risk factors for coronary heart disease (58, 90, 106), and although trials have shown that reducing cholesterol and LDL-cholesterol levels may reduce the number of coronary heart disease events and related mortality (103, 100, 99), one has to keep in mind that the reduction in the number of events has been systematically reported to reach approximately 30%. Therefore, several patients who are treated with a hypocholesterolemic medication such as an HMG-CoA reductase inhibitor may still be at very high risk for a CHD event. In this regard, we would like to propose that the optimal management of risk among patients with abdominal obesity should not only include as therapeutic targets the legitimate reduction in plasma LDL concentration, but should also aim at the loss of abdominal fat, thereby improving insulin sensitivity and several markers of CHD risk not necessarily related to plasma cholesterol or LDL-cholesterol levels.

The Visceral Obese Syndrome

It is now commonly accepted that insulin resistance and compensatory hyperinsulinemia characterize obese patients with an excess accumulation of visceral adipose tissue. These abnormalities may ultimately evolve to glucose intolerance and to type 2 diabetes among genetically susceptible patients (76, 62, 9). Furthermore, this insulin-resistance hyperinsulinemic state is associated with a dyslipidemia, which includes hypertriglyceridemia; elevated apo B concentrations; reduced HDL-cholesterol levels; and an increased proportion of small, dense LDL particles (30, 32, 27, 31, 29, 38, 40). The classic dyslipidemia of insulin resistance has been described as the lipid triad (raised triglyceride, raised LDL-cholesterol, and reduced HDL-cholesterol), and evidence suggests that this dyslipidemic profile is associated with a substantially increased risk of coronary heart disease (5). We have conducted a prospective study in Québec City; the Québec Cardiovascular Study (68) gave us the opportunity to examine the risk related to metabolic complications commonly found in insulin-resistant, visceral obese individuals. We had access to a random sample of middle-aged men with a follow-up of 5 years. Over the 5-year follow-up period, 114 cases of ischemic heart disease were noted from the initial sample of 2103 healthy men. When we first compared the prevalence of various dyslipidemic states among those who later developed ischemic heart disease, compared to those who remained healthy, about 50% of the men who remained free from clinical manifestations of ischemic heart disease had a normal lipid profile, whereas less than 1/3 of those who developed ischemic heart disease were characterized by a normal lipoprotein profile (68). Therefore, there was initially an increased prevalence of dyslipidemic states among those middle-aged men who then developed ischemic heart disease. Multiple regression analyses revealed that diabetes, apo B concentration, age, smoking, and raised systolic blood pressure were the best predictors of the risk of ischemic heart disease in this cohort (70).

In another analysis, it was also found that the cholesterol/HDL-cholesterol ratio was the traditional index commonly used in daily clinical practice showing the best relationship with ischemic heart disease, although apo B was a slightly superior predictor of IHD risk in this sample of middle-aged men. Indeed, after considering apo B in the model, no other lipoprotein variable could improve the prediction of ischemic heart disease risk in this cohort. These results emphasize the importance of measuring apo B for a more refined assessment of the risk of ischemic heart disease. From a pathophysiological standpoint, we believe that this finding of apo B being an important predictor of ischemic heart disease risk could be largely explained by the fact that apo B measurement provides a better estimate of the concentration of atherogenic lipoproteins than plasma cholesterol or LDL-cholesterol levels. Indeed, these two classic variables of the lipid profile may provide misleading information in insulin-resistant subjects, as discussed in the previous section of this chapter (see section on "The Athero-Thrombotic Metabolic Risk Profile of Visceral Obesity").

In the sample of the Québec Cardiovascular Study, we have also measured fasting plasma insulin levels as a crude marker of *in vivo* insulin resistance (36), after excluding diabetic patients from the analysis (the relationship of insulin levels to *in vivo* insulin resistance is disturbed in subjects with type 2 diabetes). We found that an increase of one standard deviation in fasting insulinemia (corresponding to an increase of approximately 30%) was associated with a 70% increase in the risk of ischemic heart disease, and this association was found to be partly independent from the variation in plasma triglyceride, apo B, and HDL-cholesterol levels (36). This finding suggests that hyperinsulinemia could be an additional marker of athero-thrombotic abnormalities that are not satisfactorily described by conventional risk factors. Furthermore, as apo B was also found to be a powerful predictor of ischemic heart disease in this cohort, we have examined the potential synergism between plasma insulin and apo B concentrations in the modulation of the risk of ischemic heart disease. In order to examine this issue, we have used the 50th percentile of the apo B distribution of our sample to identify subjects with either low or elevated apo B concentrations. We then further subdivided our sample on the basis of tertiles of fasting insulin concentrations. We have used as the reference group subjects in the first tertile of insulin levels who were also below the 50th percentile of the apo B distribution. Among subjects below the 50th percentile of the apo B distribution, being in the upper tertile of the fasting insulin levels was associated with a threefold increase in the risk of ischemic heart disease (36). However, hyperinsulinemia (upper insulin tertile) combined with an elevated apo B concentration (above the 50th percentile) was associated with more than a 10-fold increase in the risk of ischemic heart disease (36). It should be emphasized that irrespective of plasma cholesterol concentration, hyperinsulinemia with an elevated apo B concentration is precisely the combination of metabolic abnormalities found in insulin resistant visceral obese patients. Thus, the viscerally obese patients are characterized by a tremendously elevated risk of ischemic heart disease. It is suggested that the impact of this cluster of abnormalities may even be more deleterious to cardiovascular health than hypercholesterolemia.

Finally, as the hypertriglyceridemia of visceral obesity is associated with an increased proportion of small, dense particles (109), we have tested the hypothesis in our sample of middle-aged men of the Québec Cardiovascular Study that the so-called triad of nontraditional metabolic risk factors (hyperinsulinemia, elevated apo B, and small, dense LDL particles) would be a more powerful discriminator of ischemic heart disease risk than the conventional lipid triad (hypertriglyceridemia, raised LDL-cholesterol, and reduced HDL-cholesterol levels) (4, 71, 85). Whereas we found that the presence of the conventional lipid triad was associated with a 4.5-fold increase in the risk of ischemic heart disease, the simultaneous presence of hyperinsulinemia, elevated apo B, and small, dense LDL particles was associated with a 20-fold increase in the risk of ischemic heart disease; this powerful effect on risk is completely independent from the concomitant variation in plasma triglyceride, LDL-cholesterol, and HDL-cholesterol concentrations (71). This cluster of metabolic abnormalities, which are commonly found among obese patients with excess visceral adipose tissue accumulation, is associated with a remarkable increase in the risk of ischemic heart disease. We have also proposed that clinicians probably do not currently fully recognize the very high risk of ischemic heart disease that characterizes this form of obesity (figure 17.2). Indeed, it is common for these patients to be non-diabetic and to have normal cholesterol levels. If they do not smoke and they have normal blood pressure readings, the clinician may incorrectly assume that the risk of ischemic heart disease in these patients is not very high. However, based on the remarkable increase in the risk of ischemic heart disease found among our subjects with the triad of nontraditional metabolic risk variables, it is suggested that further attention should be devoted to this highly atherogenic condition.

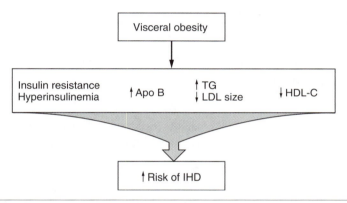

Figure 17.2   Risk of ischemic heart disease (IHD) associated with visceral obesity. Apo = apolipoproteins; TG = triglycerides; LDL = low density lipoproteins; HDL = high density lipoproteins.

Is Abdominal Obesity Due to Excess Weight or Lack of Exercise?

There is currently considerable evidence to suggest that the excess accumulation of body fat and of abdominal visceral adipose tissue are critical correlates of the insulin-resistance dyslipidemic syndrome that occurs in sedentary obese individuals. In this regard, weight loss that results from diet or exercise-induced negative energy balance is generally associated with an improvement in *in vivo* insulin sensitivity that may lead to a reduction in plasma insulin levels and to an improvement in glucose tolerance (34, 35). Furthermore, common alterations in the lipid profile associated with weight loss include a reduction in plasma triglyceride and apo B levels, an increase in HDL-cholesterol concentration (once body weight is stabilized), and a reduction in the ratio of cholesterol to HDL-cholesterol (35). However, it could also be argued that the insulin-resistant dyslipidemic state found in obese individuals may also be the consequence of the sedentary lifestyle found in these individuals. In concordance with this suggestion, studies have shown that the increased energy expenditure associated with an exercise training program could improve insulin action and the metabolic risk profile in the absence of weight loss (69, 75). Indeed, results from the Cooper Clinic in Dallas have suggested that one could be "fat and fit" and be at a reduced risk of ischemic heart disease (16, 15, 12, 13). These results suggest that the sedentary lifestyle found in obese patients may contribute to the dyslipidemia and the altered metabolic risk profile in these individuals. Thus, such a deleterious metabolic risk profile may result from a combination of excess abdominal fat and reduced physical activity. From a risk management standpoint, it will be important to better understand what are the independent contributions of body weight control and physical activity habits in the management of the metabolic risk profile of obese patients.

Effect of Weight Loss on the Metabolic Risk Profile of Obese Patients

Globally, the literature available on diet-induced weight loss suggests that reducing body weight and the amount of abdominal fat are associated with significant improvements in the cardiovascular disease risk profile (34, 50, 74); these improvements include reduced blood pressure, increased *in vivo* insulin action, and improvements in the plasma lipoprotein-lipid profile. Two remaining issues, however, need to be emphasized. First, no randomized weight-loss trial has shown that body weight loss and related metabolic improvements lead to reductions in "hard clinical endpoints" such as decreased morbidity and mortality from coronary heart disease. Second, there is controversy regarding the critical amount of weight loss necessary to substantially improve the metabolic risk profile. Although the notion that a 10% weight loss is generally associated with substantial improvements in the metabolic risk profile has recently gained greater acceptance (52), the magnitude of improvement in the metabolic risk profile may depend on the patient's initial body weight as well as his or her initial metabolic risk profile. For example, a 10% weight loss in a patient with a body mass index of 45 kg/m^2 and a body weight of 150 kg would only be associated with a 15 kg weight loss, which may not be sufficient to substantially reduce cardiovascular disease risk (although some benefits could be expected). However, a 10% weight loss in a subject weighing 100 kg may be enough to substantially reduce abdominal fat accumulation and improve his or her metabolic risk profile, particularly if this patient is insulin resistant and dyslipidemic (32). Thus, if we refer to moderate abdominal obesity, it may be legitimate to propose the concept that the normalization of body weight and of body fat content may not be necessary to substantially reduce cardiovascular disease risk.

Thus, several factors need to be taken into account when establishing the objectives of diet-induced weight loss in the obese patient. The patient's initial body weight, body-fat distribution, family history, and metabolic risk profile need to be considered.

Effect of Exercise Training on the Metabolic Risk Profile of Obese Patients

As is true for diet-induced weight loss, it should be emphasized that we have no large-scale randomized trial of exercise and its impact on cardiovascular disease risk in overweight patients. However, there is an impressive amount of literature indicating that endurance exercise training may generate some weight loss in obese patients as well as some improvements in *in vivo* insulin action and in the metabolic risk profile; these improvements would be predictive of a reduced risk of type 2 diabetes (7, 11, 88) and premature cardiovascular disease [see (34) and (35) for reviews]. These results are consistent with the notion that both physically active (87, 93, 94) and fit individuals (96, 84, 48, 16) are at reduced risk of coronary heart disease and related mortality. Although there is no consensus about the minimum amount of exercise necessary to optimally reduce CHD risk, it is generally considered that a physically active lifestyle contributes to body weight control and to improvement in the metabolic risk profile (2, 19, 20, 34, 35, 53, 54, 64).

Another critical question is whether an increase in maximal oxygen consumption ($\dot{V}O_2max$) is necessary to reduce risk in sedentary obese patients or whether increasing daily energy expenditure through a more active lifestyle would contribute to reduction in CHD risk (35). Based on the literature indicating that the $\dot{V}O_2max$ or the performance on a treadmill test are powerful predictors of coronary heart disease and related mortality (16, 48, 84, 96), one may advocate that improving fitness should be an important component of an exercise program designed for obese patients. However, results from our laboratory have suggested that one should focus on increasing energy expenditure rather than improving cardiorespiratory fitness (35). The exercise literature clearly indicates that an exercise bout of at least 20 minutes of high-intensity aerobic exercise performed three times per week may eventually improve cardiorespiratory fitness over several weeks on such an exercise program (3). However, if one calculates the increase in daily energy expenditure associated with such an exercise regimen, it is unlikely to have a substantial effect on energy balance, body weight, or metabolic risk variables (35). However, an exercise program performed at a lower intensity, such as 50% of $\dot{V}O_2max$ (which corresponds to the exercise intensity associated with a brisk walk) performed over a longer duration (up to 45 minutes) on an almost daily basis, would be more likely to have an impact on daily energy expenditure, body composition, and on the metabolic risk profile (34, 35) (table 17.1). Several exercise-training studies have shown that regular endurance exercise may eventually lead to reductions in total body fat and to the mobilization of abdominal adipose tissue, leading to improvements in the metabolic risk profile as long as the energy expenditure generated by the exercise program is sufficient (35, 53, 54, 57, 64, 81, 92, 115, 116, 117, 118, 119). However, discordant results regarding the effect of exercise on the metabolic risk profile can be found in the literature (12, 15, 57, 114). It is therefore important to consider factors that could explain such discrepancies among studies. Among these

Table 17.1   Comparison of Different Endurance Exercise Training Prescriptions on Cardiorespiratory Fitness, Energy Expenditure, and Metabolic Risk Profile

Exercise prescription	$\dot{V}O_2$max	Energy expenditure	Changes in risk profile
Short bouts (20 min) High intensity (80%)	↑ ↑	= or slight ↑	Trivial
Prolonged Low intensity (50-60%)	= or slight ↑	↑ ↑ ↑	Significant

factors, subjects' age and initial cardiovascular disease risk profile, initial body composition, level of abdominal fat, gender, and genetic factors are variables that could contribute to explain individual differences in the response to a standardized exercise training program. For example, a premenopausal woman with a normal metabolic risk profile is more likely to show trivial changes in her metabolic risk profile in response to a standardized exercise training program; an abdominal obese insulin-resistant and dyslipidemic male is likely to show substantial metabolic improvements in response to the same endurance exercise training program.

It should also be pointed out that the exercise modality that would optimally improve the metabolic risk profile for a given increase in energy expenditure related to exercise is poorly defined (15). However, it does appear that improvements in the metabolic risk profile are not closely related to changes in treadmill performance or to changes in cardiorespiratory fitness as assessed by $\dot{V}O_2$max (35). For example, a seminal study by Williams and colleagues (115) found that the improvements in the plasma lipoprotein profile induced by exercise training were not related to changes in $\dot{V}O_2$max or in treadmill test duration. However, these workers did find highly significant correlation between plasma HDL concentration and the training volume estimated by the number of miles run per week. It thus appears that the improvement in the metabolic risk profile predictive of coronary heart disease risk induced by regular exercise may be more closely related to the volume of exercise than to its intensity (35, 15). We have previously suggested (34) that the metabolic improvements generated by regular endurance exercise could be mediated by two factors. First, the chronic effect of exercise may be largely (but not entirely) resulting from the concomitant loss of body fat and of abdominal adipose tissue (34, 35, 43, 57, 118) that will be associated with improvements in insulin sensitivity and in the plasma lipoprotein profile. Therefore, on the basis of this relationship, the health professional should emphasize the importance of a moderate caloric restriction. This could be achieved by reducing the fat and sugar content of the diet rather than modifying the quantity of food consumed and by increasing the daily energy expenditure by prolonged endurance exercise. The ultimate objective will be to generate the largest possible energy deficit that could be well tolerated by the patient. Such an approach should eventually produce improvements in the metabolic risk profile (15). Second, the acute metabolic response to a single bout of exercise has been suggested to improve *in vivo* insulin action (88, 23) and the lipid profile (26) for up to 48 to 72 hours. Therefore, a prolonged endurance exercise performed on an almost daily basis (or at least every other day) may generate an acute metabolic response that could be "perpetuated" for as long as the subject is active at least three times a week. The patient may therefore benefit from both the acute metabolic response to a single bout of exercise and from the chronic response to the endurance exercise training program (figure 17.3).

Rogers and his colleagues (98) showed that a large increase in energy expenditure that resulted from daily sessions of prolonged exercise could improve *in vivo* insulin action. This led to a decreased insulinemia within a few days among subjects who were initially characterized by a disturbed metabolic risk profile. Indeed, a moderate-intensity exercise training program in which subjects exercised for about one hour per day led to rapid improvements in *in vivo* insulin action (98), which could favorably alter the plasma lipoprotein profile. These changes in the lipid profile that result from an improved *in vivo* insulin action include reductions in plasma

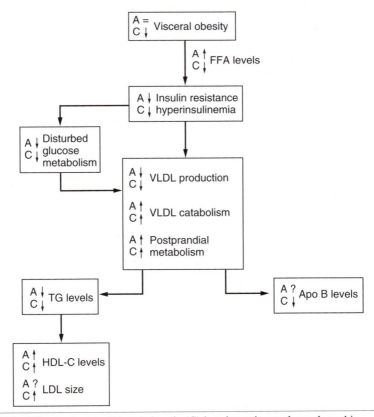

Figure 17.3 Effects of acute (A) and chronic (C) low-intensity, prolonged aerobic exercise on the metabolic complications of visceral obesity.

triglyceride and apo B levels, an increase in plasma HDL-cholesterol concentration, and changes in the activity of enzymes relevant to the regulation of plasma HDL-cholesterol levels. The latter specifically refers to lipoprotein-lipase, which has been shown to be low in abdominal obesity; and hepatic triglyceride lipase, which has been shown to be elevated in abdominal obese patients (35). Thus, regular exercise that produces improved *in vivo* insulin action leads to reciprocal changes in the activity of two enzymes that are critical determinants of plasma HDL-cholesterol levels (89, 24, 105, 47, 95, 67).

In addition to the improved *in vivo* insulin action, it is believed that the response of lipoprotein lipase and hepatic lipase to regular endurance exercise are important determinants of the increased plasma HDL-cholesterol levels found with endurance exercise training (35). It should also be emphasized that these metabolic improvements are completely dissociated from changes in cardiorespiratory fitness, seeming instead to be proportionate to the magnitude of weight loss and of abdominal fat mobilization (43). These results are concordant with the notion that the weight loss achieved through regular exercise training is an important (although not the sole component) determinant of changes in plasma HDL-cholesterol observed with endurance exercise training. However, one has to keep in mind that obesity is associated with a sedentary lifestyle and that a physically active lifestyle, in the absence of weight loss, has also been shown to improve the metabolic risk profile. We would like to suggest that the favorable alterations in the metabolic risk profile among physically active subjects who would remain obese may be related to the acute response to single bouts of prolonged exercise. Therefore, a physically active obese individual may substantially benefit from frequent sessions of exercise such as a daily 45-minute to 1-hour brisk walk. Such a large exercise prescription would acutely improve insulin action and the lipid profile for up to 48 to 72 hours, and the next exercise session would contribute to re-establish and maintain this favorable metabolic profile even in the absence of body weight loss. This notion may have important public health implications, because recent data from our laboratory have shown that one can normalize the metabolic risk profile of obese patients by diet and exercise-induced weight loss in the absence of normalization of body weight (46). Indeed, we even have found that physically active obese individuals could normalize their metabolic risk profile by regular physical activity compared to sedentary healthy normolipidemic controls while still being obese from a body-composition standpoint (46). Thus, although weight loss may further improve the metabolic response to a regular endurance exercise program, it may not be required to substantially reduce the risk of type 2 diabetes and CHD in obese patients.

Summary

The minimum exercise prescription necessary to improve $\dot{V}O_2max$ (20 to 30 minutes of endurance exercise performed three times per week at 50-70% of $\dot{V}O_2max$) may eventually improve cardiorespiratory fitness over several weeks (3).

However, the increase in energy expenditure resulting from such an exercise prescription is trivial and unlikely to induce significant weight loss and related metabolic improvements, unless a hypocaloric diet is used in combination with this exercise regimen. By using a more moderate form of exercise and by increasing its duration (such as with a daily brisk walk of 45 to 60 minutes), a substantial increase in daily energy expenditure may be generated, which may eventually lead to significant weight loss and to improvements in *in vivo* insulin action and in other metabolic risk variables. Results from metabolic studies emphasizing the importance of increasing energy expenditure rather than exercise intensity are consistent with epidemiological studies that have shown that the greatest reduction in CHD risk is obtained when comparing moderately active subjects to sedentary individuals (81, 113). Overall, it appears that increasing further the level of physical activity is associated with an additional reduction in risk, but the greatest benefit from a public health standpoint would be to transform our largely sedentary population into moderately active individuals (14, 15). However, it is critical to emphasize that no randomized mortality trial on diet-induced weight loss has been conducted. Accordingly, a large trial comparing the impact of weight loss achieved by a hypocaloric diet, by an endurance exercise program, and by a a combination of both approaches on coronary heart disease and related mortality as endpoint has never been performed. While waiting for this important evidence, it appears warranted to recommend that the level of daily physical activity should be increased in our population. Furthermore, one should expect beneficial effects on the cardiovascular disease risk profile of our population by increasing daily physical activity, although the expected benefits may vary upon subjects' initial characteristics and genetic background.

Furthermore, changes in lifestyle associated with the best compliance are likely to have the greatest long-term impact on the cardiovascular health of our population. With the sharp increase in the prevalence of obesity in affluent countries (101), we believe that it is critical to develop urban environments that would favor a physically active lifestyle. This ambitious objective goes far beyond the field of exercise physiology and medicine and will require the involvement and cooperation of a multidisciplinary team as well as greater commitment of our political leaders toward physical activity and health.

References

1. Abate, N., A. Garg, R.M. Peshock, J. Stray-Gundersen, and S.M. Grundy. 1995. Relationships of generalized and regional adiposity to insulin sensitivity in men. *Journal of Clinical Investigation* 96: 88-98.
2. Anonymous. 1990. Exercise, fitness, and health. The consensus statement. In *Exercise, fitness, and health. A consensus of current knowledge,* ed. C. Bouchard, R.J. Shephard, T. Stephens, J.R. Sutton, and R. McPherson, 3-28. Champaign, IL: Human Kinetics.

3. Anonymous. 1990. Position stand on "The recommended quantity and quality of exercise for developing and maintaining cardiorespiratory and muscular fitness in healthy adults". *Medicine and Science in Sports and Exercise* 22: 265-274.

4. Assmann, G., and H. Schulte. 1992. Relation of high-density lipoprotein cholesterol and triglycerides to incidence of atherosclerotic coronary artery disease (the PROCAM experience). *American Journal of Cardiology* 70: 733-737.

5. Assmann, G., and H. Schulte. 1994. Identification of individuals at high risk for myocardial infarction. *Atherosclerosis* 110 Suppl: S11-21.

6. Barrett-Connor, E.L. 1985. Obesity, atherosclerosis, and coronary artery disease. *Annals of Internal Medicine* 103: 1010-1019.

7. Björntorp, P. 1981. Effects of exercise on plasma insulin. *International Journal of Sports Medicine* 2: 129-139.

8. Björntorp, P. 1984. Hazards in subgroups of human obesity. *European Journal of Clinical Investigation* 14: 239-241.

9. Björntorp, P. 1988. Abdominal obesity and the development of non-insulin dependent diabetes mellitus. *Diabetes Metabolism Reviews* 4: 615-622.

10. Björntorp, P. 1990. "Portal" adipose tissue as a generator of risk factors for cardiovascular disease and diabetes. *Arteriosclerosis* 10: 493-496.

11. Björntorp, P., K. De Jounge, L. Sjöström, and L. Sullivan. 1970. The effect of physical training on insulin production in obesity. *Metabolism: Clinical & Experimental* 19: 631-638.

12. Blair, S.N., and J.C. Connelly. 1996. How much physical activity should we do? The case for moderate amounts and intensities of physical activity. *Res Q Exerc Sport* 67: 193-205.

13. Blair, S.N., and K.H. Cooper. 1997. Dose of exercise and health benefits. *Archives of Internal Medicine* 157: 153-154.

14. Blair, S.N., E. Horton, A.S. Leon, I.M. Lee, B.L. Drinkwater, R.K. Dishman, M. Mackey, and M.L. Kienholz. 1996. Physical activity, nutrition, and chronic disease. *Medicine and Science in Sports and Exercise* 28: 335-349.

15. Blair, S.N., H.W. Kohl, N.F. Gordon, and R.S.J. Paffenbarger. 1992. How much physical activity is good for health? *Annual Reviews of Public Health* 13: 99-126.

16. Blair, S.N., H.W. Kohl III, R.S. Paffenbarger, D.G. Clark, K.H. Cooper, and L.W. Gibbons. 1989. Physical fitness and all-cause mortality. A prospective study of healthy men and women. *Journal of the American Medical Association* 262: 2395-2401.

17. Bolinder, J., L. Kager, J. Ostman, and P. Arner. 1983. Differences at the receptor and postreceptor levels between human omental and subcutaneous adipose tissue in the action of insulin on lipolysis. *Diabetes* 32: 117-123.

18. Borkan, G.A., S.G. Gerzof, A.H. Robbins, D.E. Hults, C.K. Silbert, and J.E. Silbert. 1982. Assessment of abdominal fat content by computed tomography. *American Journal of Clinical Nutrition* 36: 172-177.

19. Bouchard, C., J.P. Després, and A. Tremblay. 1993. Exercise and obesity. *Obesity Research* 1: 40-54.
20. Bouchard, C., A. Tremblay, J.P. Després, A. Nadeau, P.J. Lupien, G. Thériault, J. Dussault, S. Moorjani, S. Pinault, and G. Fournier. 1990. The response to long-term overfeeding in identical twins. *New England Journal of Medicine* 322: 1477-1482.
21. Bray, G.A. 1985. Complications of obesity. *Annals of Internal Medicine* 103: 1052-1062.
22. Bray, G.A., C. Bouchard, and W. P. T. James. 1998. *Handbook of obesity.* New York: Marcel Dekker, Inc.
23. Burnstein, R., C. Polychronakos, C.J. Toews, J.D. MacDouglas, H.J. Guyda, and B.I. Pasner. 1985. Acute reversal of the enhance insulin action in trained athletes: association with insulin receptor changes. *Diabetes* 34: 756-760.
24. Clay, M.A., H.H. Newnham, and P.J. Barter. 1991. Hepatic lipase promotes a loss of apolipoprotein A-I from triglyceride-enriched human high density lipoproteins during incubation in vitro. *Arteriosclerosis & Thrombosis* 11: 415-422.
25. Couillard, C., N. Bergeron, D. Prud'homme, J. Bergeron, A. Tremblay, C. Bouchard, P. Mauriège, and J.P. Després. 1998. Postprandial triglyceride response in visceral obesity in men. *Diabetes* 47: 953-960.
26. Crouse, S.F., B.C. O'Brien, J.J. Rohack, R.C. Lowe, J.S. Green, H. Tolson, and J.L. Reed. 1995. Changes in serum lipids and apolipoproteins after exercise in men with high cholesterol: influence of intensity. *Journal of Applied Physiology* 79: 279-286.
27. Després, J.P. 1991. Lipoprotein metabolism in visceral obesity. *International Journal of Obesity & Related Metabolic Disorders* 15 Suppl 2: 45-52.
28. Després, J.P. 1991. Visceral obesity, insulin resistance, and related dyslipoproteinemias. In *Diabetes 1991,* ed. H. Riskin, J.A. Colwell, and S.I. Taylor, 95-99. Amsterdam: Excerpts Medica.
29. Després, J.P. 1991. Visceral obesity: a component of the insulin resistance-dyslipidemic syndrome. *Canadian Journal of Cardiology* 10: 17B-22B.
30. Després, J.P. 1993. Abdominal obesity as important component of insulin-resistance syndrome. *Nutrition* 9: 452-459.
31. Després, J.P. 1993. Obesity and lipid metabolism: relevance of body fat distribution. *Current Opinion in Lipidology* 2: 5-15.
32. Després, J.P. 1994. Dyslipidaemia and obesity. *Baillère's Clinical Endocrinology and Metabolism* 8: 629-660.
33. Després, J.P., M. Ferland, S. Moorjani, A. Nadeau, A. Tremblay, and P.J. Lupien. 1989. Role of hepatic-triglyceride lipase activity in the association between intra-abdominal fat and plasma HDL cholesterol in obese women. *Arteriosclerosis, Thrombosis & Vascular Biology* 9: 485-492.
34. Després, J.P., and B. Lamarche. 1993. Effects of diet and physical activity on adiposity and body fat distribution: implications for the prevention of cardiovascular disease. *Nutrition Research Reviews* 6: 137-159.

35. Després, J.P., and B. Lamarche. 1994. Low-intensity endurance exercise training, plasma lipoproteins and the risk of coronary heart disease. *Journal of Internal Medicine* 236: 7-22.

36. Després, J.P., B. Lamarche, P. Mauriège, B. Cantin, G.R. Dagenais, S. Moorjani, and P.J. Lupien. 1996. Hyperinsulinemia as an independent risk factor for ischemic heart disease. *New England Journal of Medicine* 334: 952-957.

37. Després, J.P., S. Lemieux, B. Lamarche, D. Prud'homme, S. Moorjani, L.D. Brun, C. Gagné, and P.J. Lupien. 1995. The insulin resistance-dyslipidemic syndrome: contribution of visceral obesity and therapeutic implications. *International Journal of Obesity & Related Metabolic Disorders* 19 Suppl 1: S76-86.

38. Després, J.P., and A. Marette. 1994. Relation of components of insulin resistance syndrome to coronary disease risk. *Current Opinion in Lipidology* 5: 274-289.

39. Després, J.P., S. Moorjani, M. Ferland, A. Tremblay, A. Nadeau, P.J. Lupien, S. Pinault, G. Thériault, and C. Bouchard. 1989. Adipose tissue distribution and plasma lipoprotein levels in obese women: Importance of intra-abdominal fat. *Arteriosclerosis, Thrombosis & Vascular Biology* 9: 203-210.

40. Després, J.P., S. Moorjani, P.J. Lupien, A. Tremblay, A. Nadeau, and C. Bouchard. 1990. Regional distribution of body fat, plasma lipoproteins, and cardiovascular disease. *Arteriosclerosis, Thrombosis & Vascular Biology* 10: 497-511.

41. Després, J.P., S. Moorjani, P.J. Lupien, A. Tremblay, A. Nadeau, and C. Bouchard. 1992. Genetic aspects of susceptibility to obesity and related dyslipidemias. *Molecular and Cellular Biochemistry* 113: 151-169.

42. Després, J.P., S. Moorjani, A. Tremblay, M. Ferland, P.J. Lupien, A. Nadeau, and C. Bouchard. 1989. Relation of high plasma triglyceride levels associated with obesity and regional adipose tissue distribution to plasma lipoprotein-lipid composition in premenopausal women. *Clinical Investigative Medicine* 12: 374-380.

43. Després, J.P., A. Nadeau, A. Tremblay, M. Ferland, S. Moorjani, P.J. Lupien, G. Thériault, S. Pinault, and C. Bouchard. 1989. Role of deep abdominal fat in the association between regional adipose tissue distribution and glucose tolerance in obese women. *Diabetes* 38: 304-309.

44. Després, J.P., M.C. Pouliot, S. Moorjani, A. Nadeau, A. Tremblay, P.J. Lupien, G. Thériault, and C. Bouchard. 1991. Loss of abdominal fat and metabolic response to exercise training in obese women. *American Journal of Physiology (Endocrinology and Metabolism)* 261: E159-E167.

45. Després, J.P., M.C. Pouliot, D. Prud'homme, S. Moorjani, P.J. Lupien, A. Nadeau, A. Tremblay, and C. Bouchard. 1991. Abdominal obesity, hyperinsulinaemia and related hypertriglyceridaemia are important correlates of LDL composition in men. *International Journal of Obesity & Related Metabolic Disorders* 15: O37.

46. Doucet, E., P. Imbeault, J.P. Després, and A. Tremblay. 1998. Metabolic fitness in reduced obese individuals: evidence of the robust effects of low-fat diet-exercise follow-up. *International Journal of Obesity & Related Metabolic Disorders* 22(Suppl 3): P708.

47. Eckel, R. 1989. Lipoprotein lipase: a multifunctional enzyme relevant to common metabolic diseases. *New England Journal of Medicine* 320: 1060-1068.

48. Ekelund, L.G., W.L. Haskell, J.L. Johnson, F.S. Whaley, M.H. Criqui, and D.S. Sheps. 1988. Physical fitness as a predictor of cardiovascular mortality in asymptomatic North American men: the Lipid Research Clinic Mortality Follow-up Study. *New England Journal of Medicine* 319: 1379-1384.

49. Ferland, M., J.P. Després, A. Tremblay, S. Pinault, A. Nadeau, S. Moorjani, P.J. Lupien, G. Thériault, and C. Bouchard. 1989. Assessment of adipose tissue distribution by computed axial tomography in obese women: Association with body density and anthropometric measurements. *British Journal of Nutrition* 61: 139-148.

50. Fujioka, S., Y. Matsuzawa, K. Tokunaga, T. Kawamoto, T. Kobatake, Y. Keno, K. Kotani, S. Yoshida, and S. Tarui. 1991. Improvement of glucose and lipid metabolism associated with selective reduction of intra-abdominal visceral fat in premenopausal women with visceral fat obesity. *International Journal of Obesity & Related Metabolic Disorders* 15: 853-859.

51. Fujioka, S., Y. Matsuzawa, K. Tokunaga, and S. Tarui. 1987. Contribution of intra-abdominal fat accumulation to the impairment of glucose and lipid metabolism in human obesity. *Metabolism: Clinical & Experimental* 36: 54-59.

52. Goldstein, D.J. 1992. Beneficial health effects of modest weight loss. *International Journal of Obesity & Related Metabolic Disorders* 16: 397-415.

53. Haskell, W.L. 1984. Exercise-induced changes in plasma lipids and lipoproteins. *Preventive Medicine* 13: 23-36.

54. Haskell, W.L. 1986. The influence of exercise training on plasma lipids and lipoproteins in health and disease. *Acta Medica Scandinavica* Suppl 711: 25-37.

55. Hennes, M.M., E. Shrago, and A.H. Kissebah. 1990. Receptor and postreceptor effects of free fatty acids (FFA) on hepatocyte insulin dynamics. *International Journal of Obesity & Related Metabolic Disorders* 14: 831-841.

56. Jahoor, F., S. Klein, and R. Wolfe. 1992. Mechanism of regulation of glucose production by lipolysis in humans. *American Journal of Physiology* 262: E353-E358.

57. Katzel, L.I., E.R. Bleecker, E.G. Colman, E.M. Rogus, J.D. Sorkin, and A.P. Goldberg. 1995. Effects of weight loss vs aerobic exercise training on risk factors for coronary disease in healthy, obese, middle-aged and older men. A randomized controlled trial. *Journal of the American Medical Association* 274: 1915-1921.

58. Keys, A. 1970. Coronary heart disease in seven countries. *Circulation* 41: I1-I211.
59. Kissebah, A.H., D.J. Evans, A. Peiris, and C.R. Wilson. 1985. Endocrine characteristics in regional obesities: role of sex steroids. In *Metabolic complications of human obesities,* ed. J. Vague, P. Björntorp, B. Guy-Grand, M. Rebuffé-Scrive, and P. Vague, 115-130. Amsterdam: Elsevier.
60. Kissebah, A.H., D.S. Freedman, and A.N. Peiris. 1989. Health risks of obesity. *Med Clin North Am* 73: 111-138.
61. Kissebah, A.H., and G.R. Krakower. 1994. Regional adiposity and morbidity. *American Journal of Physiology* 74: 761-811.
62. Kissebah, A.H., and A.N. Peiris. 1989. Biology of regional body fat distribution: Relationship to non-insulin-dependent diabetes mellitus. *Diabetes Metabolism Reviews* 5: 83-109.
63. Kissebah, A.H., N. Vydelingum, R. Murray, D.J. Evans, A.J. Hartz, R.K. Kalkhoff, and P.W. Adams. 1982. Relation of body fat distribution to metabolic complications of obesity. *Journal of Clinical Endocrinology & Metabolism* 54: 254-260.
64. Krauss, R.M. 1989. Exercise, lipoproteins, and coronary heart disease. *Circulation* 79: 1143-1145.
65. Krotkiewski, M., P. Björntorp, L. Sjöström, and U. Smith. 1983. Impact of obesity on metabolism in men and women. Importance of regional adipose tissue distribution. *Journal of Clinical Investigation* 72: 1150-1162.
66. Kvist, H., B. Chowdhury, U. Grangard, U. Tylen, and L. Sjöström. 1988. Total and visceral adipose-tissue volumes derived from measurements with computed tomography in adult men and women: predictive equations. *American Journal of Clinical Nutrition* 48: 1351-1361.
67. Lalouel, J.M., D.E. Wilson, and P.H. Iverius. 1992. Lipoprotein lipase and hepatic triglyceride lipase: molecular and genetic aspects. *Current Opinion in Lipidology* 3: 86-95.
68. Lamarche, B., J.P. Després, S. Moorjani, B. Cantin, G.R. Dagenais, and P.J. Lupien. 1995. Prevalence of dyslipidemic phenotypes in ischemic heart disease (prospective results from the Québec Cardiovascular Study). *American Journal of Cardiology* 75: 1189-1195.
69. Lamarche, B., J.P. Després, M.C. Pouliot, S. Moorjani, P.J. Lupien, G. Thériault, A. Tremblay, A. Nadeau, and C. Bouchard. 1992. Is body fat loss a determinant factor in the improvement of carbohydrate and lipid metabolism following aerobic exercise training in obese women? *Metabolism: Clinical & Experimental* 41: 1249-1256.
70. Lamarche, B., S. Moorjani, P.J. Lupien, B. Cantin, P.M. Bernard, G.R. Dagenais, and J.P. Després. 1996. Apolipoprotein A-I and B levels and the risk of ischemic heart disease during a five-year follow-up of men in the Québec Cardiovascular Study. *Circulation* 94: 273-278.
71. Lamarche, B., A. Tchernof, P. Mauriège, B. Cantin, G.R. Dagenais, P.J. Lupien, and J.P. Després. 1998. Fasting insulin and apolipoprotein B levels

and low-density lipoprotein particle size as risk factors for ischemic heart disease. *Journal of the American Medical Association* 279: 1955-1961.

72. Lapidus, L., C. Bengtsson, B. Larsson, K. Pennert, E. Rybo, and L. Sjöström. 1984. Distribution of adipose tissue and risk of cardiovascular disease and death: 12 year follow up of participants in the population study of women in Gothenberg, Sweden. *British Medical Journal* 289: 1257-1261.

73. Larsson, B., K. Svardsudd, L. Welin, L. Wilhelmsen, P. Björntorp, and G. Tibblin. 1984. Abdominal adipose tissue distribution, obesity, and risk of cardiovascular disease and death: 13 year follow-up of participants in the study of men born in 1913. *British Medical Journal* 288: 1401-1404.

74. Leenen, R., K. van der Kooy, P. Deurenberg, J.C. Seidell, J.A. Weststrate, F.J.M. Schouten, and J.G.A.J. Hautvast. 1992. Visceral fat accumulation in obese subjects: relation to energy expenditure and response to weight loss. *American Journal of Physiology* 263: E913-E919.

75. Lehmann, R., A. Vokac, K. Niedermann, K. Agosti, and G.A. Spinas. 1995. Loss of abdominal fat and improvement of the cardiovascular risk profile by regular moderate exercise training in patients with NIDDM. *Diabetologia* 38: 1313-1319.

76. Lemieux, S., and J.P. Després. 1994. Metabolic complications of visceral obesity: contribution to the aetiology of type 2 diabetes and implications for prevention and treatment. *Diabetes & Metabolism* 20: 375-393.

77. Lemieux, S., J.P. Després, S. Moorjani, A. Nadeau, G. Thériault, D. Prud'homme, A. Tremblay, C. Bouchard, and P.J. Lupien. 1994. Are gender differences in cardiovascular disease risk factors explained by the level of visceral adipose tissue? *Diabetologia* 37: 757-764.

78. Lemieux, S., D. Prud'homme, C. Bouchard, A. Tremblay, and J.P. Després. 1993. Sex differences in the relation of visceral adipose tissue accumulation to total body fatness. *American Journal of Clinical Nutrition* 58: 463-467.

79. Lemieux, S., D. Prud'homme, S. Moorjani, A. Tremblay, C. Bouchard, P.J. Lupien, and J.P. Després. 1995. Do elevated levels of abdominal visceral adipose tissue contribute to age-related differences in plasma lipoprotein concentrations in men? *Atherosclerosis* 118: 155-164.

80. Lemieux, S., D. Prud'homme, A. Tremblay, C. Bouchard, and J.P. Després. 1996. Anthropometric correlates to changes in visceral adipose tissue over 7 years in women. *International Journal of Obesity and Related Metabolic Disorders* 20: 618-624.

81. Leon, A.S., J. Conrad, D.B. Hunninghake, and R. Serfass. 1979. Effects of a vigorous walking program on body composition, and carbohydrate and lipid metabolism of obese young men. *American Journal of Clinical Nutrition* 32: 1776-1787.

82. Lewis, G.F. 1997. Fatty acid regulation of very low density lipoprotein (VLDL) production. *Current Opinion in Lipidology* 8: 146-153.

83. Lewis, G.F., and G. Steiner. 1996. Hypertriglyceridemia and its metabolic consequences as a risk factor for atherosclerotic cardiovascular disease in

non-insulin- dependent diabetes mellitus. *Diabetes-Metabolism Reviews* 12: 37-56.

84. Lie, H., R. Mundal, and J. Erikssen. 1985. Coronary risk factors and incidence of coronary death in relation to physical fitness: seven-year follow-up study of middle-aged and elderly men. *European Heart Journal* 6: 147-157.

85. Manninen, V., L. Tenkanen, P. Koshinen, J.K. Huttunen, M. Mänttäri, O.P. Heinonen, and M.H. Frick. 1992. Joint effects of serum triglyceride and LDL cholesterol and HDL cholesterol concentrations on coronary heart disease risk in the Helsinki Heart Study: implications for treatment. *Circulation* 85: 37-45.

86. Mauriège, P., A. Marette, C. Atgie, C. Bouchard, G. Thériault, L.K. Bukowiecki, P. Marceau, S. Biron, A. Nadeau, and J.P. Després. 1995. Regional variation in adipose tissue metabolism of severely obese premeno-pausal women. *Journal of Lipid Research* 36: 672-684.

87. Morris, J.N., S.P.W. Chave, C. Adam, C. Sirey, L.H. Epstein, and D.J. Sheehan. 1973. Vigorous exercise in leisure-time and the incidence of coronary heart disease. *Lancet* 1: 333-339.

88. National Institutes of Health. 1987. Consensus development conference on diet and exercise in non-insulin-dependent diabetes mellitus. *Diabetes Care* 10: 639-644.

89. Newnham, H.H., G.J. Hopkins, S. Devlin, and P.J. Barter. 1990. Lipoprotein lipase prevents the hepatic lipase-induced reduction in particle size of high density lipoproteins during incubation of human plasma. *Atherosclerosis* 82: 167-176.

90. NIH Consensus Conference. 1985. Lowering cholesterol to prevent heart disease. *Journal of the American Medical Association* 253: 2080-2086.

91. Ohlson, L.O., B. Larsson, K. Svardsudd, L. Welin, H. Eriksson, L. Wilhelmsen, P. Björntorp, and G. Tibblin. 1985. The influence of body fat distribution on the incidence of diabetes mellitus: 13.5 years of follow-up of the participants in the study of men born in 1913. *Diabetes* 34: 1055-1058.

92. Oshida, Y., K. Yamanouchi, S. Hayamizu, and Y. Sato. 1989. Long-term mild jogging increases insulin action despite no influence on body mass index or $\dot{V}O_2$max. *Journal of Applied Physiology* 66: 2206-2210.

93. Paffenbarger, R.S., and W.E. Hale. 1975. Work activity and coronary heart mortality. *New England Journal of Medicine* 292: 545-550.

94. Paffenbarger, R.S., A.L. Wing, and R.T. Hyde. 1978. Physical activity as an index for heart attack risk in college alumni. *American Journal of Epidemiology* 108: 161-175.

95. Patsch, J.R., S. Prasad, A.M. Gotto Jr., and W. Patsch. 1987. High density lipoprotein. Relationship of the plasma levels of this lipoprotein species to its composition, to the magnitude of post-prandial lipemia, and to the activities of lipoprotein lipase and hepatic lipase. *Journal of Clinical Investigation* 80: 341-347.

96. Peters, R.K., L.D. Cady Jr., D.P. Bischoff, L. Bernstein, and M.C. Pike. 1983. Physical fitness and subsequent myocardial infarction in healthy workers. *Journal of the American Medical Association* 249: 3052-3056.
97. Pouliot, M.C., J.P. Després, A. Nadeau, S. Moorjani, D. Prud'homme, P.J. Lupien, A. Tremblay, and C. Bouchard. 1992. Visceral obesity in men. Associations with glucose tolerance, plasma insulin and lipoprotein levels. *Diabetes* 41: 826-834.
98. Rogers, M.A., C. Yamamoto, D.S. King, J.M. Hagberg, A.A. Ehsani, and J.O. Holloszy. 1988. Improvement in glucose tolerance after 1 wk of exercise in patients with mild NIDDM. *Diabetes Care* 11: 613-618.
99. Sacks, F.M., M.A. Pfeffer, L.A. Moye, J.L. Rouleau, J.D. Rutherford, T.G. Cole, L. Brown, J.W. Warnica, J.M. Arnold, C.C. Wun, B.R. Davis, and E. Braunwald. 1996. The effect of pravastatin on coronary events after myocardial infarction in patients with average cholesterol levels. Cholesterol and Recurrent Events Trial investigators. *New England Journal of Medicine* 335: 1001-1009.
100. Scandinavian Simvastatin Survival Study Group. 1994. Randomized trial of cholesterol lowering in 4444 patients with coronary heart disease: the Scandinavian Simvastatin Survival Study (4S). *Lancet* 344: 1383-1389.
101. Seidell, J.C. 1995. Obesity in Europe: scaling an epidemic. *International Journal of Obesity and Related Metabolic Disorders* 19 Suppl 3: S1-S4.
102. Seidell, J.C., and K.M. Flegal. 1997. Assessing obesity: classification and epidemiology. *British Medical Bulletin* 53: 238-252.
103. Shepherd, J., S.M. Cobbe, I. Ford, C.G. Isles, A.R. Lorimer, P.W. MacFarlane, J.H. McKillop, and C.J. Packard. 1995. Prevention of coronary heart disease with pravastatin in men with hypercholesterolemia. West of Scotland Coronary Prevention Study Group. *New England Journal of Medicine* 333: 1301-1307.
104. Sjöström, L., H. Kvist, A. Cederblad, and U. Tylen. 1986. Determination of total adipose tissue and body fat in women by computed tomography, 40K, and tritium. *American Journal of Physiology* 250: E736-E745.
105. Skinner, E.R. 1994. High-density lipoprotein subclasses. *Current Opinion in Lipidology* 5: 241-247.
106. Stamler, J., D. Wentworth, and J.D. Neaton. 1986. Is the relationship between serum cholesterol and risk of premature death from coronary heart disease continuous and graded? Findings in 356,222 primary screens of the Multiple Risk Factor Intervention Trial (MRFIT). *Journal of the American Medical Association* 256: 2823-2828.
107. Svedberg, F., P. Björntorp, U. Smith, and P. Lönnroth. 1990. Free fatty acid inhibition of insulin binding, degradation, and action in isolated rat hepatocytes. *Diabetes* 39: 570-574.
108. Tchernof, A., J. Calles-Escandon, C.K. Setes, and E.T. Poehlman. 1998. Menopause, central body fatness, and insulin resistance: effects of hormone replacement therapy. *Coronary Artery Disease* 9: 503-511.

109. Tchernof, A., B. Lamarche, A. Nadeau, S. Moorjani, F. Labrie, P.J. Lupien, and J.P. Després. 1996. The dense LDL phenotype: association with plasma lipoprotein levels, visceral obesity and hyperinsulinemia in men. *Diabetes Care* 19: 629-637.

110. Tchernof, A., and E.T. Poehlman. 1998. Effects of the menopause transition on body fatness and body fat distribution. *Obesity Research* 6: 246-254.

111. Tokunaga, K., Y. Matsuzawa, K. Ishikawa, and S. Tarui. 1983. A novel technique for the determination of body fat by computed tomography. *International Journal of Obesity & Related Metabolic Disorders* 7: 437-445.

112. Vague, J. 1947. Sexual differentiation, a factor affecting the forms of obesity. *Presse Médicale* 30: 339-340.

113. Wei, M., C.A. Macera, C.A. Hornung, and S.N. Blair. 1997. Changes in lipids associated with change in regular exercise in free-living men. *Journal of Clinical Epidemiology* 50: 1137-1142.

114. Whaley, M.H., and S.N. Blair. 1995. Epidemiology of physical activity, physical fitness and coronary heart disease. *J Cardiovasc Risk* 2: 289-295.

115. Williams, P.T., P.D. Wood, R.M. Krauss, and W.L. Haskell. 1982. The effect of running mileage and duration on plasma lipoprotein levels. *Journal of the American Medical Association* 247: 2674-2679.

116. Wood, P., and W.L. Haskell. 1979. The effect of exercise on plasma high-density lipoproteins. *Lipids* 14: 417-427.

117. Wood, P.D., and M.L. Stefanick. 1988. Exercise, fitness and atherosclerosis. In *Exercise, fitness and health. A consensus of current knowledge,* ed. C. Bouchard, R.J. Shephard, T. Stephens, J.R. Sutton, and R. McPherson, 409-420. Champaign, IL: Human Kinetics.

118. Wood, P.D., M.L. Stefanick, D.M. Dreon, B. Frey-Hewitt, S.C. Gray, and P.T. Williams. 1988. Changes in plasma lipids and lipoproteins in overweight men during weight loss through dieting as compared with exercise. *New England Journal of Medicine* 319: 1173-1179.

119. Wood, P.D., P.T. Williams, and W.L. Haskell. 1984. Physical activity and high-density lipoproteins. In *Clinical and metabolic aspects of high-density lipoproteins,* ed. N.E. Miller and G.J. Miller, 133-165. Amsterdam: Elsevier Science Publishers.

120. World Health Organization. 1988. Obesity. Preventing and managing the global epidemic. Report of a WHO Consultation on Obesity. 1-276.

CHAPTER 18

Is It Possible to Be Overweight or Obese and Fit and Healthy?

Suzanne Brodney, Steven N. Blair, and Chong Do Lee
Cooper Institute for Aerobics Research, Dallas, Texas, U.S.A.

There is an abundant body of literature describing higher rates of morbidity and mortality in those who are overweight or obese when compared to normal-weight individuals. However, most of these studies have not addressed the possible role of physical activity. A sedentary lifestyle, which is a major risk factor for weight gain, is a potential confounder in the relationship between overweight and obesity and morbidity and mortality. In this chapter we discuss the relationship of overweight and obesity to morbidity and mortality after taking physical activity or cardiorespiratory fitness into account.

Overweight, Obesity, and Health

Overweight and obesity have increased steadily over the past several decades. Current estimates are that 50% of US adults are overweight or obese as characterized by a body mass index (BMI) of > 25 kg/m^2. During this same time period, participation in physical activity by US adults has remained well below recommended levels despite evidence supporting the protective effect of physical activity on morbidity and mortality. This section describes the relationship between overweight, obesity, and health and explores the importance of an active lifestyle in relation to excess weight, morbidity, and mortality.

Long-Term Studies of Overweight, Obesity, and Health

A large number of long-term studies on the relationship between overweight or obesity and morbidity and mortality have been published. The long-term studies reviewed in this section are prospective cohort studies. This type of epidemiologic study design follows a group of disease-free individuals, who are categorized by exposure status at the beginning of the study, for a period of several years to determine who will develop the outcome of interest in the different exposure groups. Cohort studies, though time consuming and expensive to conduct, provide valuable data because they can establish a temporal relationship between exposure and disease.

The Framingham Heart Study is an investigation of 2,873 women and 2,336 men residing in the town of Framingham, MA. One of the best reports of this study is given by Hubert et al. (23). These study participants, who were between the ages of 28 and 62 when initially examined in 1948, were reexamined periodically over the next 26 years. Data from the initial examination were used to classify individuals as obese using the Metropolitan Relative Weight (ratio of actual weight to desirable weight × 100). The authors concluded that the degree of obesity independently predicted the incidence of cardiovascular disease (CVD) after controlling for known cardiovascular risk factors.

Investigators at Harvard Medical School have been following more than 115,000 women enrolled in the Nurses' Health Study (31, 46). The nurses were 30 to 55 years of age at baseline in 1976. The relationship between mortality and BMI was examined after excluding women with known CVD or cancer. There were 4,726 deaths during 16 years of follow-up, 881 from CVD. After adjusting for age, a J-shaped relationship between BMI and mortality was observed. After accounting for cigarette smoking and early mortality, the association between BMI and all-cause mortality was linear. Another analysis of this population examined the relationship between body weight and incident coronary heart disease (CHD) in women without a history of CHD at baseline. A total of 1,292 cases of CHD were observed over 14 years of follow-up. This study observed a linear relationship between body weight and incident CHD. In general, data from these studies reflect a strong direct relationship between BMI and CHD, CVD, and all-cause mortality. The Nurses' Health Study data show the lowest morbidity and mortality rates in the leanest women, even in those with a BMI of < 19 kg/m².

The Harvard Alumni Health Study, which is a cohort of men who were undergraduates at Harvard College between 1916 and 1950, includes assessments of all-cause mortality and body weight (35). A total of 4,370 deaths and 425,718 person-years were accumulated in alumni who responded to a questionnaire in 1962 or 1966 on their medical history, body habits, and health practices. Body mass index, which was calculated in 1962 and 1966 using self-reported height and weight, was stratified as < 22.5, 22.5-< 23.5, 23.5-< 24.5, 24.5-< 26.0, or ≥ 26.0 kg/m². After 27 years of follow-up, the lowest risk of mortality was observed for those who never smoked and had a BMI between 23.5 and 24.5 kg/m², after omitting the first five years of follow-up to account for those with underlying disease at the beginning of the study and adjusting for age and physical activity. Men with BMIs of > 26.0 kg/m² had the highest risk of mortality.

To explain further the relationship between obesity and health, several studies describe the level of BMI associated with minimum mortality. The National Health and Nutrition Examination Survey (NHANES) I Epidemiologic Follow-Up Study includes data on BMI and all-cause mortality in a random, biethnic sample of the US population (15). For each of the four race-sex categories, the BMI of minimum mortality was 27.1 kg/m² in black men, 26.8 kg/m² in black women, 24.8 kg/m² in white men, and 24.3 kg/m² in white women. The authors concluded that the lower end of the BMI distribution for each subgroup was not necessarily associated with the optimal BMI (15).

Using a different BMI classification, investigators at the British Regional Heart Study followed 7,735 men aged 40-59 years at screening in 1978-1980 to determine the BMI associated with the lowest risk of morbidity and mortality (39). BMI was divided into seven categories: < 20, 20.0-21.9, 22.0-23.9, 24.0-25.9, 26.0-27.9, 28.0-29.9, and ≥ 30. After 14.8 years of follow-up, all-cause mortality was elevated in men with BMI < 20 and ≥ 30.

A meta-analysis of 19 prospective cohort studies of white males examined the relationship between body weight and mortality, concluding a U-shaped relationship (42). Mortality risk was higher for BMI levels < 23 or > 28 kg/m². In contrast to the Nurses' Health Study, but in agreement with the NHANES I Epidemiologic Follow-Up Study, the range of minimum mortality in this analysis was a BMI between 24 and 27 kg/m² for nonsmoking, white men followed for 30 years. These results indicated that moderately overweight individuals did not have an increased risk of all-cause mortality.

Interestingly, a prospective study of 6,193 obese patients in Germany with a mean of 14 years of follow-up had similar but more liberal findings (3). Participants were categorized by moderate obesity (BMI: 25-< 32 kg/m²), gross obesity (BMI: 32-< 40 kg/m²), and morbid obesity (BMI: > 40 kg/m²). Individuals classified as grossly obese did not have a significant increase in excess mortality as compared to those with moderate obesity [men (RR = 1.02; 95% CI = 0.76-1.37) and women (RR = 1.23; 95% CI = 0.96-1.58)]. As expected, the highest mortality was seen in the morbidly obese with an elevated risk of 3.05 (95% CI = 2.47-3.73) for men and 2.31 (95% CI = 2.04-2.60) for women when compared to those in the lowest BMI category.

Weight Change

There is recent evidence that weight gain during the adult years may contribute to the incidence of morbidity and mortality (19). Weight change in Japanese-American men participating in the Honolulu Heart Program was calculated for two time periods: (1) recalled weight at age 25 and weight at Examination I (1965-1968), and (2) weight at Examination I and weight at Examination III (1971-1974). Weight gains earlier in life of 5.1 to 10 kg and > 10 kg were associated with a 1.60 (95% CI = 1.22-2.11) and 1.75 (95% CI = 1.32-2.33) elevated risk of CHD incidence, respectively. Weight loss later in life also was associated with an increased risk of CHD. In the Multiple Risk Factor Intervention Trial, investigators found an increased risk of CVD and all-cause mortality associated with high weight variability [RR=1.64, (95% CI= 1.21-2.23)] compared with those who had stable weights (9). This increased risk associated with weight change was not observed for the heaviest men. A positive association between weight gain and risk of CVD was also supported by two large cohort studies (23, 46).

There is clear consensus that higher BMIs are associated with elevated rates of CVD and decreased longevity (16, 30). BMI is also directly associated with numerous other health problems, including type II diabetes (39) and osteoarthritis of the knee (17). Most authorities view these associations as causal, and much attention is given to the treatment of obesity. A potential problem with studies on obesity and health, and one which we will return to later in more detail, is that few published studies took physical activity into account in the analyses. In cases where the investigators indicated that they adjusted for physical activity, little detail was given regarding how physical activity was assessed, or if the measure had acceptable reliability and validity.

Physical Activity, Cardiorespiratory Fitness, and Health

There is agreement that a sedentary way of life is hazardous to health. Recent reports from the Centers for Disease Control and Prevention and the American College of Sports Medicine (CDC/ACSM) (37); American Heart Association (AHA) (18); the U.S. Surgeon General (44); and a National Institutes of Health (NIH) consensus development conference (34) indicate that physical inactivity increases risk of atherosclerotic, metabolic, and hypertensive diseases; increases risk of some cancers; decreases longevity; reduces quality of life; and increases functional limitations, especially in older individuals.

The protective effects of physical activity are extensive, as documented in recent large prospective studies (22, 25, 26, 38, 41). The benefits of physical activity extend to both men (22, 38) and women (26, 41); remain after adjustment for potential confounding variables, including genetic factors (25); and apply to older and younger individuals (22, 26).

One of the important recent findings from physical activity research is that moderate-intensity physical activity, typically defined as activities requiring an increase from 3 to 6 times the resting metabolic rate, provides substantial health benefits. For example, Kushi et al. report on physical activity and mortality in a cohort of 27,974 postmenopausal women (26). When compared with women who rarely or never engaged in moderate physical activity, a strong inverse relationship was observed between total mortality and the increasing frequency of participation in moderate physical activity. Multivariate-adjusted relative risks for total mortality by frequency categories were 0.73 (once a week to a few times per month); 0.69 (2 to 4 times a week); and 0.61 (> 4 times a week) (p for trend < 0.001). Hakim et al. (22) report a similar inverse gradient across categories of miles walked per week in older, nonsmoking, retired men participating in the Honolulu Heart Study.

Leisure-time physical activity is linked to a decreased risk of premature all-cause and CVD mortality in middle-aged men (21). The Finnish men in this cohort were followed for 10 years and 10 months. Sedentary men [energy expenditure (EE) < 800 kcals/week] had an elevated risk of 2.74 (95% CI = 1.46-5.14) for all-cause mortality and 3.58 (95% CI = 1.45-8.85) for CVD mortality, as compared to the most active men (EE > 2,100 kcals/week) after adjustment for several factors thought to be related to both mortality and physical activity. Lack of participation in leisure-time activities, such as gardening or engine repair at least once per week, contributed to an elevated risk of all-cause and CVD mortality as compared to those who participated in these activities at least once per week (likelihood ratio = 1.75 and 2.71, respectively).

A stronger test of the causal hypothesis for the protective effect of physical activity is provided from Paffenbarger and colleagues on changes in physical activity and mortality in Harvard College alumni (36). They report on 14,786 men who had their physical activity assessed two times, in either 1962 or 1966 and again in 1977. The alumni were then followed for mortality until 1988, during which time 2,343 men died. Compared with men whose physical activity remained stable (changed < 250 kcal per week), men who increased their activity by ≥ 1,250 kcal per week had a 24% lower risk of dying (p < 0.001), and men who decreased their activity by ≥ 1,250 kcal per week had a 43% higher risk of dying (p < 0.001).

The British Regional Heart Study, which was conducted to examine the relationship between physical activity, all-cause mortality, and the incidence of CHD in middle- and older-aged men, reported similar results in 4,311 men examined in 1978-80 and again in 1992 (45). Men who were sedentary at baseline and who began light activity by 1992 had a lower risk of morbidity and mortality as compared with those who remained sedentary [RR = 0.55, (95% CI = 0.36-0.84)].

The protective effects of physical activity or fitness appear to apply in various subgroups of the population (5). For example, activity benefits both healthy and unhealthy individuals, smokers and nonsmokers, hypertensives and normotensives, those with a positive family history of early CHD death and those without such a history, and those with elevated cholesterol or glucose and those with normal values on these measurements (8, 11, 26, 36, 38, 40).

Higher levels of activity or fitness also may protect against mortality in overweight individuals (6, 35), although this association was of marginal significance (p = 0.08) for diabetes incidence in Japanese-American men (40), and not significant for breast cancer in a recent prospective study (41). In published reports on the relationship of activity or fitness to mortality in BMI strata, the investigators typically performed the analyses in BMI strata defined by tertiles (thirds of the population) or quartiles (quarters of the population). This often resulted in the highest BMI category having a lower bound of a BMI of 24 or 25 kg/m^2, so that it is not possible to determine if the inverse association between activity or fitness and mortality pertains in those with mild or moderate obesity. This issue needs further investigation.

Activity, Body Composition, and Weight Control

The majority of overweight and obese individuals is less likely to be physically active or fit than normal-weight individuals. In cross-sectional analyses of fitness level and coronary risk factors of participants in the Aerobics Center Longitudinal Study (ACLS), cardiorespiratory fitness was inversely associated with body weight in men (13) and women (20). Kushi et al. (26) observed a similar pattern between BMI and activity level in 40,417 postmenopausal Iowa women (26). BMI and subscapular-tricep skinfold ratio were significantly and inversely related to physical activity quintile (fifths of the population) in the Honolulu Heart Program (11).

Physical activity is a crucial component of weight control (4, 12). Cross-sectional studies show a strong inverse gradient between weight or BMI across categories of physical activity or cardiorespiratory fitness; exercise training studies show weight loss with increased physical activity; and maintenance of weight loss, once it is achieved, is strongly determined by physical activity (32).

A fit and active way of life also may help prevent weight gain and subsequent obesity, although this has been difficult to confirm because of a lack of population-based studies with assessment of physical activity or fitness and weight over multiple times. DiPietro et al. recently reported on weight change in a cohort of 4,599 men and 724 women who had their weights and cardiorespiratory fitness measured at least three times over a maximal interval of 24 years (average follow-up = 7.5 years) (14). An increase of one minute in treadmill time from the first to the second examination was associated with reduced odds of a ≥ 5 kg weight gain by 14% in men and 9% in women; the odds of having a ≥ 10 kg weight gain was reduced by 21% in both men and women. This is the first study with three measurements of fitness and weight, and it provides strong evidence that an active life may help prevent age-related weight gain.

Physical Activity and Mortality as a Function of BMI

As mentioned earlier, many of the large cohort studies that examined obesity and health either did not account for physical activity or little detail was provided on the physical activity measurement tool. Of the studies identified that examined physical activity and morbidity or mortality, many did not examine the data by categories of BMI, and those that did utilized different BMI cutpoints making it difficult to compare results across studies. This section is a review of the studies on physical activity and mortality that stratified the results by BMI.

Paffenbarger and colleagues examined the relationship between physical activity and all-cause mortality in the Harvard Alumni Health Study (35). As previously described, the men completed a self-administered questionnaire in 1962 or 1966 and reported the amount of weekly energy expenditure and other lifestyle characteristics. The 16,936 alumni were categorized by BMI and amount of physical activity in kilocalories per week (< 500, 500-2000, > 2000). Those who were least active and had the lowest BMI had the highest mortality. For all BMI categories, the risk of death was lower in men who were most active.

Incidences of diabetes and physical activity were examined in the Honolulu Heart Study over a six-year period in 8,006 middle-aged Japanese-American men (11). A physical activity index was created based on the hours spent in specific activities. The age-adjusted cumulative incidence of diabetes was inversely related to the quintile of physical activity index. Investigators observed a significant trend for diabetes across physical activity quintile within the mean of the lower four quintiles of BMI (p = 0.0004). This pattern was evident for all men and for men with glucose < 225 mg/dl (nonfasting measurement taken one hour after a 50-gm load). This same trend was not observed for either group of men in the upper BMI quintile.

The Multifactor Primary Prevention Study in Goteborg began in 1970 when researchers recruited all men in that city born between 1915 and 1925, excluding 1923, and followed them for a mean of 20 years to determine incidence of disease (38). Subjects completed a mailed questionnaire on several risk factors, including physical activity. The men were stratified by BMI (< 24.1, 24.1-26.6, and > 26.6) and quartile of occupational and leisure-time physical activity. The most physically active men during leisure time had a significantly lower risk of CHD death as compared with sedentary men, after adjusting for various cardiovascular risk factors [RR = 0.72, (95% CI = 0.56-0.92)]. This same finding was not observed for occupational physical activity.

In a study of 9,376 male civil servants aged 45-64 at baseline in 1976, Morris et al. examined the association between leisure-time exercise and coronary attack and death rates (33). BMI was stratified by < 24, 24-26.9, and ≥ 27 and amount of vigorous aerobic exercise by quartiles (frequent/intense; next lesser degree; residual, little; none). In each BMI category there was an inverse gradient of CHD across physical activity groups.

Summary of Overweight, Obesity, and Health

An active way of life protects against early death, and this appears to hold true in various subgroups, perhaps including those who are overweight or obese. Studies on the relationship of cardiorespiratory fitness and health outcomes generally have stronger associations with morbidity or mortality and fitness than is seen in similar studies with physical activity as the exposure (7). This presumably results from less misclassification when fitness is the exposure, due to its more objective assessment as compared with physical activity self-report. Thus, physical activity and health studies may underestimate the effect of physical inactivity on morbidity and mortality.

A Hypothesis for the Relationship Between Inactivity and Morbidity in the Obese

Findings from the preceding sections lead to several conclusions: (1) There is a strong direct relation of BMI to morbidity and mortality, especially from CVD and diabetes; (2) sedentary habits and low levels of cardiorespiratory fitness are important predictors of the incidence of chronic atherosclerotic, hypertensive, and metabolic diseases; some cancers; and premature CVD and all-cause mortality; (3) overweight and obese individuals are more likely to be sedentary and unfit than their active and fit peers; and, (4) physical activity appears to protect against chronic disease morbidity in various subgroups of the population. These findings led us to develop the hypothesis that some, and perhaps much, of the increased morbidity and mortality observed in overweight and obese individuals may be due to the lower levels of activity and cardiorespiratory fitness seen in this group. In other words, the fundamental problem is physical inactivity, which leads to higher disease incidence; and also to weight gain, which may further exacerbate disease processes.

Cardiorespiratory Fitness, Obesity, and Health: Evidence from the Aerobics Center Longitudinal Study (ACLS)

We have the opportunity to follow, since 1970, a large population of men and women who have been examined at the Cooper Clinic in Dallas, Texas. These individuals are primarily from the middle- to upper-socioeconomic strata. More than 75% are college graduates and are employed as business executives or professionals. Most are married, and approximately 97% are non-Hispanic whites. At the time of their baseline examinations at the Cooper Clinic, they ranged in age from 20 to 88 years. More than 7,000 women and 25,000 men comprise the study group. Participants are followed from the day of their baseline examination until the day they die, or currently up to 12/31/89 for the survivors. Follow-up for mortality resulted in detection of 89 deaths in more than 52,000 woman-years and 601 deaths in more than 211,000 man-years of observation. The principal method of mortality follow-up was the National Death Index, which is a record of the majority of deaths in the US maintained by the National Center for Health Statistics.

The baseline examination is thorough and includes a physical examination, medical and family history, anthropometry, clinical assessments, blood chemistry evaluations, and a maximal exercise test on a treadmill. Further details of examination content and procedures are available in prior reports (6-8). The two principal exposure variables discussed here are body composition and cardiorespiratory fitness. We measured body composition in several ways. Height and weight were obtained on periodically calibrated physician's balance beam scales and stadiometers;

BMI was calculated as kg/m^2 from this data. We estimated % body fat by either hydrostatic weighing or from a seven-site skinfold test (24).

We measured cardiorespiratory fitness on a treadmill using a modified Balke protocol (1). The treadmill was set initially at 0 grade, with a beginning speed of 88 m/min. After the first minute, the grade was increased to 2%, and was further increased by 1% each minute until the 25th minute whereupon the speed was increased by 5.4 m/min until the test was terminated when the patient reached exhaustion. In most analyses we analyze cardiorespiratory fitness as a categorical variable where the least fit 20% of men and women in each age group are labeled as lowly fit, the next 40% of the distribution as moderately fit, and the top 40% as highly fit.

Preliminary Results on the Relationship of Cardiorespiratory Fitness to Mortality in BMI

In our first report on fitness and mortality in this cohort, we presented results for all-cause mortality across low-, moderate-, and high-fitness levels in BMI strata (8). Body mass index strata for these analyses were < 20, 20-25, and > 25. These strata were arbitrarily chosen to provide adequate numbers of deaths in each cell, although the number of deaths in women, especially in the highest BMI stratum, was small. There was an inverse gradient for all-cause mortality across fitness categories in all three BMI strata, with the least fit individuals having mortality rates at least twice as high as the highly fit men and women in all BMI strata. These findings suggested the hypothesis that fitness might protect against obesity, but the highest BMI category (> 25 kg/m^2) included both overweight and obese individuals, so the results could not be considered definitive.

In order to test further the hypothesis that cardiorespiratory fitness provides protection against the hazards of obesity, we performed additional cross-tabulation analyses for all-cause mortality by strata of BMI and fitness (2). This second set of analyses was done after extending mortality surveillance for an additional four years, during which there were 433 additional deaths in men. This allowed for creating different BMI strata and focusing attention on the relationship of fitness to mortality in men with higher BMIs. At this time, there are still too few deaths in our smaller cohort of women to conduct analyses in women who meet conventional criteria for obesity. The BMI categories used in these analyses were < 27, 27-30, and > 30, and results are presented in table 18.1. Overweight and obese men were more likely than normal-weight men to be unfit, with 63.4 % of man-years of observation in the > 30 BMI group classified as unfit, and 17.4 % of man-years in the < 27 BMI category in this bottom fitness group. Conversely, the percentages of man-years of observation in highly fit men were 38.9 and 17.0 in the two lower BMI groups, respectively. Men with a BMI > 30 were less likely than other men to be highly fit, and we combined the moderately and highly fit groups for analysis in this category. This resulted in 36.6% of man-years in those with a BMI > 30 occurring in moderately and highly fit men.

Table 18.1   Age-Adjusted All-Cause Death Rates for Lowly, Moderately, and Highly Fit Men by BMI Strata

Fitness category	Number of deaths	Age-adjusted death rate/10,000	Age-adjusted RR	95% CI
BMI < 27 n = 17,178				
Low	133	52.1	1.0	
Moderate	180	28.6	0.49	0.31-0.76
High	119	20.0	0.34	0.21-0.57
BMI 27 to 30 n = 5,277				
Low	63	49.1	1.0	
Moderate	67	29.8	0.61	0.38-0.96
High	17	19.7	0.40	0.24-0.68
BMI > 30 n = 2,934				
Low	75	62.1	1.0	
Moderate-High	19	18.0	0.29	0.17-0.49

RR = relative risk; CI = confidence interval.

Adapted, by permission, from C.E. Barlow, H.W. Kohl, III, L.W. Gibbons, and S.N. Blair, 1995, "Physical fitness, mortality and obesity," *International Journal of Obesity and Related Metabolic Disorders* 19: S41-S44.

There was a steep inverse gradient of mortality risk across fitness categories in each of the BMI strata. Highly fit men in BMI groups < 27 and 27-30 had comparable death rates. Obese men who were at least moderately fit did not have an elevated mortality risk, and in fact this group had a much lower death rate than unfit men in the < 27 BMI category (18.0 compared with 52.1 deaths per 10,000 man-years). We also calculated the mortality risk associated per one minute of treadmill time for men with a BMI ≥ 27. A 9% lower risk was observed for each minute higher treadmill time after adjustment for age, health status, BMI, smoking status, systolic blood pressure, cholesterol, and fasting glucose. The percent difference in all-cause mortality risk per minute of greater treadmill time was greatest (15%) for those with a BMI ≥ 30.

Cardiovascular Fitness and Body Weight Standards

Cardiovascular fitness should be considered when defining body weight standards. The 1995 weight guidelines (43) classify a healthy weight as a BMI between 19-25

kg/m^2. To examine the validity of the 1995 weight guidelines while controlling for cardiorespiratory fitness, 21,586 men aged 30-83 years participating in the ACLS cohort were examined. All-cause and CVD mortality was calculated in fit versus unfit men stratified by BMI level (19.0-< 25.0, 25.0-27.8, ≥ 27.8) (table 18.2). In all three BMI strata, the relative risk of all-cause and CVD mortality was higher in the unfit versus the fit men. Fit, overweight men and fit, normal-weight men had similar risks of mortality. Using fit, normal-weight men as the referent group, unfit, normal-weight men had a relative risk of 2.2 (95% CI = 1.24-3.83, p = 0.007) after excluding smokers and the first five years of follow-up data, which may have included men with underlying disease at baseline. This study provides evidence that fitness may be more important than BMI as a mortality predictor (29).

Body composition and fat distribution measurements may better predict CVD than BMI, because BMI measures overweight only for height. The effects of body fatness, % body fat, fat distribution, waist circumference, and cardiorespiratory fitness in relation to all-cause and CVD mortality also were examined in the ACLS cohort. Stratification by body fatness (% body fat) [< 16.7 (lean), 16.7-< 25.0 (normal), and ≥ 25.0 (obese)] and waist circumference [< 87 (low), 87-< 99 (moderate), and ≥ 99 (obese)] for the fit and unfit men are presented in tables 18.3 and 18.4, respectively. Men who were unfit had a higher relative risk for all-cause mortality than their fit peers at all body fatness and waist circumference categories (28).

Table 18.2   US Weight Guidelines and Relative Risk of All-Cause Mortality by Cardiorespiratory Fitness Levels in Men: Aerobics Center Longitudinal Study

Fitness	BMI[a]	Deaths (n)	Subjects (n)	Man-years of follow-up (%)	Multivariate RR of death (95% CI)[b]
Fit	19.0-< 25.0	121	8123	64233 (37)	1.00
Unfit	19.0-< 25.0	45	939	10412 (6)	2.25 (1.59-3.17)
Fit	25.0-< 27.8	89	6073	45877 (26)	0.96 (0.73-1.26)
Unfit	25.0-< 27.8	46	1296	14358 (8)	1.68 (1.19-2.37)
Fit	≥ 27.8	49	3307	21683 (12)	1.08 (0.77-1.50)
Unfit	≥ 27.8	77	2118	19626 (11)	2.24 (1.68-2.98)

[a]US weight guidelines: US Department of Agriculture.

[b]Adjusted for age (single year), examination year, smoking habit (never, former, or current)(< 20, 20-40, or > 40 cigarettes/day), and alcohol intake (none, light, moderate, heavy).

Adapted, by permission, from C.D. Lee, A.S. Jackson, and S.N. Blair, 1998, "U.S. weight guidelines: Is it also important to consider cardiorespiratory fitness?" *International Journal of Obesity and Related Metabolic Disorders* 22: S2-S7.

Table 18.3 Body Fatness and Relative Risks of All-Cause and Cardiovascular Disease Mortality by Cardiorespiratory Fitness Levels in Men

Cardiorespiratory fitness and % body fat		Number of subjects	Man-years of follow-up	Number of deaths	All-cause multivariate RR of deaths (95% CI)[a]
Fit	Lean (< 16.7)	5093	41854	68	1.00
Unfit	Lean (< 16.7)	327	3883	14	2.07 (1.16-3.69)
Fit	Normal (16.7-< 25.0)	9255	68546	127	0.80 (0.80-1.08)
Unfit	Normal (16.7-< 25.0)	1851	19669	60	1.62 (1.15-2.30)
Fit	Obese (≥ 25)	3217	21874	65	0.92 (0.65-1.31)
Unfit	Obese (≥ 25)	2182	20916	94	1.90 (1.39-2.60)

[a]Adjusted for age (single year), examination year, smoking habit (never, former, or current [< 20, 20-40, or > 40 cigarettes/day]), alcohol intake (none, light, moderate, heavy), and parental history of coronary heart disease (either parent died of coronary heart disease). Adapted, by permission, from C.D. Lee, S.N. Blair, and A.S. Jackson, 1999, "Cardiorespiratory fitness, body composition, and all-cause and cardiovascular disease mortality in men," *American Journal of Clinical Nutrition* 69: 373-380. © Am. J. Clin. Nutr. American Society for Clinical Nutrition.

Table 18.4 Waist Circumference and Relative Risks of All-Cause Mortality by Cardiorespiratory Fitness Levels in 14,043 Men

Cardiorespiratory fitness and waist circumference (cm)		Number of subjects	Man-years of follow-up	Number of deaths	All-cause multivariate RR of deaths (95% CI)[a]
Fit	Lean (< 87)	3247	18579	26	1.00
Unfit	Lean (< 87)	136	1022	8	4.88 (2.20-10.83)
Fit	Normal (87-< 99)	6237	34189	60	1.05 (0.66-1.67)
Unfit	Normal (87-< 99)	616	4211	15	2.05 (1.08-3.87)
Fit	Obese (≥ 99)	2645	12994	24	0.95 (0.54-1.66)
Unfit	Obese (≥ 99)	1162	7013	29	2.40 (1.41-4.07)

[a]Adjusted for age (single year), examination year, smoking habit (never, former, or current [< 20, 20-40, or > 40 cigarettes/day]), alcohol intake (none, light, moderate, heavy), and parental history of coronary heart disease (either parent died of coronary heart disease). Adapted, by permission, from C.D. Lee, S.N. Blair, and A.S. Jackson, 1999, "Cardiorespiratory fitness, body composition, and all-cause and cardiovascular disease mortality in men," *American Journal of Clinical Nutrition* 69: 373-380. © Am. J. Clin. Nutr. American Society for Clinical Nutrition.

Summary and Conclusions

There is evidence from several large cohort studies to support the following conclusions: (1) There is a strong direct relationship of BMI to morbidity and mortality; (2) sedentary habits and low levels of cardiorespiratory fitness are important predictors of disease incidence and all-cause mortality; (3) overweight and obese individuals are more likely to be sedentary and unfit than their active and fit peers; and (4) physical activity appears to protect against chronic disease morbidity and mortality in various subgroups of the population.

Results from the ACLS cohort indicate that fitness protects against mortality, even in individuals who are overweight or obese. When this same cohort was stratified by the US weight guidelines or BMI category, the risk of all-cause and CVD mortality was higher in the unfit versus the fit men for all three BMI strata (2, 29). A similar pattern of lower risk was observed for men who were physically fit, even if they were classified as obese by either waist circumference or percent body fat (28).

An individual's quality of life depends on mental, physical, and social well-being. Realistically, overweight or obese men, women, and children, physically fit or not, will still encounter the social stigmatism and prejudice attached to fatness. Even though fitness may protect against mortality in overweight and obese individuals, there is less known about the role of physical fitness on some of the other morbidities associated with obesity. A report from the Women's Health Australia Project shows an association between BMI and indicators of health and well-being in 13,431 women aged 45-59 (10). Summary scores for the physical functioning component were incrementally lower across BMI categories (< 20; ≥ 20- ≤ 25; > 25- ≤ 30; > 30- ≤ 40; > 40). Lean et al. (27) reported decreases in physical functioning in men and women at higher waist circumference categories. Weight loss was shown to reduce the risk of symptomatic knee osteoarthritis in overweight, middle-, and older-aged women (17). Activity and fitness might ameliorate some of the negative associations of obesity on morbidity outcomes.

Physical inactivity is a modifiable risk factor for premature morbidity and mortality. Physical activity plays a crucial role in weight loss, maintenance of weight loss, and prevention of weight gain. Clinicians are in a unique role to encourage a physically active lifestyle for all patients by prescribing moderate-intensity activities that are enjoyable and in keeping with the recommendations of the CDC/ACSM (37), AHA (18), U.S. Surgeon General (44), and the NIH consensus development conference (34). In addition, public policy must support these recommendations through community-wide interventions and environmental changes to assure safety and accessibility.

Obesity is a serious public health problem in the US and the current strategies to prevent and treat obesity, which focus mainly on classifying individuals as overweight or obese and on prescribing dietary changes, are not working. We believe that the public health would be better served with more comprehensive attempts to increase population levels of physical activity, rather than emphasizing ideal weight ranges and raising an alarm about increasing prevalence rates of obesity.

Acknowledgments

This research was supported by U.S. Public Health Service research grant AG06945 from the National Institute on Aging, Bethesda, MD; Polar Electro Oy, Kempele, Finland; and by a grant from Kellogg.

References

1. Balke, B., and R.W. Ware. 1959. An experimental study of physical fitness in Air Force personnel. *US Armed Forces Medical Journal* 10:675-688.
2. Barlow, C.E., H.W. Kohl III, L.W. Gibbons, and S.N. Blair. 1995. Physical fitness, mortality and obesity. *International Journal of Obesity and Related Metabolic Disorders* 19(Suppl. 2):S41-S44.
3. Bender, R., C. Trautner, M. Spraul, and M. Berger. 1998. Assessment of excess mortality in obesity. *American Journal of Epidemiology* 147:42-48.
4. Blair, S.N. 1993. Evidence for success of exercise in weight loss and control. *Annals of Internal Medicine* 119:702-706.
5. Blair, S.N. 1997. Effects of physical activity on cardiovascular disease mortality independent of risk factors. In *Physical activity and cardiovascular health: A national consensus,* ed. A.S. Leon, 127-136. Champaign, IL: Human Kinetics.
6. Blair, S.N., J.B. Kampert, H.W. Kohl III, C.E. Barlow, C.A. Macera, R.S. Paffenbarger Jr., and L.W. Gibbons. 1996. Influences of cardiorespiratory fitness and other precursors on cardiovascular disease and all-cause mortality in men and women. *Journal of the American Medical Association* 276:205-210.
7. Blair, S.N., H.W. Kohl III, C.E. Barlow, R.S. Paffenbarger Jr., L.W. Gibbons, and C.A. Macera. 1995. Changes in physical fitness and all-cause mortality: A prospective study of healthy and unhealthy men. *Journal of the American Medical Association* 273:1093-1098.
8. Blair, S.N., H.W. Kohl III, R.S. Paffenbarger Jr., D.G. Clark, K.H. Cooper, and L.W. Gibbons. 1989. Physical fitness and all-cause mortality: A prospective study of healthy men and women. *Journal of the American Medical Association* 262:2395-2401.
9. Blair, S.N., J. Shaten, K. Brownell, G. Collins, and L. Lissner. 1993. Body weight change, all-cause mortality, and cause-specific mortality in the Multiple Risk Factor Intervention Trial. *Annals of Internal Medicine* 119:749-757.

10. Brown, W.J., A.J. Dobson, and G. Mishra. 1998. What is a healthy weight for middle aged women? *International Journal of Obesity and Related Metabolic Disorders* 22:520-528.
11. Burchfiel, C.M., D.S. Sharp, J.D. Curb, B.L. Rodriguez, L.-J. Hwang, E.B. Marcus, and K. Yano. 1995. Physical activity and incidence of diabetes: The Honolulu Heart Program. *American Journal of Epidemiology* 141:360-368.
12. Byers, T. 1995. Body weight and mortality. *New England Journal of Medicine* 333:723-724.
13. Cooper, K.H., M.L. Pollock, R.P. Martin, S.R. White, A.C. Linnerud, and A. Jackson. 1976. Physical fitness levels vs selected coronary risk factors: A cross-sectional study. *Journal of the American Medical Association* 236:166-169.
14. DiPietro, L., H.W. Kohl III, C.E. Barlow, and S.N. Blair. 1997. Physical fitness and risk of weight gain in men and women: The Aerobics Center Longitudinal Study. *Medicine and Science in Sports and Exercise* 29(Suppl 5):S115.
15. Durazo-Arvizu, R.A., D.L. McGee, R.S. Cooper, Y. Liao, and A. Luke. 1998. Mortality and optimal body mass index in a sample of the US population. *American Journal of Epidemiology* 147:739-749.
16. Eckel, R.H. 1997. Obesity and heart disease: A statement for healthcare professionals from the Nutrition Committee, American Heart Association. *Circulation* 96:3248-3250.
17. Felson, D.T., Y. Zhang, J.M. Anthony, A. Naimark, and J.J. Anderson. 1992. Weight loss reduces the risk for symptomatic knee osteoarthritis in women. The Framingham Study. *Annals of Internal Medicine* 116:535-539.
18. Fletcher, G.F., G. Balady, S.N. Blair, J. Blumenthal, C. Caspersen, B. Chaitman, S. Epstein, E.S.S. Froelicher, V.F. Froelicher, I.L. Pina, and M.L. Pollock. 1996. Statement on exercise: Benefits and recommendations for physical activity programs for all Americans: A statement for health professionals by the Committee on Exercise and Cardiac Rehabilitation of the Council on Clinical Cardiology, American Heart Association. *Circulation* 94:857-862.
19. Galanis, D.J., T. Harris, D.S. Sharp, and H. Petrovitch. 1998. Relative weight, weight change, and risk of coronary heart disease in the Honolulu Heart Program. *American Journal of Epidemiology* 147:379-386.
20. Gibbons, L.W., S.N. Blair, K.H. Cooper, and M. Smith. 1983. Association between coronary heart disease risk factors and physical fitness in healthy adult women. *Circulation* 67:977-983.
21. Haapanen, N., S. Miilunpalo, I. Vuori, P. Oja, and M. Pasanen. 1996. Characteristics of leisure time physical activity associated with decreased risk of premature all-cause and cardiovascular disease in mortality in middle-aged men. *American Journal of Epidemiology* 143:870-880.
22. Hakim, A.A., H. Petrovitch, C.M. Burchfiel, G.W. Ross, B.L. Rodriguez, L.R. White, K. Yano, J.D. Curb, and R.D. Abbott. 1998. Effects of walking on mortality among nonsmoking retired men. *New England Journal of Medicine* 338:94-99.

23. Hubert, H.B., M. Feinleib, P.M. McNamara, and W.P. Castelli. 1983. Obesity as an independent risk factor for cardiovascular disease: A 26-year follow-up of participants in the Framingham Heart Study. *Circulation* 67:968-977.

24. Jackson, A.S., and M.L. Pollock. 1978. Generalized equations for predicting body density of men. *British Journal of Nutrition* 40:496-504.

25. Kujala, U.M., J. Kaprio, S. Sarna, and M. Koskenvuo. 1998. Relationship of leisure-time physical activity and mortality: the Finnish twin cohort. *Journal of the American Medical Association* 279:440-444.

26. Kushi, L.H., R.M. Fee, A.R. Folsom, P.J. Mink, K.E. Anderson, and T.A. Sellers. 1997. Physical activity and mortality in postmenopausal women. *Journal of the American Medical Association* 277:1287-1292.

27. Lean, M.E., T.S. Han, and J.C. Seidell. 1998. Impairment of health and quality of life in people with large waist circumference. *Lancet* 351:853-856.

28. Lee, C.D., S.N. Blair, and A.S. Jackson. 1999. Cardiorespiratory fitness, body composition, and all-cause and cardiovascular disease mortality in men. *American Journal of Clinical Nutrition* 691:373-380.

29. Lee, C.D., A.S. Jackson, and S.N. Blair. 1998. US weight guidelines: Is it also important to consider cardiorespiratory fitness? *International Journal of Obesity and Related Metabolic Disorders* 22:S2-S7.

30. Manson, J.E., M.J. Stampfer, C.H. Hennekens, and W.C. Willett. 1987. Body weight and longevity: A reassessment. *Journal of the American Medical Association* 257:353-358.

31. Manson, J.E., W.C. Willett, M.J. Stampfer, G.A. Colditz, D.J. Hunter, S.E. Hankinson, C.H. Hennekens, and F.E. Speizer. 1995. The Nurses' Health Study: Body weight and mortality among women. *New England Journal of Medicine* 333:677-685.

32. McGuire, M.T., R.R. Wing, M.L. Klem, H.M. Seagle, and J.O. Hill. 1998. Long-term maintenance of weight loss: Do people who lose weight through various weight loss methods use different behaviors to maintain their weight? *International Journal of Obesity and Related Metabolic Disorders* 22:572-577.

33. Morris, J.N., D.G. Clayton, M.G. Everitt, A.M. Semmence, and E.H. Burgess. 1990. Exercise in leisure time: Coronary attack and death rates. *British Heart Journal* 63:325-334.

34. National Institutes of Health Consensus Development Panel on the Health Implications of Obesity. 1985. Health implications of obesity: National Institutes of Health Consensus Development Conference Statement. *Annals of Internal Medicine* 103:1073-1077.

35. Paffenbarger, R.S. Jr., R.T. Hyde, A.L. Wing, and C.C. Hsieh. 1986. Physical activity, all-cause mortality, and longevity of college alumni. *New England Journal of Medicine* 314:605-613.

36. Paffenbarger, R.S. Jr., R.T. Hyde, A.L. Wing, I.-M. Lee, D.L. Jung, and J.B. Kampert. 1993. The association of changes in physical-activity level and

other lifestyle characteristics with mortality among men. *New England Journal of Medicine* 328:538-545.

37. Pate, R.R., M. Pratt, S.N. Blair, W.L. Haskell, C.A. Macera, C. Bouchard, D. Buchner, W. Ettinger, G.W. Heath, A.C. King, A. Kriska, A.S. Leon, B.H. Marcus, J. Morris, R.S. Paffenbarger Jr., K. Patrick, M.L. Pollock, J.M. Rippe, J. Sallis, and J.H. Wilmore. 1995. Physical activity and public health: A recommendation from the Centers for Disease Control and Prevention and the American College of Sports Medicine. *Journal of the American Medical Association* 273:402-407.

38. Rosengren, A., and L. Wilhelmsen. 1997. Physical activity protects against coronary death and deaths from all causes in middle-aged men: Evidence from a 20-year follow-up of the Primary Prevention Study in Goteborg. *Annals of Epidemiology* 7:69-75.

39. Shaper, A.G., S.G. Wannamethee, and M. Walker. 1997. Body weight: Implications for the prevention of coronary heart disease, stroke, and diabetes mellitus in a cohort study of middle aged men. *British Medical Journal* 314:1311-1317.

40. Stewart, A.L., A.C. King, and W.L. Haskell. 1993. Endurance exercise and health-related quality of life in 50-65-year-old adults. *The Gerontologist* 33:782-789.

41. Thune, I., T. Brenn, E. Lund, and M. Gaard. 1997. Physical activity and the risk of breast cancer. *New England Journal of Medicine* 336:1269-1275.

42. Troiano, R.P., E.A.J. Frongillo, J. Sobal, and D.A. Levitsky. 1996. The relationship between body weight and mortality: a quantitative analysis of combined information from existing studies. *International Journal of Obesity and Related Metabolic Disorders* 20:63-75.

43. U.S. Department of Agriculture and U.S. Department of Health and Human Services. 1995. *Nutrition and your health: Dietary guidelines for Americans*, 4th ed. Washington, D.C.: U.S. Department of Agriculture and U.S. Department of Health and Human Services.

44. U.S. Department of Health and Human Servces. 1996. *Physical activity and health: A report of the Surgeon General*. Atlanta, Georgia: U.S. Department of Health and Human Services, Centers for Disease Control and Prevention, National Center for Chronic Disease Prevention and Health Promotion.

45. Wannamethee, S.G., A.G. Shaper, and M. Walker. 1998. Changes in physical activity, mortality, and incidence of coronary heart disease in older men. *Lancet* 351:1603-1608.

46. Willett, W.C., J.E. Manson, M.J. Stampfer, G.A. Colditz, B. Rosner, F.E. Speizer, and C.H. Hennekens. 1995. Weight, weight change, and coronary heart disease in women. Risk within the 'normal' weight range. *Journal of the American Medical Association* 273:461-465.

CHAPTER 19

Changing Lifestyle: Moving From Sedentary to Active

Rena R. Wing, PhD, and John M. Jakicic, PhD
University of Pittsburgh School of Medicine, Western Psychiatric Institute and Clinic, Pittsburgh, Pennsylvania, U.S.A.

Exercise has been shown to be an important component of behavioral weight control programs. Randomized trials suggest that the combination of diet plus exercise produces better short- and long-term weight losses than either of these interventions used alone (19, 20, 45, 60). Likewise, correlational studies show that participation in a regular exercise program is one of the strongest predictors of long-term maintenance of weight loss (49). Kuczmarski et al. (35) recently reported that low levels of physical activity are predictive of subsequent weight gain in both men and women. Thus, increased exercise may be important in both the prevention and treatment of obesity.

Despite the important role of exercise in the management of body weight, overweight individuals appear to have particular difficulty maintaining their exercise participation. The review of the literature conducted by Dishman (8) concluded that obesity is a barrier to physical activity, and that the obese have dropout rates from exercise programs that are as high as 70% within a 6-12 month period. This chapter will discuss strategies that might be used to improve adherence to exercise in overweight individuals. Much of the research on exercise adherence has been done with sedentary individuals, regardless of their body weight. Because the strategies that have been successful in improving exercise in these studies are clearly applicable to overweight patients, we will review these findings as well as studies directly focused on obese patients seeking weight control.

Encouraging Individuals to Initiate Exercise

Until recently, much of the research on exercise adherence has focused on people who have already entered an exercise program or a weight control program, and research has been designed to improve their long-term adherence to such an intervention. Recently, however, it has been recognized that efforts should also be focused on the large population who have not made the decision to start exercising. The goal for these individuals is to get them to initiate an exercise program. Although this research has not focused on overweight individuals per se, the research findings are extremely relevant to these patients.

Using the stages-of-change model (48), Marcus et al. (38) hypothesized that individuals who were given self-help manuals about exercise that were tailored to their stage of motivational readiness for exercise adoption would be more likely to increase their exercise than individuals who received standard booklets. Thus, individuals at the lowest stage of readiness to exercise (precontemplators) were given a self-help manual ("Do I Need This?") that focused on increasing awareness of the benefits of exercise and encouraged these individuals to think about the barriers that prevented activity. Similarly, individuals at other stages of readiness for change received booklets tailored to their level. Contemplators, who are a little further along the motivational readiness scale, received a manual ("Try It, You'll Like It") focusing on reasons for becoming more active vs. staying inactive (decision making), learning to reward oneself for setting realistic goals, and social support. Those in the preparation stage, who were exercising occasionally, received a manual ("I'm on My Way") that reviewed the benefits of activity and goal setting, provided tips on safe and enjoyable exercise, and discussed obstacles to regular activity. A booklet entitled "Keeping It Going" was given to those who were already active. This booklet focused on the benefits of regular activity, staying motivated, self-rewards, enhancing confidence, and decreasing obstacles. Finally, those individuals already in maintenance received a booklet ("I Won't Stop Now") that emphasized the benefits of regular activity, varying activities, rewarding oneself, and planning ahead to stay active.

These tailored booklets were compared to standard booklets in a randomized controlled study conducted at 11 worksites. The booklet and cover letter were sent to employees at entry into the study, with a second booklet sent one month later. The main dependent measure was self-reported exercise stage and minutes of exercise at 3 months. Thirty- seven percent of subjects receiving tailored booklets moved ahead in their stage of motivational readiness for exercise compared to 27% of those in the standard booklet condition. The tailored booklets were particularly effective for those subjects in the contemplation and precontemplation stage at baseline. Among subjects in the contemplation stage, for example, 61% of those who received tailored pamphlets moved forward a stage, compared to 52% of those in the standard condition.

Physicians may be in a particularly good position to encourage people to embark on an exercise program. Because overweight individuals frequently visit physicians for the treatment of obesity-related co-morbidities, this approach may be particularly relevant for obese patients. Structured counseling protocols and physician training have been shown to be effective for smoking cessation, and similar positive results have been reported recently for physical activity (5, 36). The largest study of such physician training was conducted using the Physician-Based Assessment and Counseling for Exercise (PACE) protocol, which was also based on the stages-of-change model (5).

In the PACE protocol, patients complete a short questionnaire that assesses their current stage of motivational readiness for physical activity. Then, based on their responses, patients are given a stage-matched written protocol, which they partially complete before seeing the physician. The physician then reviews the protocol with the patient, provides 3-5 minutes of stage-relevant information for the patient, and completes the recommendation section of the protocol (giving a copy to the patient and keeping one for the patient's file).

This PACE intervention was evaluated in a quasi-experimental design, where physicians interested in physical activity were recruited for the intervention condition and taught to implement the PACE curriculum. These physicians were matched for ethnicity, practice type, practice specialty, and geographic location with control physicians who were provided training on hepatitis B, rather than physical activity. Data were presented on 98 patients who were seen by the intervention physicians and on 164 control patients. All patients in this study were in the contemplation stage of exercise at baseline, and received in addition to the PACE protocol, a booster phone call from staff and a mailing with exercise tips, if desired. The effectiveness of the PACE intervention was evaluated by obtaining self-reported exercise information 4 to 6 weeks after the intervention and through use of Caltrac accelerometers on a subgroup of the subjects. The intervention patients reported an increase in exercise of almost 40 min/week, compared to 10 min/week in control patients (5), and a 22% increase in Caltrac activity counts/hour compared to a slight decrease in control subjects. Similarly, a larger number of intervention patients moved from the "contemplation" to the "active" stage than control patients (52% vs. 12%, $p < 0.0001$). Other measures also supported the effectiveness of the intervention.

Despite the limitations of this study (its quasi-experimental design and short follow-up interval), the positive effects were noteworthy. Further research is needed to determine whether such brief physician counseling can also be effective with patients in other stages of motivational readiness, and to determine which components of the intervention (e.g., the stage-based message; the booster phone calls) were responsible for its success.

Evidence that the written recommendations may contribute to the effectiveness of the physician intervention is apparent in another recent study (56). This study evaluated the effect of verbal advice from a general practitioner vs. a written prescription for physical activity. A total of 491 subjects were randomized and 456 of these were recontacted at 6 weeks. These patients averaged 49 years of age, and 55% of them had a medical condition related to inactivity. The physician spent approximately 5 minutes with the patient, assessing activity and setting a goal, and then gave an activity prescription either verbally or in writing. In most cases, the advice given was to increase walking. Patients given the written prescription were more likely to engage in some physical activity at follow-up

(86% vs. 77%, p = 0.004) and were more likely to report an increase in physical activity (73% vs. 63%, p = 0.02). Differences in duration of activity were not significant.

Thus, a written prescription may be a simple, effective component of a physician-delivered exercise intervention. The prescription may serve as a cue to exercise, but may also emphasize the importance of the physician's advice to exercise, elevating it to the same degree of importance usually given to medications. Because physicians are most likely to recommend physical activity for their overweight patients (34), this approach may be particularly useful in this population.

Maintaining Exercise Adherence Over Time

In order for exercise to be helpful in the long-term maintenance of weight loss, it is necessary that obese individuals not only adopt an exercise program, but also maintain it over long periods of time. Unfortunately, long-term adherence to exercise remains a major problem. Prior studies suggest that over 50% of individuals will drop out of exercise programs within the first 6 months of starting the program. Poor adherence has been noted particularly in the obese, in smokers, in blue collar employees, and in low socioeconomic status individuals (8). In order to maximize adherence, it is important to consider the reasons people give for dropping out of exercise programs and to develop strategies that address each of these barriers. The most commonly cited barriers include lack of time, inconvenience, lack of interest, and lack of social support. The following sections describe behavioral strategies that address each of these specific barriers.

Lack of Time

Lack of time is often cited as a major reason for failure to adopt a pattern of regular exercise. Previous studies suggest that an exercise expenditure of between 1000 and 2500 calories per week may be necessary to enhance long-term weight loss (1, 32, 53). If the exercise is performed at a moderate intensity, this level of expenditure would require approximately 200 to 500 minutes per week (30-70 minutes per day). Strategies are therefore needed to help overweight adults deal with this large time commitment.

Exercise Intensity

Because there is an inverse relationship between exercise intensity and duration when attempting to achieve a given level of energy expenditure, one strategy to reduce the amount of time required for exercise is to exercise at a higher intensity. For example, in a study of overweight women conducted by Duncan et al. (10), the time required to complete a 4.8 kilometer (3.0 miles) distance was 60 minutes for those individuals walking at a 4.8 km/hr pace compared to 36 minutes for those individuals walking at an 8.0 km/hr pace, a 40% decrease in the time required to complete the exercise.

Studies of the effect of intensity on exercise adherence have produced a mixture of results. A review conducted by Dishman and Sallis (9) concluded that exercise intensity does not have a clear effect on exercise adherence and dropout rates. For example, in a study conducted by King et al. (30), individuals assigned to a higher intensity exercise group (75-85% $\dot{V}O_2$max) reported completing more exercise than individuals assigned to a lower intensity exercise group (60-73% $\dot{V}O_2$max) over the 2-year trial. In contrast, in a community sample, Sallis et al. (52) showed that the dropout rate was 25-35% for individuals performing moderate intensity exercise compared to a dropout rate of 50% for individuals performing vigorous intensity exercise. The increased dropout rate in higher intensity exercise programs may be due to increased frequency of injuries. A gradual progression in exercise intensity and the use of lower impact exercise such as walking may minimize this injury rate and thus lead to better adherence to this type of regimen (47).

Exercise Duration
An alternative approach to overcoming lack of time as a barrier to exercise is to encourage individuals to try to accumulate short periods of moderate intensity exercise over the course of the day. This strategy has been recommended by the Centers for Disease Control and Prevention and the American College of Sports Medicine (44), yet this recommendation was based on little empirical data. However, in recent years, data are emerging that support the use of this strategy for improving exercise adherence. Exercise performed in this manner has also been shown to positively impact fitness levels and other health-related factors.

Studies have shown that the incorporation of multiple (2-4) short bouts of exercise (10-15 minutes in length) performed each day can significantly improve cardiorespiratory fitness (7, 12, 23, 55). In addition, some studies have shown that short bouts of exercise may also have an impact on high-density lipoproteins (HDL-C) (12), insulin, and body composition (55). However, only one published study has examined whether the use of multiple short bouts of exercise will improve exercise adherence.

Jakicic et al. (23) studied 56 overweight women enrolled in a 20-week behavioral weight-loss program. The women were randomly assigned to either a long-bout or short-bout exercise condition. Both groups received a comparable behavioral program, with strategies to facilitate adoption of healthy eating and exercise behavior. Both groups were encouraged to exercise on their own and to gradually increase their exercise to 40 minutes/day on 5 days in the week. The exercise prescription for the long-bout group was to progress to 40 minutes of continuous exercise on each of the 5 days, whereas the prescription for the short-bout group was to divide the 40 minutes into four 10-minute bouts performed throughout the day. Women in the short-bout exercise condition reported exercising for more total minutes per week (224 min/week) and reported at least one bout of exercise on more total days (87 days) than women in the long-bout exercise condition (188 min/week, 69 total days). Moreover, this increase in exercise participation may have contributed to the trend for greater weight loss in the short-bout exercise condition (-9 kg) compared to the long-bout exercise condition (-6 kg). We are currently examining the long-term impact of this strategy.

Inconvenience

The belief that exercise is inconvenient may be another barrier to exercise participation. In a study of men in a supervised cardiac rehabilitation program, lack of convenience (time of day sessions were scheduled) was identified as a major reason for individuals to discontinue their participation in the program (43). These data suggest that strategies to increase convenience of exercise may improve compliance.

Exercise Facilities and Equipment

A strategy to increase the convenience of exercise is to increase access to exercise facilities and/or exercise equipment. Using cross-sectional data, Sallis et al. (51) found a significant correlation between proximity to exercise facilities and the level of self-reported physical activity. However, because these are correlational data, it is difficult to interpret the direction of the correlation: Does living close to an exercise facility promote physical activity, or do active individuals select to live close to exercise facilities? A partial answer to this question is found in a laboratory study conducted by Raynor et al. (50). These investigators showed that when access to physical activity equipment was made easier for subjects and sedentary activities were moved further away, the chances that the individual would be active were increased. Therefore, both the development of parks, bike paths, and community recreation centers, and the removal of TVs from many rooms in the house may be effective approaches to increasing physical activity on a large-scale basis.

Another approach to increasing the convenience of exercise is to increase the options available within the individuals' immediate environment, primarily within their home. The home-exercise equipment industry is a $5 billion per year industry, yet little is known about whether having access to this equipment actually increases exercise participation rates. In a cross-sectional study of healthy adults, Jakicic et al. (22) found a weak but significant correlation between the amount of exercise equipment in the home and level of physical activity. Based on the results of this cross-sectional study, Jakicic and colleagues are currently investigating whether providing a treadmill to overweight women in a behavioral weight loss program enhances exercise participation and long-term weight loss. Preliminary results indicate that this may be an effective strategy when coupled with a behavioral intervention program (24). In addition to making exercise convenient, home exercise equipment may also act as a cue to exercise. Despite these promising results, additional research in this area is necessary before drawing definitive conclusions from these few studies.

Home-Based Versus Supervised Exercise

Many exercise programs have traditionally been conducted in supervised settings. The advantage of this type of intervention is that patients typically will have access to state-of-the-art facilities and well-trained staff. In addition, from a research perspective, it becomes much easier to quantify the volume of exercise being performed. However, the necessity of traveling to a facility during specific times of the day may make exercise participation inconvenient, which may act as a barrier to continued participation. In addition, membership in these types of facilities is typically expensive, and this may create another barrier for many individuals. Thus, it is important to examine other options and strategies that are available to individuals to promote long-term exercise adherence.

One possible option would be to encourage individuals to participate in home-based exercise and physical activity rather than supervised exercise. King et al. (29) examined exercise adherence rates in over 350 men and women 50 to 65 years of age. Individuals were assigned to one of two home-based exercise groups (a higher intensity exercise group or a lower intensity exercise group) or a supervised group-based exercise condition. After completion of the 12-month program, results showed that exercise adherence was greater in the home-based groups (higher intensity = 79%, lower intensity = 75%) compared to the supervised exercise group (53%). In general, this trend continued throughout the two years of treatment (30); however, as described earlier, the higher intensity home-based exercise group maintained the highest level of adherence over this 2-year period.

In a study of overweight women in a behavioral weight loss program, Perri et al. (46) also showed better exercise participation in a home-based exercise compared to group-based exercise. Over the 12-month study, women in the home-based program attended 84% of their prescribed sessions versus 62% of prescribed sessions in the group-based exercise condition. Exercise participation was similar in the groups during months 0-6 (~104 min/week). However, during months 7-12, the home-based group was able to maintain a significantly higher level of exercise participation than the group-based condition (69 min/week vs. 58 min/week).

The improved exercise participation in the home-based group may have increased weight loss during months 7-12. All the women were prescribed the same dietary intervention, and weight change during months 0-6 was similar between the treatment groups (home-based = -10 kg vs. group-based = -9 kg). However, individuals in the home-based condition lost an additional 2 kg during months 7-12, whereas the group-based condition gained 1 kg during this same time period. Therefore, home-based exercise may be an effective strategy for improving both exercise participation and long-term weight loss compared to group-based exercise.

Other studies (2, 11) have also shown improvements in health-related variables from home exercise that appear comparable to those achieved using supervised exercise. For example, Juneau et al. (26) showed significant increases in aerobic capacity following 6 months of home-based exercise. Therefore, home-based exercise may provide a viable alternative to supervised exercise, which may make exercise participation more appealing to otherwise sedentary individuals.

Lack of Interest

Another commonly cited barrier to exercise is lack of interest. This barrier has been addressed by attempting to develop more flexible exercise regimens, by trying to increase environmental control over exercise through increased cues for exercise, and by increased tangible rewards for exercise. Each of these approaches is discussed below.

Increasing Flexibility by Providing Choice of Exercise

One approach to this problem is to provide participants with a choice in the type of exercise they are prescribed, allowing them to choose types of physical activity they find most enjoyable and to vary the type of exercise over time. Epstein et al. (14, 15) evaluated the effectiveness of providing such flexibility in a 10-year study with overweight children, aged 8 to 12. The children were randomly assigned to a structured exercise program, where they had to select a time of day and a type of exercise to perform; or to a lifestyle exercise, where they could continually select different types of exercise and convenient times for exercise. A calisthenics control condition was also included. The lifestyle exercise condition led to the greatest change in percent overweight. The difference between the structured and lifestyle exercise condition gradually increased over the 10 years of follow-up.

Prompts for Exercise

Another approach to the problem of lack of interest in exercise is to use stimulus-control techniques to rearrange the environment to prompt or cue physical activity. Several studies have shown the effectiveness of simple cueing or prompting strategies. For example, Brownell and colleagues (4) posted a sign at the point where people choose to use the stairs or elevator; the sign encouraged people to use the stairs to promote heart health. This simple manipulation was effective in markedly increasing the number of people who chose to use the stairs.

Other types of prompting have been shown to be effective in encouraging individuals to maintain their exercise regimen over time. King, Taylor, Haskell, and DeBush (31) randomized 52 individuals to either receive 10 staff-initiated phone calls over a 6-month interval or to receive no phone contact. Subjects in both conditions received instructions on a home-based exercise regimen, information about strategies to promote adherence, and relapse-prevention instruction. At the end of 6 months, the group given the phone contact had greater increases in cardiorespiratory fitness. This finding suggests that they adhered more to the exercise regimen, although such data were not presented.

Likewise, mail and phone contact may be helpful in maintaining an exercise regimen in overweight individuals who have lost weight through exercise. Following a year of diet or exercise intervention, subjects were randomized to receive contact by mail and phone or to receive no contact (28). The major focus of the contact was on preventing relapse, with mailings sent monthly, and brief calls during the first 3 months and at 6, 9, and 12 months of follow-up. This mail/phone contact had no effect on the dieters, but was effective in the exercise condition. In the exercise condition subjects given phone contact experienced less weight regain over the year of follow-up. Interestingly no data are presented in whether this cueing strategy increased maintenance of the exercise behavior.

More recently, Lombard and colleagues (37) began to investigate the effect of various components of prompts used to increase physical activity. They varied the frequency of the prompt (once a week vs. once every 3 weeks) and the structure of the prompt (highly structured vs. touching base). The highly structured prompt included feedback to the subject on their past exercise, support for future exercise, and goal setting. The goal of the prompting was to increase the number of participants in a walking program who met the ACSM goals of exercising 20 minutes/day for 3 days a week. Compared to a no-prompt control condition, the prompting strategies increased the number of participants who met the exercise goal during the 12 weeks of intervention; there was also a significant effect of frequency of prompting, but no evidence that the nature of the message affected outcome.

To date, there has been no comparison of prompting by mail vs. phone. If the effectiveness of these strategies is due to the social support provided, then phone contact should be more effective than mail contact.

Tangible Rewards for Exercise

One of the most studied strategies for increasing interest in exercise and consequently improving exercise adherence is the use of a tangible reward, including lottery systems or direct payment for exercise. Epstein et al. (16) found positive effects of a lottery procedure on attendance at exercise sessions, but most studies have failed to confirm this. Martin et al. (41) found no effect of a lottery in which attendance at exercise sessions resulted in chances for weekly lotteries for jogging apparel (T-shirts, sweat bands, etc.), or for midpoint and endpoint lotteries (gift certificates to sporting goods store). Similarly, Marcus and Stanton (39) and Wing et al. (61) found no improvements in attendance at exercise sessions when attendance was reinforced by lottery tickets.

In contrast, direct reward systems have had quite positive effects on exercise adherence. Epstein et al. found that money deposited at the start of a program that was returned contingent on attendance at exercise sessions improved attendance at these sessions (16). Similar results have been shown when participants could earn clothing (33) or tokens that could be redeemed for a reward of the participant's choice (42), contingent on attendance at exercise sessions or when participants were paid small amounts for attending exercise sessions (25).

Lack of Social Support

Finally, another commonly cited barrier to exercise is lack of social support. Several studies have suggested an association between the level of social support for exercise and the amount of exercise performed (21, 57). Moreover, in one report (59), individuals who joined an exercise program with their spouse had better attendance at the exercise program and lower levels of dropping out than married individuals who joined the program alone.

Few studies have attempted to manipulate support as a way to improve exercise adherence, and the results of these studies have been mixed. King and Frederiksen (27) found that a social support program, where 3 to 4 individuals who previously did not know each other were grouped together and given group cohesiveness activities (encouraging intragroup cooperation and intergroup competition) had better exercise adherence than the usual care control condition. In contrast, Kravitz and Furst (33) reported attendance rates for subjects in aerobic dance classes who self-selected to either work individually toward rewards for class attendance, to work with a group of 2 or 3 other individuals for comparable rewards, or to receive no rewards or social support. Subjects in both the reward groups had better adherence than those given no rewards; however, there were no differences between the individual and group reward conditions.

Social support from exercise leaders, through personalized feedback, has also been shown to improve exercise adherence (41). As noted above, cueing strategies may also be effective because they provide ongoing social support for the new behavior.

Relapse Prevention and Coping Skills

Finally, another approach to improving exercise adherence is to try to develop ways to keep exercise lapses from becoming relapses. This approach, which is based on Marlatt and Gordon's theory of relapse prevention (40), suggests that all people will at times have lapses from exercise, for example due to illness or injury. The key is to teach participants to anticipate high-risk situations that might lead to lapses; to develop strategies to cope with these situations effectively; and to teach participants to quickly return to their exercise routine after a lapse.

Based on this theory, Simkin and Gross (54) evaluated the association between coping responses to hypothetical high-risk situations for exercise and subsequent patterns of relapse from exercise. Subjects were 29 healthy women who were adopting exercise without formal intervention. At baseline, the subjects were asked to listen to an audiotape with vignettes of 10 high-risk situations for lapsing from exercise (negative mood, lack of time, etc.). In response to these vignettes, subjects indicated what they would say or do to help themselves exercise as planned. Lapses, defined as a period of 1 week during the 14-week project in which the subject did not exercise, occurred in 41% of the women. Those who lapsed were more likely to report a prior history of adopting and quitting exercise and had lower self-motivation to exercise at baseline. Lapsers reported significantly fewer behavioral and cognitive coping strategies in response to the vignettes. The effect of coping responses on chances of lapsing was highly significant even after adjusting for baseline self-motivation level. Thus, this study suggests that teaching participants how to cope with high-risk situations may decrease the chances that they will lapse from exercise.

Given data such as these, it would appear logical to try to improve exercise adherence by providing training in relapse-prevention skills. This training has been provided in several exercise studies (27, 41) with a focus on helping subjects identify high-risk situations. The fact that the relapse-prevention training was used as only one component of a multifactorial program (27) and small sample sizes (41) have made it difficult to assess the effectiveness of the relapse-prevention training.

Changing From Sedentary to Active Lifestyle

Many of the studies discussed above utilized behavioral strategies to increase attendance at exercise sessions. However, several studies, such as those of Goran and Poehlman (17) and Jeffery et al. (25), found that increasing attendance at supervised exercise sessions did not necessarily increase overall physical activity level. For example, Jeffery et al. (25) studied two new approaches to increasing attendance at supervised exercise sessions. The investigators hypothesized that increased attendance at such sessions would increase overall exercise levels and therefore lead to improved maintenance of weight loss. One approach used to increase attendance at exercise sessions was a "personal trainer," who called subjects to schedule and cue exercise, and then exercised with the participants at scheduled sessions. Thus the personal trainer served not only as a prompt, but also as a source of social support for exercise. The second procedure was to provide small financial incentives for attending each supervised exercise session.

To test these procedures, Jeffery et al. (25) recruited 196 overweight subjects. These subjects were randomly assigned to one of five groups: (1) A standard behavior therapy (SBT) group received a state-of-the-art behavioral weight loss program that included six months of weekly meetings followed by a full year of monthly meetings. Participants were given goals for calories and fat and encouraged to expend 1000 kcal/week in home-based exercise. (2) Another group was given the same SBT program and the same diet and exercise goals, but invited to 3 supervised exercise sessions/week. (3) A third group was given the same SBT program, the supervised walks, and personal trainers to cue their exercise. (4) A fourth group was given SBT, walks, and financial incentives for attending the walks. (5) The fifth group was given SBT, walks, personal trainers, and financial incentives. Subjects in the SBT group (Group 1) completed 35 walks out of a possible total of 222 walks. Both the use of the personal trainer and the use of incentives markedly increased attendance at walks; the average attendance of these groups was 80 walks in the personal trainer group and 66 walks in the incentive group. Attendance was best when the two strategies were utilized in combination (average walks attended = 103). However, in contrast to the investigator's expectations, the increased attendance at supervised walks did not lead to better overall exercise levels or to better weight loss and maintenance. All five groups reported expending 1000-1200 kcal/week in exercise at both 6 months and 18 months. Thus all groups met, but none really exceeded, the exercise goal that was prescribed. Surprisingly, weight losses at both times were best in the standard behavioral program. This study thus suggests that behavioral strategies can indeed increase attendance at exercise sessions. However, such attendance does not automatically affect overall exercise expenditure. Increasing attendance at exercise sessions may consequently not be the appropriate target to use in behavioral programs. Moreover, in a secondary analysis these investigators found that long-term weight loss was significantly better only in those in the top quartile of overall energy expenditure, who reported expending approximately 2500 kcal/day. Thus, this study also raises the question of whether the exercise target used in the behavioral weight loss program (typically 1000 kcal/week) is too low to promote significant long-term weight loss, and whether a more appropriate target would be 2500 kcal/week.

Another issue is whether it is most effective to focus on increasing physical activity or on decreasing sedentary activity. Epidemiological studies have shown a positive association between amount of sedentary activity (such as TV watching) and obesity; i.e., people who watch the most television have the highest levels of body mass index (3, 6, 18). Sedentary activity may be associated with obesity because it reduces the amount of time that is spent in physical activity (58). However, the fact that the association of sedentary activity with obesity is independent of the association with physical activity (6) suggests that other mechanisms may also be operative. Activities such as television watching may increase the amount of snacking and total calories consumed, and lead to increased obesity in this manner.

These findings suggest the possibility that interventions should focus on decreasing sedentary behavior instead of, or in addition to, focusing on increasing physical activity. Epstein and colleagues (13) have investigated this issue in the study of overweight children, age 8 to 12. Sixty-one families (including an overweight child and the parent) were randomly assigned to one of three groups: reinforce children for increasing their physical activity; reinforce children for decreasing their sedentary activities; or the combination of both. All three groups received comparable diet and behavioral information. Children who were reinforced for decreasing sedentary activities showed the greatest reductions in percent overweight at 4 and 12 months. In addition, children reinforced for decreasing sedentary behavior reported the greatest increases in preferences for high-intensity physical activities. Although the study did not include direct measures of physical activity, and changes in fitness over time were comparable in the three groups, the investigators suggest that children who were reinforced for reducing sedentary behaviors may have substituted higher expenditure activities for these sedentary pursuits. The fact that the children were in control of the choice to substitute higher intensity activities and could select whatever activities they wished may have led to better adherence.

Summary

It is clear that exercise can be beneficial to overweight individuals who are attempting to control their body weight, and yet limited numbers of overweight individuals are able to maintain their exercise behavior beyond 6 to 12 months. The purpose of this chapter was to review various strategies for increasing exercise adherence in this population of typically sedentary individuals. A number of strategies have shown promise for increasing the adoption and maintenance of a physically active lifestyle; however, only a few of these strategies have specifically focused on increasing physical activity in obese individuals. Therefore, additional studies need to be conducted to further evaluate the effectiveness of these strategies for increasing physical activity in overweight adults. Moreover, it is unlikely that any one strategy will be effective for all individuals, and it is possible that physical activity lapses/relapses are a result of a periodic change in the barriers that affect physical activity participation. For example, weather may be the barrier for physical activity during the winter months, whereas lack of time due to family commitments or lack of motivation may be the barriers that the individual is confronted with during other times of the year. Therefore, different strategies may be needed at different times, depending on the barrier the individual is currently confronting. Little is known about factors that affect physical activity participation beyond 6 to 12 months or that would help overweight adults achieve the relatively high level of physical activity that appears characteristic of successful long-term weight losers.

References

1. American College of Sports Medicine. 1995. *ACSM's guidelines for exercise testing and prescription.* Philadelphia: Williams and Wilkins.
2. Andersen, R.E., S.J. Bartlett, C.D. Moser, M.I. Evangelisti, and T.J. Verde. 1997. Lifestyle or aerobic exercise to treat obesity in dieting women. *Medicine and Science in Sports and Exercise* 29: S46.

3. Andersen, R.E., C.J. Crespo, S.J. Bartlett, L.J. Cheskin, and M. Pratt. 1998. Relationship of physical activity and television watching with body weight and level of fatness among children. *Journal of the American Medical Association* 279: 938-942.

4. Brownell, K.D., A.J. Stunkard, and J.M. Albaum. 1980. Evaluation and modification of exercise patterns in the natural environment. *American Journal of Psychiatry* 137: 1540-1545.

5. Calfas, K.J., B.J. Long, J.F. Sallis, W.J. Wooten, M. Pratt, and K. Patrick. 1996. A controlled trial of physician counseling to promote the adoption of physical activity. *Preventive Medicine* 25: 225-233.

6. Ching, P.I.Y.H., W.C. Willett, E.B. Rimm, G.A. Colditz, S.L. Gortmaker, and M.J. Stampfer. 1996. Activity level and risk of overweight in male health professionals. *American Journal of Public Health* 86: 25-30.

7. DeBusk, R.F., U. Stenestrand, M. Sheehan, and W.L. Haskell. 1990. Training effects of long versus short bouts of exercise in healthy subjects. *American Journal of Cardiology* 65: 1010-1013.

8. Dishman, R.K. 1990. Determinants of participation in physical activity. In *Exercise, fitness, and health,* eds. C. Bouchard, R.J. Shephard, T. Stephens, J.R. Sutton, and B.D. McPherson, 75-102. Champaign, IL: Human Kinetics.

9. Dishman, R.K., and J.F. Sallis. 1994. Determinants and interventions for physical activity and exercise. In *Physical activity, fitness, and health,* eds C. Bouchard, R.J. Shephard, and T. Stephens, 214-238. Champaign, IL: Human Kinetics.

10. Duncan, J.J., N.F. Gordon, and C.B. Scott. 1991. Women walking for health and fitness: How much is enough? *Journal of the American Medical Association* 266: 3295-3299.

11. Dunn, A.L., B.H. Marcus, J.B. Kampert, M.E. Garcia, H.W. Kohl, and S.N. Blair. 1997. Reduction in cardiovascular disease risk factors: 6-month results from Project Active. *Preventive Medicine* 26: 883-892.

12. Ebisu, T. 1985. Splitting the distance of endurance running: On cardiovascular endurance and blood lipids. *Japanese Journal of Physical Education* 30: 37-43.

13. Epstein, L.H., A.M. Valoski, L.S. Vara, J. McCurley, L. Wisniewski, M.A. Kalarchian, K.R. Klein, and L.R. Shrager. 1995. Effects of decreasing sedentary behavior and increasing activity on weight change in obese children. *Health Psychology* 14: 109-115.

14. Epstein, L.H., A. Valoski, R.R. Wing, and J. McCurley. 1994. Ten year outcomes of behavioral family based treatment for childhood obesity. *Health Psychology* 13: 373-383.

15. Epstein, L.H., R.R. Wing, R. Koeske, and A. Valoski. 1985. A comparison of lifestyle exercise, aerobic exercise, and calisthenics on weight loss in obese children. *Behavior Therapy* 16: 345-356.

16. Epstein, L.H., R.R. Wing, J.K. Thompson, and W. Griffin. 1980. Attendance and fitness in aerobics exercise: The effects of contract and lottery procedures. *Behavior Modification* 4: 465-479.

17. Goran, M.I., and E.T. Poehlman. 1992. Total energy expenditure and energy requirements in healthy elderly persons. *Metabolism* 41: 744-753.

18. Gortmaker, S.L., A. Must, A.M. Sobol, K. Peterson, G.A. Colditz, and W.H. Dietz. 1996. Television viewing as a cause of increasing obesity among children in the United States, 1986-1990. *Archives of Pediatrics & Adolescent Medicine* 150: 356-362.

19. Hagan, R.D., S.J. Upton, L. Wong, and J. Whittam. 1986. The effects of aerobic conditioning and/or caloric restriction in overweight men and women. *Medicine and Science in Sports and Exercise* 18: 87-94.

20. Hammer, R.L., C.A. Barrier, E.S. Roundy, J.M. Bradford, and A.G. Fisher. 1989. Calorie-restricted low-fat diet and exercise in obese women. *American Journal of Clinical Nutrition* 49: 77-85.

21. Heinzelmann, F., and R.W. Bagley. 1970. Response to physical activity programs and their effects on health behavior. *Public Health Report* 86: 905-911.

22. Jakicic, J.M., R.R. Wing, B.A. Butler, and R.W. Jeffery. 1997. The relationship between presence of exercise equipment in the home and physical activity level. *American Journal of Health Promotion* 11: 363-365.

23. Jakicic, J.M., R.R. Wing, B.A. Butler, and R.J. Robertson. 1995. Prescribing exercise in multiple short bouts versus one continuous bout: Effects on adherence, cardiorespiratory fitness, and weight loss in overweight women. *International Journal of Obesity* 19: 893-901.

24. Jakicic, J.M., R.R. Wing, C. Winters, and L. Clifford. 1998. Exercise adherence in overweight women: Effect of short-bouts of exercise and exercise equipment. *Annals of Behavioral Medicine* 20 (Supp): S068(abstract).

25. Jeffery, R.W., R.R. Wing, C. Thorson, and L.C. Burton. 1998. Use of personal trainers and financial incentives to increase exercise in a behavioral weight-loss program. *Journal of Consulting and Clinical Psychology* 66: 777-783.

26. Juneau, M., F. Rogers, V. DeSantos, M. Yee, A. Evans, A. Bohn, W.L. Haskell, C.B. Taylor, and R.F. DeBusk. 1987. Effectiveness of self-monitored, home-based, moderate-intensity exercise training in middle-aged men and women. *American Journal of Cardiology* 60: 66-70.

27. King, A.C., and L.W. Frederiksen. 1984. Low cost strategies for increasing exercise behavior: Relapse prevention and social support. *Behavior Modification* 8: 13-21.

28. King, A.C., B. Frey-Hewitt, D.M. Dreon, and P.D. Wood. 1989. The effects of minimal intervention strategies on long-term outcomes in men. *Archives of Internal Medicine* 149: 2741-2746.

29. King, A.C., W.L. Haskell, C.B. Taylor, H.C. Kraemer, and R.F. DeBusk. 1991. Group- vs home-based exercise training in healthy older men and women: A community-based clinical trial. *Journal of the American Medical Association* 266: 1535-1542.

30. King, A.C., W.L. Haskell, C.B. Taylor, H.C. Kraemer, and R.F. DeBusk. 1995. Long-term effects of varying intensities and formats of physical activity on participation rates, fitness, and lipoproteins in men and women aged 50-65 years. *Circulation* 91: 2596-2604.

31. King, A.C., C.B. Taylor, W.L. Haskell, and R.F. DeBusk. 1988. Strategies for increasing early adherence to and long-term maintenance of home-based exercise training in healthy middle-aged men and women. *American Journal of Cardiology* 61: 628-632.

32. Klem, M.L., R.R. Wing, M.T. McGuire, H.M. Seagle, and J.O. Hill. 1997. A descriptive study of individuals successful at long-term maintenance of substantial weight loss. *American Journal of Clinical Nutrition* 66: 239-246.

33. Kravitz, L., and D. Furst. 1991. Influence of reward and social support on exercise adherence in aerobic dance classes. *Psychological Reports* 69: 423-426.

34. Kreuter, M.W., D.P. Scharff, L.K. Brennan, and S.N. Lukwago. 1997. Physician recommendations for diet and physical activity: Which patients get advised to change? *Preventive Medicine* 26: 825-833.

35. Kuczmarski, R.J., K.M. Flegal, S.M. Campbell, and C.L. Johnson. 1994. Increasing prevalence of overweight among U.S. adults. The National Health and Nutrition Examination Surveys, 1960 to 1991. *Journal of the American Medical Association* 272: 205-211.

36. Lewis, B.S., and W.D. Lynch. 1993. The effect of physician advice on exercise behavior. *Preventive Medicine* 22: 110-121.

37. Lombard, D.N., T.N. Lombard, and R.A. Winett. 1995. Walking to meet health guidelines—the effect of prompting. *Health Psychology* 14: 164-170.

38. Marcus, B.H., K.M. Emmons, L. Simkin-Silverman, L.A. Linnan, E.R. Taylor, B.C. Bock, M.B. Roberts, J.S. Rossi, and D.B. Abrams. 1998. Evaluation of motivationally tailored vs. standard self-help physical activity interventions at the workplace. *American Journal of Health Promotion* 12: 246-253.

39. Marcus, B.H., and A.L. Stanton. 1993. Evaluation of relapse prevention and reinforcement interventions to promote exercise adherence in sedentary females. *Research Quarterly for Exercise and Sport* 64: 447-452.

40. Marlatt, G.A., and J.R. Gordon. 1985. *Relapse prevention: Maintenance strategies in addictive behavior change.* New York: Guilford.

41. Martin, J.E., P.M. Dubbert, A.D. Katell, J.K. Thompson, J.R. Raczynski, M. Lake, P.O. Smith, J.S. Webster, T. Sikora, and R.E. Cohen. 1984. Behavioral control of exercise in sedentary adults: Studies 1 through 6. *Journal of Consulting and Clinical Psychology* 52: 795-811.

42. Noland, M.P. 1989. The effects of self-monitoring and reinforcement on exercise adherence. *Research Quarterly for Exercise and Sport* 60: 216-224.

43. Oldridge, N.B., A.P. Donner, C.W. Buck, N.L. Jones, G.M. Andrew, J.O. Parker, D.A. Cunningham, T. Kavanaugh, P.A. Rechnitzer, and J.R. Sutton. 1983. Predictors of dropout from cardiac exercise rehabilitation: Ontario Exercise-Heart Collaborative Study. *American Journal of Cardiology* 51: 70-74.

44. Pate, R.R., M. Pratt, S.N. Blair, W.L. Haskell, C.A. Macera, C. Bouchard, D. Buchner, W. Ettinger, G.W. Heath, A.C. King, A. Kriska, A.S. Leon, B.H. Marcus, J. Morris, R.S. Paffenbarger, K. Patrick, M.L. Pollock, J.M. Rippe, J. Sallis, and J.H. Wilmore. 1995. Physical activity and public health: a recommendation from the Centers for Disease Control and Prevention and the American College of Sports Medicine. *Journal of the American Medical Association* 273: 402-407.

45. Pavlou, K.N., S. Krey, and W.P. Steffee. 1989. Exercise as an adjunct to weight loss and maintenance in moderately obese subjects. *American Journal of Clinical Nutrition* 49: 1115-1123.

46. Perri, M.G., A.D. Martin, E.A. Leermakers, S.F. Sears, and M. Notelovitz. 1997. Effects of group- versus home-based exercise in the treatment of obesity. *Journal of Consulting and Clinical Psychology* 65: 278-285.

47. Pollock, M.L., J.F. Carroll, J.E. Graves, S.H. Leggett, R.W. Braith, M. Limacher, and J.M. Hagberg. 1991. Injuries and adherence to walk/jog and resistance training programs in the elderly. *Medicine and Science in Sports and Exercise* 23: 1194-1200.

48. Prochaska, J.O., and C.C. DiClemente. 1984. *The transtheoretical approach: Crossing traditional boundaries of therapy.* Homewood, IL: Dow Jones/Irwin.

49. Pronk, N.P., and R.R. Wing. 1994. Physical activity and long-term maintenance of weight loss. *Obesity Research* 2: 587-599.

50. Raynor, D.A., K.J. Coleman, and L.H. Epstein. 1998. Effects of proximity on the choice to be physically active or sedentary. *Research Quarterly for Exercise and Sport* 69: 99-103.

51. Sallis, J.F., M.F. Hovell, C.R. Hofstetter, J.P. Elder, M. Hackley, C.J. Caspersen, and K.E. Powell. 1990. Distance between homes and exercise facilities related to frequency of exercise among San Diego residents. *Public Health Report* 105: 179-185.

52. Sallis, J.F., W.L. Haskell, S.P. Fortmann, K.M. Vranizan, C.B. Taylor, and D.S. Solomon. 1986. Predictors of adoption and maintenance of physical activity in a community sample. *Preventive Medicine* 15: 331-346.

53. Schoeller, D.A., K. Shay, and R.F. Kushner. 1997. How much physical activity is needed to minimize weight gain in previously obese women? *American Journal of Clinical Nutrition* 66: 551-556.

54. Simkin, L.R., and A.M. Gross. 1994. Assessment of coping with high-risk situations for exercise relapse among healthy women. *Health Psychology* 13: 274-277.
55. Snyder, K.A., J.E. Donnelly, D.J. Jacobsen, G. Hertner, and J.M. Jakicic. 1997. The effects of long-term, moderate intensity, intermittent exercise on aerobic capacity, body composition, blood lipids, insulin and glucose in overweight females. *International Journal of Obesity* 21: 1180-1189.
56. Swinburn, B.A., L.G. Walter, B. Arroll, M.W. Tilyard, and D.G. Russell. 1998. The green prescription study: A randomized controlled trial of written exercise advice provided by general practitioners. *American Journal of Public Health* 88: 288-291.
57. Treiber, F.A., T. Baranowski, D.S. Braden, W.B. Strong, M. Levy, and W. Knox. 1991. Social support for exercise: Relationship to physical activity in young adults. *Preventive Medicine* 20: 737-750.
58. Tucker, L.A., and G.M. Friedman. 1989. Television viewing and obesity in adult males. *American Journal of Public Health* 79: 516-518.
59. Wallace, J.P., J.S. Raglin, and C.A. Jastremski. 1995. Twelve month adherence of adults who joined a fitness program with a spouse vs without a spouse. *The Journal of Sports Medicine and Physical Fitness* 35: 206-213.
60. Wing, R.R., L.H. Epstein, M. Paternostro-Bayles, A. Kriska, M.P. Nowalk, and W. Gooding. 1988. Exercise in a behavioural weight control programme for obese patients with type 2 (non-insulin-dependent) diabetes. *Diabetologia* 31: 902-909.
61. Wing, R.R., R.W. Jeffery, N. Pronk, and W.L. Hellerstedt. 1996. Effects of a personal trainer and financial incentives on exercise adherence in overweight women in a behavioral weight loss program. *Obesity Research* 4: 457-462.

Index

About the Editor

Claude Bouchard, one of the world's leaders in the study of genetics and physical activity, is the executive director of the Pennington Biomedical Research Center and the George A. Bray Chair in Nutrition at Louisiana State University. Previously, he held the Donald B. Brown Research Chair on Obesity.

Dr. Bouchard's research has focused on the genetics of adaptation to exercise, nutritional interventions, and the genetics of obesity and its co-morbidities. His efforts have been funded from various agencies including the National Institute of Health and the Medical Research Council of Canada.

Dr. Bouchard has authored or coauthored several books and has published more than 500 professional and scientific papers. Former president of the North American Association for the Study of Obesity, Dr. Bouchard is the president-elect of the International Association for the Study of Obesity.

Among other awards, he has received the Honor Award from the Canadian Association of Sport Sciences, a Citation Award from the Americian College of Sports Medicine, the Benjamin Delessert Award in Nutrition from France, the Willendorf Award from the International Association for the Study of Obesity, the Sandoz Award from the Canadian Atherosclerosis Society, the Albert Creff Award in Nutrition of the National Academy of Medicine of France, and an Honoris Causa Doctorate in Science from the Katholieke Universiteit Leuven.

Dr. Bouchard holds an MSc in Exercise Physiology from the University of Oregon and a PhD in Anthropological Genetics from the University of Texas.